A Guide to Thermal Physics

Chris McMullen, Ph.D.
Northwestern State University of Louisiana

from the Fundamentals thru Callen-Level Equilibrium Thermodynamics

A Guide to Thermal Physics from the Fundamentals thru Callen-Level Equilibrium Thermodynamics

Copyright © 2010 Chris McMullen, Ph.D.

All rights reserved. This includes the right to reproduce any portion of this book in any form.

Chris McMullen, Ph.D.
Physics Instructor
Department of Mathematics and Physical Sciences
Northwestern State University of Louisiana
http://mathematics.nsula.edu/faculty/
mcmullenc@nsula.edu

CreateSpace

Textbooks / science / physics / thermodynamics
Nonfiction / science / physics / thermodynamics

ISBN: 1453772804

EAN-13: 9781453772805

Contents

Introduction	5
0 Basic Terminology and Measurements of Thermal Physics	
0.1 Basic Measureable Quantities that Relate to Thermal Physics	7
0.2 Basic Constants that Relate to Thermal Physics	14
0.3 Basic Units and Conversions of Thermal Physics	16
0.4 Basic Measurements that Relate to Thermal Physics	17
0.5 Basic Thermodynamics States and Processes	19
1 Basic Relationships of Thermal Physics	
1.1 The Four Laws of Thermodynamics	26
1.2 Basic Mathematical Requisites of Thermal Physics	33
1.3 Basic Quantitative Relationships of Thermal Physics	41
1.4 Work: Interpreting Pressure-Volume (P-V) Diagrams	46
1.5 Phase Transitions: Interpreting Pressure-Temperature (P-T) Diagrams	51
2 Basic States and Processes of Thermal Physics	
2.1 Thermal Expansion of Liquids and Solids	58
2.2 Thermal Equilibrium: Calorimetry	60
2.3 A Steady State: Heat Conduction	62
2.4 Radiative Equilibrium: Electromagnetic Radiation	68
2.5 Conservation of Energy: The First Law	70
3 Ideal Gases in Thermal Physics	
3.1 Definition of an Ideal Gas	79
3.2 The Ideal Gas Law	81
3.3 Pressure on an Ideal Gas	83
3.4 Energy of an Ideal Gas	86
3.5 Heat Capacities for an Ideal Gas	88
3.6 Isothermal Expansion of an Ideal Gas	90
3.7 Adiabatic Expansion of an Ideal Gas	92
4 Probability Distribution Functions	
4.1 Probability and Statistics	100
4.2 Discrete Probability Distributions	104
4.3 Continuous Probability Distributions	110
4.4 Most Probable, Average, and Root-Mean-Square Values	116
4.5 Maxwell-Boltzmann Distribution of Speeds for an Ideal Gas	121
4.6 Mean-Free Path and Molecular Flux	131

5 Heat Engines and Refrigerators
- 5.1 Entropy and Reversibility … 140
- 5.2 Heat Engines and Refrigerators … 155
- 5.3 The Carnot Cycle … 163
- 5.4 Maximum Efficiency of a Heat Engine … 174
- 5.5 Practical Heat Engines … 182

6 An Introduction to Thermodynamics
- 6.1 Formulation of Thermodynamics … 202
- 6.2 Equilibrium Thermodynamics … 214
- 6.3 Thermodynamic Potentials … 222
- 6.4 Simple Thermodynamic Systems … 228
- 6.5 Extremum Principles … 241
- 6.6 The Maxwell Relations … 249
- 6.7 Phase Transitions … 260

Hints to Conceptual Questions and Practice Problems … 281
Answers to Conceptual Questions and Selected Practice Problems … 298
References … 311
Index … 312

Introduction

The primary goal of this text is to help students pursue a better understanding of both the concepts and mathematics of thermal physics from the basic principles through the more advanced formulation of equilibrium thermodynamics. This includes students who are taking a course in thermodynamics – such as third-semester calculus-based physics, an advanced undergraduate course in thermodynamics, or an introductory graduate course in thermodynamics – as well as students who are reviewing thermodynamics and even self-learners.

There is often a significant disparity between the level of rigor, level of mathematics, conceptual approach, and even the perception of what thermodynamics is between the third semester of calculus-based physics and the first undergraduate course devoted toward thermodynamics – in terms of the textbook from which these separate courses are taught. In order to facilitate this transition, this text covers material from both worlds, attempting to present a more unifying picture. Therefore, this text does not aim to simply reproduce the same material at the same level and with the same organization as the commonly used textbooks on this subject. Rather, more depth has been given to the introductory material in order to aid in the transition to the more advanced formulation of thermodynamics, and the material has been organized such that it may seem somewhat more unified. For example, one concern was that third-semester calculus-based physics often appears as a set of disparate empirical observations, whereas the formulation of equilibrium thermodynamics is derived from a minimum number of postulates.

The introductory material is also presented with a more advanced mathematical formulation in order to help bridge the gap between the introductory material and the advanced formulation of thermodynamics. To go along with this, an effort has been made to make the concepts behind the mathematics as well as the mathematical techniques clear to help students fully grasp the inherently mathematical component of the course. Some sections of this text have even been dedicated solely toward the mathematical techniques of thermal physics.

As the mathematics and concepts are inherently intertwined, much effort has also been made to elucidate the concepts and help students understand the marriage between the mathematics and the concepts. The introductory chapters are largely geared toward providing a solid foundation of the fundamental concepts and their relationship with the mathematics. The material from these chapters is intended to serve as a valuable introduction for beginning students and also as a useful review for advanced students.

There is not a universal set of notation in thermal physics, so there will indubitably be differences between the notation of this text and other textbooks. Effort has been made to make the notation of this text clear, and notes have even been included to discuss important notational differences. For example, many calculus-based textbooks use n for mole number and N for the number of molecules, while Callen's popular textbook uses N for mole number. In this text, the choice was made to use n for mole number, which is consistent with the common habit of using lowercase symbols for molar quantities.

Introduction

The format of this text has been designed to accommodate both students who wish to read straight through as well as those who are selectively reviewing or searching for specific concepts. Nearly 300 footnotes have been included in order to provide more detail for those students who want more information without interrupting the main ideas with long diversions. Boxed paragraphs highlight important distinctions between fundamental concepts – such as the distinction between temperature and heat or the distinction between an adiabatic and an isentropic process. Each chapter includes a selection of conceptual and mathematical examples that are deemed to be most instructive and to illustrate the main problem-solving strategies. At the end of each chapter is a selection of conceptual questions and practice problems, with hints and answers at the end of the book – to help students learn the concepts, master the mathematical techniques, and learn to apply the material.

This text is organized as follows. Chapter 0 begins with precise definitions of the most common fundamental concepts of thermal physics. This is crucial for beginning students, and even advanced students should browse through this to gauge – and possibly perfect – their understanding of these basics. Also included in this chapter is a tabulation of the most handy constants and conversions (along with references to online tables of standards). Chapter 1 introduces the requisite mathematical techniques and elementary quantitative relationships of thermal physics. It begins with a conceptual introduction to each of the four laws of thermodynamics (which are developed more thoroughly, and more mathematically, later in the text). The sections on mathematical techniques and quantitative relationships should be valuable to beginning students, and are also detailed enough to be worth skimming through by advanced students; and if the math seems challenging later in the text, it may be worth returning here to review these techniques. The section on *P-V* diagrams is especially fundamental to thermal physics, as it relates directly to applications of the first law of thermodynamics. A handful of fundamental empirical observations from introductory thermal physics – such as heat conduction, thermal expansion, and the first law of thermodynamics – are discussed conceptually and also with a rigorous level of mathematics in Chapter 2. For example, heat conduction is compared in detail with electrical conduction – with which students are generally apt to be more familiar – and the integrals are cast in a variety of forms, from the elegant to the practical. The titles of these sections aim to plant in the student's mind that some of these seemingly different ideas are at least related in that they apply to systems either in equilibrium or in a steady state. It would be a good exercise for more advanced students who are studying Chapter 6 to look back at Chapter 2 and try to figure out where each of these sections fits in with the formulation of thermodynamics.

Chapter 3 is dedicated solely to the ideal gas, which for its simplicity and wide applicability is the most common system to be encountered in textbook problems. The subject of Chapter 4 is probability distribution functions, which provides some background on introductory probability theory. This extra material on the basics of probability should be helpful in grasping the abstract and mathematically challenging Maxwell-Boltzmann distribution and the framework of kinetic theory. Chapter 5 is focused on heat engines and refrigerators, and the details of this chapter should be particularly useful for all students of thermal physics. Here the student will find a detailed conceptual introduction to the main thermodynamic heat engine cycles, applications to heat pumps and refrigerators, detailed mathematical calculations of the efficiencies of heat engines, and a thorough introduction to entropy and the second law of thermodynamics. Chapter 6 is the most advanced chapter, covering the more advanced formulation of thermodynamics, and so is the lengthiest and most detailed chapter. This is the main chapter that more advanced students will be studying; yet the earlier chapters should prove quite helpful, as they help to bridge the gap between introductory thermal physics and thermodynamics.

0 Basic Terminology and Measurements of Thermal Physics

0.1 Basic Measureable Quantities that Relate to Thermal Physics

Temperature (T): Our experience intuitively suggests that temperature relates to our perceptions of how hot or cold an object is. However, when we touch an object, what we "feel" relates to how much heat is transferred from the object during the thermal contact. Of course, this also depends on the temperature of the object, but this intuitive notion of temperature does not distinguish it from the similar term, heat. Temperature is more appropriately defined in terms of the average kinetic energy of the molecules of a substance. An object with low temperature consists of molecules that have little kinetic energy on average; if the molecules move faster on average, the temperature rises. This definition is called absolute temperature in the sense that the average kinetic energy of the molecules has an absolute minimum.[1] Absolute temperature is measured in Kelvin. Many formulas in thermal physics are only valid using absolute temperature (i.e. not Celsius or Fahrenheit).

> The average kinetic energy of the molecules of a substance provides a conceptually useful measure of absolute temperature.[2] Absolute zero thus establishes a lower limit.[3]

Heat (Q): The thermal energy transferred from one substance to another is called heat. If two objects are in thermal contact, heat tends to flow from the higher-temperature object to the lower-temperature object. Again we see a connection between heat and temperature. We also see that each is essentially a measure of energy. Yet, heat and temperature are fundamentally different, so it is important to distinguish between the two. Indeed, it is possible for a system to absorb heat without increasing temperature, and it is also possible for a system to change temperature without exchanging heat with its surroundings. Heat is only one form of energy that affects temperature.

> A transfer of thermal energy two objects in thermal contact is an exchange of heat. Heat is positive ($dQ > 0$) if it is added to the system.[4]

---

[1] As the absolute temperature approaches absolute zero, average kinetic energy approaches a finite zero-point energy. That is, the minimum kinetic energy of the molecules is not actually zero.
[2] Technically, molecules are the smallest neutral arrangement of atoms that form the object. Some objects are composed of monatomic, diatomic, or polyatomic molecules, but some actually do not consist of molecules at all – they consist of groups of ions that attract one another. However, we'll use the term molecule more loosely here.
[3] Absolute zero can be approached, but not attained. On the other hand, it is possible to achieve negative absolute temperatures, but this is not achieved by passing through absolute zero: Counterintuitively, negative absolute temperatures are obtained by increasing the temperature past infinity, rather than decreasing it below zero!
[4] These imperfect (or inexact) differentials remind us that heat and work are path-dependent.

Important Distinction. Temperature relates to the average kinetic energy of the molecules, whereas heat refers to thermal energy transferred to or from a system. When a system absorbs heat from its surroundings, this thermal energy can be used to increase the overall kinetic energy of the system's molecules, thereby increasing the temperature of the system. However, heat absorbed does not necessarily increase the kinetic energy of the system's molecules. The system may also use part or all of the thermal energy gained as heat to do work (i.e. to expand in volume). This is how it is possible for a system to gain or lose heat without changing temperature. Similarly, work done by or on the system can result in a change in temperature without any exchange of heat between the system and the surroundings.[5] There is yet another way that a system can exchange heat with its surroundings without changing temperature: Molecules in the system may undergo a phase transition.

Conceptual Examples. Suppose that you pour hot coffee from a thermos into a cup. Over time, the coffee cools. One obvious reason is that the hot coffee is in thermal contact with the cup and the air: The coffee loses heat, which the cup and air gain. The coffee thus cools, while the cup warms; there is so much air, that it probably does not warm noticeably (it is essentially a large heat reservoir). There is also a subtle way that the coffee loses heat. The faster coffee molecules could escape. If a substance loses its most energetic molecules, its average kinetic energy – i.e. its temperature – decreases. Looking for evidence that the coffee is missing some molecules? Try using your nose.

Consider two pots of water boiling on a stovetop, one of which has twice the volume of the other. The pot with greater volume has more molecules, yet both pots are at the same temperature – since water boils at $100°\,C$ at household pressure. Thus, the water molecules of both pots have the same average kinetic energy. However, one pot has twice as much total kinetic energy; as a result, it has the potential to transfer more heat.

Latent heat (L): A substance absorbs or releases heat when it undergoes a phase change. For example, heat is absorbed when a solid melts into liquid, whereas heat is released when a gas condenses into liquid. The heat per unit mass can be measured for a given substance, and this is called the latent heat. Latent heat is the general term; the actual transformation is specified when applied to a particular phase change (e.g. heat of fusion). The heat of fusion corresponds to a phase transition between solid and liquid, and is the same for freezing or melting except for a minus sign. The heat of vaporization corresponds to a phase transition between liquid and gas, and is the same as the heat of condensation apart from a minus sign. The heat of sublimation corresponds to a phase transition between solid and gas (skipping the liquid phase), and is the same as the heat of deposition apart from a minus sign.

> Latent heat[6] is the heat per unit mass absorbed or gained when a substance undergoes a phase transition ($Q = ML$).[7]

---

[5] But you won't have to guess at it: Once we learn the first law of thermodynamics, you will see precisely how heat and work are related to one another and will be able to use it to calculate what effects they have on a system.

[6] Don't be fooled by its name: It does not have the same dimensionality as heat. While the SI unit of heat is the Joule (J), the SI units for latent heat are Joules per kilogram (J/kg). The same amount of heat is involved in freezing and melting (and the same for vaporization and condensation and also for sublimation and deposition).

[7] The National Institute of Standards and Technology at www.nist.gov gives the latent heats for many substances.

Phase transition	Process	Latent heat
solid to liquid	melting	heat of fusion (+)
liquid to solid	freezing	heat of fusion (−)
liquid to gas	boiling	heat of vaporization (+)
gas to liquid	condensation	heat of vaporization (−)
solid to gas	sublimation	heat of sublimation (+)
gas to solid	deposition	heat of sublimation (−)

Heat capacity (C): The heat that an object exchanges with its surroundings has the capacity to change the object's temperature. (It can also impact work done by or on the system, and can supply a phase transition without changing temperature.) For a given substance, the amount of heat that must be absorbed or released to result in a given temperature change is called heat capacity. The heat capacity per unit mass is called specific heat, which in practice is either measured at constant volume (C_V) or constant pressure (C_P). The heat capacity per mole is called the molar specific heat (c_V or c_P).

The heat gain required to supply a temperature increase for a given substance mathematically defines heat capacity on the differential scale ($đQ = CdT$). In the case of constant heat capacity, this simplifies to the ratio of the heat absorbed or emitted to the temperature change ($Q = C\Delta T$). Specific heat is heat capacity per unit mass ($đQ = MC_V dT$ or $đQ = MC_P dT$). Molar specific heat is heat capacity per mole ($đQ = nc_V dT$ or $đQ = nc_P dT$).[8]

Mass (M): Conceptually, mass is best defined as a measure of inertia, according to Newton's laws: The more massive an object, the more net external force is needed to give it a certain acceleration. This is a very practical definition to have in mind in most branches of physics. In chemistry and thermodynamics, however, where it is often useful to regard an object as a very large sum of molecules, it is more useful to think of the mass of the object in terms of the sum of the masses of its parts.[9]

The mass of an object is a measure of the object's inertia, which very approximately equals the sum of the masses of its molecules for most problems in thermal physics.

Mole number (n): An object can be thought of as being composed of N molecules or n moles, where one mole is equal to Avogadro's number of molecules. The molar mass of a substance is the mass of one mole of the substance.

---

[8] Material Property Data at www.matweb.com is a convenient online resource for values of specific heat capacities for elements, compounds, and common materials.

[9] Inertia provides the more precise definition, even though it is not as practical in thermal physics and chemistry. The total mass of an object generally does not equal the sum of its parts – since according to Einstein's famous equation, $E = m_0 c^2$, binding energy actually contributes toward the measured mass of an object (nuclear fission is one way to make use of this energy). Thus, the mass of the most abundant isotope of helium does not actually equal the mass of two protons plus two neutrons plus two electrons. However, the sum of the masses of the molecules of an object generally provides a very good approximation for the mass of the object.

The number of molecules of a substance, N, divided by Avogadro's number equals the mole number, n, of the substance. The mass of the substance, M, divided by the mole number equals the molar mass, m, of the substance.

Mole fraction (x_k): If a substance is not pure, but a mixture of chemical components (elements, ions, or molecules), it is useful to work with mole fractions. The mole fraction equals the ratio of the presence of a particular chemical component to that of all of the chemical components combined together. The molar volume is the volume of the substance divided by the sum of the mole numbers of its chemical components.

Mole fraction, x_k, is defined as the number, n_k, of moles of chemical substance k divided by the total number, n, of all of the moles of all of the r chemical substances composing the chemical substance: $x_k = n_k/n$. Note that $\sum_{k=1}^{r} n_k = n$. It follows that the sum of x_k for all r chemical components of a substance is $\sum_{k=1}^{r} x_k = 1$. The molar volume, v, of the substance equals its volume, V, divided by the sum of n_k: $v = V/\sum_{k=1}^{r} n_k = V/n$.

Example. A gas mixture contains 16.8 g of helium and 35.6 g of diatomic oxygen (O_2). Determine the mole numbers and mole fractions for each component of the mixture.

The mole numbers for each substance can be found by dividing its mass by its molar mass (noting that the molar mass of O_2 is twice the molar mass of oxygen):

$$n_{He} = \frac{M_{He}}{m_{He}} = \frac{16.8 \text{ g}}{4.00 \text{ g/mol}} = 4.20 \text{ mol}$$

$$n_{O_2} = \frac{M_{O_2}}{m_{O_2}} = \frac{35.6 \text{ g}}{32.0 \text{ g/mol}} = 1.11 \text{ mol}$$

The mole fractions equal the mole numbers divided by the sum of the mole numbers:

$$x_{He} = n_{He}/n = 4.20 \text{ mol}/5.31 \text{ mol} = 0.791$$
$$x_{O_2} = n_{O_2}/n = 1.11 \text{ mol}/5.31 \text{ mol} = 0.209$$

Volume (V): The space a substance occupies is commonly referred to as volume. Microscopically, however, every substance is almost entirely empty space: The electrons surrounding the nuclei are evidently pointlike elementary particles, and similarly with the quarks that make up the protons and neutrons in the nuclei. More precisely, the molecules at the surface of a substance define a geometric object, and the region of space bounded by this surface represents the volume of space that the object occupies.[10] The shape of the object defines our intuitive sense of volume.

The geometric shape of the surface of an object bounds a three-dimensional region of space which we call volume. A sphere has a volume of $V = 4\pi R^3/3$ and a right-circular cylinder has a volume of $V = \pi R^2 h$.

---

[10] If you really want to be meticulous, this definition needs to be further expanded to account for any macroscopic hollow regions in the interior of the object.

Surface area (A): The geometric shape defined by the surface molecules of a three-dimensional object defines a shape that encloses its volume. The area of this shape is called surface area. The volume enclosed is three-dimensional, yet the surface area is two-dimensional.[11] An open surface also has a surface area (an open surface, though, does not enclose a volume). A cross-sectional area is one type of open surface. A cylinder sliced perpendicular to its axis has circular cross section, for example.

> An object has a geometric shape that is bounded by a two-dimensional surface area. A sphere has a surface area of $A = 4\pi R^2$ and a right-circular cylinder has a surface area of $A = 2\pi R h + 2\pi R^2$. A slice of the object – i.e. the intersection of the object and a plane – represents a cross-sectional area. The circular cross section of a cylinder has area $A = \pi R^2$.

Pressure (P): A measure of how a force is distributed over a surface is provided by pressure.[12] There is greater pressure if the force is perpendicular to the surface, and none if the force is tangential to the surface; also, pressure is greater if more force is exerted per unit area of the surface.

> The pressure exerted by a differential force $d\vec{\mathbf{F}}$ over a differential area dA is defined mathematically as $dF_n = PdA$, where dF_n is the component of the force perpendicular to the surface. If the force is applied evenly and everywhere perpendicular to the surface, this reduces to force per unit area ($P = F/A$).

Conceptual Examples. You can comfortably stomp on the floor with a bare foot, but if there happens to be a nail protruding upward through the floor, even a light step can be painful. The force of the stomp is spread over the sole of the foot, equating to a moderate pressure; a small force applied to the nail, on the other hand, can equate to a very large pressure, since the area of contact is very small. However, it can be comfortable to lie on a bed of nails (yes, on the points of the nails protruding through a board!), since a large number of closely spaced nails can amount to a large surface area. You might not want to stand on the bed of nails with bare feet, though, as in this case the force (your weight) is distributed over a much smaller surface area (just your feet, instead of the length of your body).

The hydraulic press is a simple machine that makes practical use of the concept of pressure. A hydraulic press looks like a U-tube – two columns of liquid connected at the bottom – where one column is much narrower than the other. A piston is placed at the top of each column. According to Pascal's principle, the pressure is the same throughout the liquid. A force pushing the large piston downward thus results in an equal pressure pushing the small piston upward. The pressures are equal, but the forces differ since the areas differ: $P_1 = P_2$ implies that $F_1/A_1 = F_2/A_2$. A small downward push on the small piston therefore results in a large upward push on the large piston. This is very useful for lifting a heavy object like a car.

---

[11] Only two independent generalized coordinates are needed to map out the surface. For example, every point on the surface of a sphere can be specified in terms of just two independent coordinates – longitude and latitude – whereas three coordinates are needed to map out its volume.

[12] The common units for tire pressure, pounds per square inch (PSI), can help you remember this definition. But PSI is not SI! The SI unit of pressure is the Pascal: $1 \text{ Pa} = 1 \text{ N/m}^2$.

$$P_1 = P_2$$

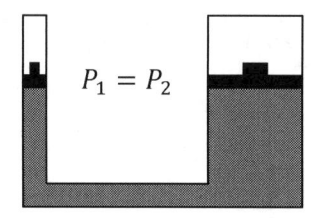

Work (W): A force does work if it contributes toward the displacement of an object. In the case of a gas, pressure exerted on the walls of a container creates the force (equal to the force that the walls exert on the gas, according to Newton's third law), and this pressure does work if the volume changes.

In thermodynamics, work is done by a system when the volume changes under pressure ($đW = PdV$). Work is positive ($đW > 0$) when work is done by the system.

Power (P): Power is the instantaneous rate at which work is being done, or – since energy represents capacity to do work – the rate at which energy is being transferred. Since heat corresponds to a transfer of energy, the rate at which a system is absorbing or releasing heat is also power. It is important not to confuse power (P) with pressure (P), which are both represented by the same symbol.[13]

Power can be expressed as the instantaneous rate at which a system is doing work ($P = đW/dt$), internal energy is increasing ($P = dU/dt$), heat is being added to a system ($P = đQ/dt$), or any other form of energy is being transformed ($P = dE/dt$).[14]

Internal energy (U): A substance can have potential energy associated with the position of its center of mass in an external gravitational or electric field, translational kinetic energy associated with the motion of its center of mass, and rotational kinetic energy for its rotation as a whole about a given axis. Additionally, a substance has internal energy associated with the total kinetic energy and potential energy of its molecules. Thus, a ball at rest on the ground, which would have zero mechanical energy for a mechanics problem, does have internal energy. For example, molecules have chemical energy in the bonds between atoms or atom groups, and the atoms have nuclear energy holding their nuclei together. Internal energy cannot be measured directly, as potential energy is defined through an indefinite integral, $\int \vec{F} \cdot d\vec{s}$, that inherently includes an arbitrary additive constant (conceptually, this corresponds to the choice of reference position); however, change in internal energy can be measured, as this relates to a definite integral, in which that additive constant cancels out.

The internal energy of a substance equals the total energy – potential energy plus translational, rotational, and vibrational kinetic energy – of its molecules, excluding their interactions with external fields (e.g. gravity). The internal energy of a system can be changed by exchanging heat, doing work (or having work done on it), or by exchanging particles with the surroundings (including another system).

---

[13] The distinction should be clear from the context, and seldom do both show up in the same equation.
[14] So for each of these cases, if work is done on the system, the internal energy of the system decreases, or heat is lost by the system, then the power is negative.

Important Distinction. Temperature is actually more directly connected to internal energy than it is to heat. The total internal energy of a system equals the sum of the potential and kinetic energies of all of its molecules, and the average kinetic energy of the molecules is proportional to the absolute temperature. There are two important differences between absolute temperature and internal energy: Internal energy includes potential energies of the molecules, and internal energy involves total, not average, kinetic energy. If the number of molecules remains constant and the sum of their potential energies does not change, then the change in internal energy is proportional to the change in temperature. However, it is possible for one to change and not the other.[15] For example, molecules have latent potential – i.e. potential energy associated with a phase change. During a phase change, the potential energy of the molecules changes, permitting internal energy to change without changing temperature. On the other hand, a change in temperature is usually associated with a change in internal energy, since a change in the average kinetic energy of the molecules generally implies a change in the total kinetic energy of the molecules.[16] Also, if the system gains or loses molecules, this affects the internal energy directly – but only affects temperature if they impact the average kinetic energy.

Conceptual Example. Reconsider the example of two pots of water boiling on a stovetop – where one pot has twice the volume of the other and both pots have the same temperature. The molecules of both pots have the same kinetic energy, on average, while one pot has twice the total kinetic energy. The pot with twice the total kinetic energy has more internal energy.

Enthalpy (H): Heat may be absorbed or released during a chemical reaction. When this occurs at constant pressure, the heat absorbed or released is called the enthalpy. In this context, it is often referred to as the heat content. The enthalpy of a system equals the internal energy of the system plus the product of the pressure and volume. The additional energy term PV may be thought of as energy stored in the surroundings – e.g. objects near the surface of the earth are pushing against atmospheric pressure. Thermodynamics is often concerned with the internal properties of a system, which relate to the internal energy, whereas chemical reactions involve the total energy, including the surroundings, where enthalpy is more useful.

> Heat absorbed or released during a chemical reaction that occurs under constant pressure equals the change in enthalpy of the system. The enthalpy of a system is related to the internal energy of the system by $H = U + PV$.

Entropy (S): The statistical disorder of the molecules of a substance provides a useful thermodynamic measure called entropy. This is not limited to just the geometric arrangement of the molecules, but also relates to physical quantities such as spin angular momentum (some may be spin up, others spin down, and the arrangement of spins can be ordered to some extent or completely random).

---

[15] When we study ideal gases, we will see that their molecules are assumed to have zero potential energy (or, equivalently, that the potential energy of their molecules does not change). Thus, in the case of an ideal gas with constant mole numbers, a change in internal energy implies a change in temperature, and vice-versa.

[16] In principle, the change in the kinetic energy of the molecules can be accompanied by a change in the potential energy of the molecules, allowing temperature to change without changing internal energy.

Entropy provides a quantitative measure of the statistical disorder of a system.

Density (ρ): If the molecules of a substance are distributed uniformly, on average, then the density of the substance equals its mass divided by its volume. If the density is non-uniform – i.e. parts of an object are more dense than others – then the ratio of the mass to the volume is only equal to the density at the infinitesimal level.

Most generally, the density of an object is mathematically expressed as a derivative of its mass with respect to its volume ($dM = \rho dV$). If the density is constant – i.e. the mass is distributed uniformly throughout the volume of the object – then integration of both sides of the equation leads to the more common notion that density equals mass over volume ($M = \rho V$).

Chemical potential (μ_k): If the number, N_k, of atoms (or ions or molecules) of chemical substance k in a system increases or decreases, this affects the energy of the system. Chemical potential refers to the energy gained or lost associated with this change in particle number.[17]

The change in energy of a system that results from gaining or losing a given type of atom is referred to as chemical potential.

Efficiency (e): A device that transforms heat input into a useful form of work output, called a heat engine, inevitably produces some exhaust. The efficiency of such a device equals the ratio of the work output to the heat input from a heat reservoir.

The efficiency of a heat engine equals the ratio of the net work output (W) to the heat input (Q_h) from a heat reservoir during each cycle of operation: $e = W/Q_h$. The efficiency is always less than one because there is always some heat expelled (exhaust) to a cold reservoir: $e = 1 - Q_c/Q_h$.

0.2 Basic Constants that Relate to Thermal Physics

Avogadro's number (N_A): Avogadro's number equals the number of carbon atoms in 12 g of $^{12}_{6}$C. One mole of atoms, molecules, or any entities consists of Avogadro's number of those entities.

Avogadro's number equals $N_A = 6.022 \times 10^{23}$ mol^{-1}.

---

[17] Technically, there is an important distinction between chemical potential, which is internal, and electrochemical potential, which includes external potential energies. The external fields are more relevant for chemistry, whereas thermodynamics is mostly concerned with the internal energies. For a similar reason, internal energy is used more in thermodynamics, while enthalpy is much more common in chemistry.

Universal gas constant (R): The universal gas constant is the constant of proportionality in the ideal gas law, $PV = nRT$. For an ideal gas, it turns out to be the difference between the molar specific heats at constant pressure and volume: $R = c_P - c_V$.

The ideal gas constant equals $R = 8.314$ J/mol/K.[18]

Boltzmann constant (k_B): The Boltzmann constant is the constant of proportionality in the ideal gas law in terms of the number of molecules rather than the number of moles: $PV = Nk_BT$. Therefore, the Boltzmann constant is related to the universal gas constant by $R = N_A k_B$. The Boltzmann constant often shows up as the coefficient of temperature in the arguments of functions in thermodynamics and statistical mechanics; observe that the combination $k_B T$ is dimensionless.

The Boltzmann constant equals $k_B = 1.381 \times 10^{-23}$ J/K.

Coefficient of linear expansion (α): Most substances expand when heated. However, there are notable exceptions: For example, water expands when freezing, which poses a danger of pipes bursting during cold winter nights. A long slender object, such as a rod, expands most significantly along its length. The coefficient of linear expansion (α) quantifies how much it expands for a given temperature change. A two- or three-dimensional expansion can also be expressed in terms of α.

The relative increase (or decrease) in length that a substance expands (or contracts[19]) along its length per temperature change is quantified as the coefficient of linear expansion ($dL = \alpha L_0 dT$ for relatively small expansions).[20]

Thermal conductivity (k): Metals and some other substances are good conductors of heat. Thermal conductivity provides a measure of how well (or poorly, if this value is small) a substance conducts heat.

Thermal conductivity quantifies how well a substance conducts heat.[20]

Emissivity (ϵ): Objects can absorb and emit thermal radiation. The emissivity is a measure of how efficiently a given substance absorbs and emits thermal radiation.

Emissivity quantifies how efficiently a substance absorbs or emits thermal radiation.[20]

Stefan-Boltzmann constant (σ): The Stefan-Boltzmann constant is the constant of proportionality in Stefan's law for the power radiated by a perfect emitter of thermal radiation, called a blackbody. It is related to other fundamental constants by $\sigma = 2\pi^5 k_B^4/(15c^2 h^3)$, where c is the speed of light in vacuum and h is Planck's constant for the energy of a photon of frequency ν ($E = h\nu$). The emissivity of a blackbody is unity: $e_{bb} = 1$.

---

[18] You might know this as a different value from chemistry, where different units (i.e. not SI) are used.
[19] Substances, like water, which expand when freezing, have negative α for some pressure-temperature ranges.
[20] Material Property Data at www.matweb.com is a convenient online resource for values of thermal conductivity, emissivity, and coefficients of linear expansion for elements, compounds, and common materials.

The Stefan-Boltzmann constant equals $\sigma = 5.670 \times 10^{-8}$ W/m^2/K^4.

0.3 Basic Units and Conversions of Thermal Physics

Kelvin (K): The Kelvin is the SI unit for temperature. Expressed in Kelvin, it is referred to as absolute temperature. The Kelvin scale is a simple shift of the Celsius scale to make absolute zero temperature correspond to the zero-point average kinetic energy of the molecules of a substance. The Fahrenheit scale is particularly common in the United States.[21,22]

$$T_K = T_C + 273.15 \text{ K}$$
$$T_F = 9T_C/5 + 32° \text{ F}$$

Cubic meter (m^3): The SI unit for volume is the cubic meter – not a liter (L), milliliter, or cubic centimeter.[23] Cubic feet and gallons are more common in the United States.

$$1 \text{ m}^3 = 10^6 \text{ cm}^3 = 10^3 \text{ L} = 35.31 \text{ ft.}^3 = 264.2 \text{ gal}$$

Kilogram (kg): The SI unit for mass is the kilogram. The kilogram is not a dimensionally viable unit for weight, however, which is a force (measured instead in Newtons).[24] Nor is the pound (lb.) a dimensionally viable unit for mass in British units. Rather, the slug is the British (or English or imperial) unit for mass: One slug equals a pound times a second-squared per foot.

$$1 \text{ slug} = 14.59 \text{ kg}$$

Newton (N): The SI unit of force is the Newton (1 N = 1 kg m/s^2). The British equivalent is the pound.

$$1 \text{ lb.} = 4.448 \text{ N}$$

---

[21] Unlike Celsius and Fahrenheit, Kelvin is not referred to in degrees: Cf. 273.16 K and 0.01° C.
[22] If you know that 0° C corresponds to 32° F and 100° C to 212° F (freezing and boiling points for water), it's easy to work out that the temperature in Fahrenheit equals 9/5 the temperature in Celsius plus 32° F.
[23] When relating cubic meters to cubic centimeters, it is important to realize that the conversion factor is cubed – i.e. there are a million (10^6) cc in a m^3, not a hundred (10^2). To see this, consider a square centimeter, which is a grid 10 mm across and 10 mm high, so there are 100 (or 10^2) mm^2 in a cm^2, for example.
[24] Strictly speaking, it is impossible to 'convert' kilograms to Newtons or pounds, since one is a unit of inertia (a scalar quantity) and the other a unit of force (a vector). The units pertain to fundamentally different physical quantities with different dimensionalities. It is, however, possible to determine the weight of an object in Newtons (or pounds) from its mass in kilograms (or slugs) through the formula $W = mg$. Many conversion tables apply this formula to offer an equivalent weight for a given mass assuming gravitational acceleration near earth's surface. Just remember, if you visit a popular science exhibit with scales that tell you your weight on other planets or moons, if the scale claims to tell you your 'weight' in 'kilograms,' just shake your head and groan.

Joule (J): The SI unit for work or energy, including heat, is the Joule (1 J = 1 N m = 1 kg m^2/s^2). Other common units for energy include ergs, calories (cal), and British thermal units (BTU).[25]

$$1\text{ J} = 10^7 \text{ ergs} = 0.2389 \text{ cal} = 0.9478 \text{ BTU}$$

Watt (W): The SI unit for power, or the rate at which work is done or energy is transferred, is the Watt. Motors in the United States generally use units of horsepower (hp).

$$1 \text{ hp} = 745.7 \text{ W}$$

Pascal (Pa): The SI unit for pressure is the Pascal (Pa), which is shorthand for a Newton per square meter. When working with pressures comparable to atmospheric pressure, or when concerned with the pressure exerted on an object by the atmosphere, atmospheres (atm) are practical units for pressure. The tire pressure of cars in the United States is often measured in pounds per square inch (PSI).[26] Millibars (mbar) and millimeters of mercury (mmHg, or torr) are other common units of pressure.

$$1 \text{ Pa} = 1 \text{ N/m}^2 = 9.869 \times 10^{-6} \text{ atm} = 10^{-5} \text{ bar} = 7.501 \times 10^{-3} \text{ mmHg} = 1.450 \times 10^{-4} \text{ psi}$$

Mole (mol): The mole is the SI unit for the number of entities (such as atoms, ions, or molecules) in a statistically large sample (usually a substance composed of molecules). One mole of a substance equals Avogadro's number of entities.

$$1 \text{ mole} = 6.022 \times 10^{23} \text{ entities}$$

0.4 Basic Measurements that Relate to Thermal Physics

Melting and freezing points: The melting point of a substance refers to the temperature at which it changes phase from solid to liquid, for a given pressure, by absorption of heat.[27] The freezing point is the same except for changing phase from liquid to solid by releasing heat. The substance coexists in the solid and liquid phases at its melting point (and similarly for its freezing point if it is instead being cooled). The melting and freezing points are approximately the same for most, but not all, substances.

---

[25] Work and all forms of energy, including heat, are measured in Joules. However, torque has units of Newtons times meters, as do work and energy, and yet torque is not expressed in Joules because it is not a measure of work or energy. Torque is an angular quantity, and there are some radians that get 'swept under the rug' in the context of meters, as with the arc length formula ($ds = rd\theta$ along a circular arc), where literally a meter equals a radian times a meter (a radian expresses a fraction of the distance traveled around a circle). The work done by a torque, $W = \int \tau d\theta$, also has units of Newtons times meters, and this quantity, unlike torque itself, is expressed in Joules.
[26] So why isn't it common to use the Greek symbol ψ (psi) to represent the unit PSI?
[27] First, the solid substance must absorb enough heat to reach the melting point, and then it needs heat according to its latent heat of fusion to make the phase transition; and similarly for boiling or sublimation.

The freezing point of pure water is 273.15 K at atmospheric pressure.[28]

Boiling and condensation points: The boiling point of a substance refers to the temperature at which it changes phase from liquid to gas, for a given pressure, by absorption of heat.[27] The condensation point similarly corresponds to a phase transition from gas to liquid by releasing heat.

The boiling point of pure water is 373.17 K at atmospheric pressure.[28]

> **Important Distinction.** At the boiling point, the process of boiling occurs, in which molecules throughout the volume of the liquid change into gas. You can see this, for example, when water boils and water vapor bubbles rise to the top. The process of evaporation can occur at temperatures below the boiling point, in which molecules at the surface escape in the gaseous phase. Thus water in a cup on a hot day may slowly decrease in volume by evaporation.

Sublimation and deposition points: The sublimation point of a substance refers to the temperature at which it changes phase from solid to gas, for a given pressure, by absorption of heat.[27] The deposition point similarly corresponds to a phase transition from gas to solid by releasing heat.

Triple points: The triple point of a substance refers to the combination of temperature and pressure for which the substance coexists in the solid, liquid, and gaseous phases. At the melting point, the substance coexists in the solid and liquid phases as it lies on the brink of a phase transition; similarly, at the boiling point, the substance coexists in the liquid and gaseous phases, and at the sublimation point, it coexists in the solid and gaseous phases. On a pressure-temperature diagram (P-T), the three phase transitions (melting/freezing, boiling/condensation, and sublimation/deposition) are each represented as curves, since the temperature at which a given phase transition occurs depends upon the pressure. These three phase transition curves, corresponding to boundaries between two phases, intersect at the triple point. The triple point thus serves as a fundamental characteristic of a given substance.

The triple point of pure water is 273.16 K at a pressure of 0.61173 kPa.[28]

Dew point: The dew point refers to the temperature at which water vapor in the air condenses into liquid form, for a given pressure. The dew point depends not only upon the pressure, but also on the relative humidity of the air.

Standard temperature and pressure: The values for Standard Temperature and Pressure (STP) are not actually standard: The National Institute of Standards and Technology works with room temperature (20° C)[29] and atmospheric pressure (1 atm), while international systems often adopt the freezing point of water (0° C) and 100 kPa (which is close to one atmosphere).

---

[28] The melting points, boiling points, and triple points of elements and a variety of compounds are available from the National Institute of Standards and Technology at www.nist.gov.
[29] Room temperature is itself sometimes taken to be 25° C in different references.

Molar masses: The molar mass of a substance is the mass of one mole of that substance, which equals the mass of the substance divided by the mole number: $m = M/n$.

Chemical composition: The chemical make-up of a substance that obeys the law of definite proportions can be represented by its chemical formula. For example, pure water consists of two-parts hydrogen to each part oxygen (H_2O). Here is another important example: At STP, helium gas consists of individual helium atoms (He), while oxygen consists of pairs of atoms bound together (O_2): Helium is monatomic, while oxygen is diatomic. The diatomic gases are easy to find: There is hydrogen, and the others form a seven on the periodic table: nitrogen, oxygen, fluorine, chlorine, bromine, and iodine.[30]

Substance	Chemical composition[31]
pure water	H_2O
ammonia	NH_3
baking soda	$NaHCO_3$
methane	CH_4
table salt	$NaCl$
table sugar	$C_{12}H_{22}O_{11}$

0.5 Basic Thermodynamics States and Processes

Microstate: A state of a system that is specified in terms of the properties of the individual molecules of the system is called a microstate. For example, a tabulation of the spins of the atoms in an array or the velocities of molecules in a gas are microscopic viewpoints.

Macrostate: A state of a system that is specified only in terms of coarse (compared to the microscopic behavior of the individual molecules) spatial and temporal averages is called a macrostate. For example, measurements of pressure, temperature, and volume pertain to a macroscopic viewpoint.

Thermal equilibrium: Two systems in thermal contact with one another achieve thermal equilibrium at such time as their temperatures become equal.

Mechanical equilibrium: Two systems achieve mechanical equilibrium when their pressures equalize. The net external force and torque are both zero for a system of particles in mechanical equilibrium.

Diffusive equilibrium: Two systems achieve diffusive equilibrium when their chemical potentials become equal. This is also referred to as equilibrium with respect to matter flow.

---

[30] The last halogen, astatine, is very rare and short-lived.
[31] These chemical formulas correspond to the most abundant forms at STP. Oxygen, for example, can also form ozone O_3.

Phase equilibrium: A system can coexist in two phases at once – if the temperature and pressure are just right that the system lies at a boundary between two phases. In this case, phase equilibrium is reached when their chemical potentials equalize.

Chemical equilibrium: A system may undergo chemical reactions. The reactants and products dynamically stabilize (the net concentrations and chemical activities of the reactants and products become constant, or the forward and reverse reaction rates balance) when chemical equilibrium is reached.

Radiative equilibrium: Objects continuously emit electromagnetic radiation as a result of possessing temperature (as it causes vibrations of the molecules of the object). The higher the temperature of the object, the greater the rate at which it radiates heat. Objects also continuously absorb electromagnetic radiation from their surroundings. This absorbed radiation is a form of heat gain that has the effect of raising the object's temperature, while the radiation it emits is a form of heat loss that has the effect of lowering the object's temperature. The power for the radiation an object absorbs depends on the temperature of the surroundings, whereas the power for the radiation an object emits depends on the temperature of the object. When the rates at which thermal radiation is emitted and absorbed by an system are equal, the system is in radiative equilibrium.

Thermodynamic equilibrium: A system that is in thermal equilibrium, mechanical equilibrium, diffusive equilibrium, and radiative equilibrium is said to be in thermodynamic equilibrium.

Non-equilibrium state: A system passes through a series of non-equilibrium states prior to reaching equilibrium. Systems in non-equilibrium states tend have time-dependent properties and tend to be non-uniform. Thus, in contrast to equilibrium states, some macroscopic variables, such as temperature and pressure, are not well-defined as they may not apply to the system as a whole.

Steady state: Similar to various types of equilibrium, a system that is in a steady state has multiple quantities that remain constant as time increases. Yet a system that is in a steady state is not necessarily in equilibrium. In particular, the net external force may be nonzero, matter may flow, and entropy may increase over time.

Transient state: A system for which equilibrium is disturbed and later reaches equilibrium (not necessarily the same equilibrium conditions) is said to be in a transient state in the interim period.

Ground state: The state of a system with the absolute minimum energy level is its ground state.

Excited state: A system where the energy level exceeds the ground state energy is in an excited state.

Adiabatic: There is no heat absorbed or lost during an adiabatic process ($đQ = 0$). It follows that the change in internal energy for a system with constant mole numbers equals the negative of the work done by the system along an adiabat: $dU = -đW$.[32]

Isothermal: The temperature of the system remains constant during an isothermal process ($dT = 0$).

Important Distinction. It may seem very intuitive to want to associate absorption of heat with an increase in temperature and heat loss with a decrease in temperature, as suggested by the equation for heat capacity. However, the heat that an object exchanges with its surroundings does not necessarily go toward changing the temperature of the object – it is also related to the work done and to latent heat of transformation. For example, a system may absorb heat and use this thermal energy exclusively to do work, without changing temperature, or a substance may absorb heat and use this thermal energy to undergo a phase transition rather than change temperature.[33]

The important point is that a process that is isothermal may not be adiabatic, and a process that is adiabatic may not be isothermal. That is, a system can absorb heat without changing temperature, and a system can change temperature without exchanging any heat.[34]

Isobaric: The pressure remains constant for a system during an isobaric process ($dP = 0$). Thus, pressure can come out of the work integral along an isobar: $\int đW = \int P dV = P\Delta V$.

Isochoric: The volume occupied by a system is constant during an isochoric process ($dV = 0$). This implies that the system does no work along an isochor: $đW = PdV = 0$. Furthermore, the change in internal energy for a system with constant mole numbers equals the heat added to the system for an isochoric process: $dU = đQ$.[35] A constant-volume process is also called isovolumetric.

Steady-state:[36] The internal energy of a system remains constant over the course of a steady-state process ($dU = 0$). It follows that the heat absorbed by a system with constant mole numbers equals the work done by the system for a steady-state process: $đQ = đW$.[37]

---

[32] As we shall learn in the last section of this chapter, for a system with constant mole numbers, the internal energy of the system, the heat absorbed by the system, and the work done by the system are related through the first law of thermodynamics as $dU = đQ - đW$.

[33] You don't have to guess. The first law of thermodynamics helps you relate heat to other forms of energy.

[34] In fact, when we get to ideal gases, we will see that an isothermal process means that internal energy, but not heat, is constant.

[35] We often speak of adiabatic and isothermal expansions or compressions, but be careful not to add the words expansion or contraction after isochoric, as you would create an oxymoron!

[36] A note of mathematical English: The hyphen can be used to combine an adjective (descriptive word) and noun (being, object, or place) together to serve as an adjective; hence the distinction between steady state (a thermodynamic state) and a steady-state process (a thermodynamic process). Compare with a monkey that has a mass of 40 kg and a 40-kg monkey.

[37] Note the distinction between a steady-state process and a cycle. The internal energy may increase and decrease over parts of a cycle, so long as these increases and decreases balance out – i.e. the net internal energy change is zero for a cycle. For a steady-state process, the internal energy is constant throughout.

Cyclic: The internal energy of a system returns to its initial value at the completion of a thermodynamic cycle – or a cyclic process.[38] A cycle corresponds to a closed path on a pressure-volume (P-V) diagram. The change in internal energy for a complete cycle (which is a closed integral) equals zero. Hence, for a system with constant mole numbers, the heat added to the system equals the work done by the system over a complete cycle: $\Delta Q = \Delta W$.[39]

Free expansion: If a gas is allowed to expand freely, by definition the pressure that would ordinarily contain the gas has been removed; hence, no work is done during a free expansion ($đW = 0$). This implies that the change in internal energy for a system with constant mole numbers equals the heat added to the system for a free expansion: $dU = đQ$.[40]

Isentropic: The entropy of a system remains constant over the course of an isentrope ($dS = 0$).

Isenthalpic: The enthalpy of a system is constant along an isenthalp ($dH = 0$).

Spontaneous: A process that occurs without being driven by an energy source is spontaneous. Spontaneous processes result in an increase in total entropy. For example, sugar dissolves spontaneously when it is poured into tea and heat flows spontaneously when two objects at different temperatures are placed in thermal contact.

Quasistatic: A process that is carried out through a series of infinitesimal (in practice, this means slight) disturbances from equilibrium and slowly enough that the system establishes equilibrium (not necessarily the exact same equilibrium conditions) between infinitesimal disturbances is termed quasistatic. This series of differential changes eventually leads to finite changes in some of its thermodynamic properties. A quasistatic process is special in that it can change a system from one equilibrium state to another without passing through a series of non-equilibrium states. Thus, the macroscopic observables remain well-defined over the course of a quasistatic process.

Reversible: A quasistatic process that can be carried out in reverse is called a reversible process. Not all quasistatic processes are reversible: For example, there may be frictional or dissipative forces during a quasistatic process that prevent the identical process from occurring in reverse.

Irreversible: Any process that is not quasistatic (and thus includes a series of non-equilibrium states) or features frictional or dissipative forces is irreversible.

---

[38] Whereas pressure is constant for an isobar and temperature is constant for an isotherm, for example, it would be incorrect to state that internal energy is constant for a cycle. Indeed, internal energy generally increases and decreases over the course of the cycle. It is the sum of these increases and decreases that totals to zero for a complete thermodynamic cycle. Thus, it is most precise to state that the final internal energy equals the initial internal energy for the complete cycle.

[39] This relation holds at the finite level, in general, but not in differential form.

[40] A free expansion is unique in that the volume changes, yet no work is done.

Conceptual Questions

The selection of conceptual questions is intended to enhance the conceptual understanding of students who spend time reasoning through them. You will receive the most benefit if you first try it yourself, then consult the hints, and finally check your answer after reasoning through it again. The hints and answers can be found, separately, toward the back of the book.

1. Two pots contain boiling water. One pot has twice the volume of water as the other. Would it be safer to dip your finger into the water of one pot compared to the other? Explain. Would either be more effective in melting ice from stairs on a cold winter day by pouring the water onto the stairs?

2. A washer is placed on a driveway and then struck with a sledgehammer. What happens to the temperature of the washer? If it changes, explain why.

3. Compare both the temperature and internal energy of an iceberg to that of a pot of boiling water. Explain. Which has the potential to transfer more heat?

4. For a given substance, how do you expect its latent heat of fusion to compare with its latent heat of vaporization? Explain.

5. Why is it more dangerous to receive a burn from steam at $100°$ C than from water at $100°$ C?

6. From everyday experience, what evidence do you have that ice, wood, and oil are less dense than water? What is a simple way to find out if an object is more dense or less dense than water? From the same line of reasoning, how dense must a fish be compared to water? What might a fish do to rise toward the surface or sink toward the bottom?

7. Would an object weigh less, the same, or more in air or in a vacuum? Explain. In both cases, assume that the object is near the surface of the earth.

8. An ice cube that is 95% as dense as water is placed in a cup of water. Why does the ice cube float at the top, and why does part of the ice cube stick out into the air above? What fraction of the ice cube sticks out of the water? What – if anything – happens to the water level as the ice cube melts? Explain.

9. Consider a quantity similar to density that is defined as mass per unit surface area instead of mass per unit volume. Why would this similar physical quantity be very impractical compared to density? How could an object have finite density, yet infinitesimal mass per unit surface area? Is it possible for an object to have infinitesimal density, but finite mass per unit surface area? Explain.

10. Is it better to crack open an egg by gently banging it against the top of a table or the edge of a table? Explain.

11. Consider a quantity similar to pressure that is defined as force per unit volume instead of force per unit area. Why would this similar physical quantity be very impractical compared to pressure? Would it be possible for an object to have finite force per unit volume, yet infinitesimal pressure? Explain.

12. In chemistry, the molarity of a solution is defined as the number of moles of solute per volume (usually, per liter) of solution. Its common chemistry units are mol/L. The molarity thus provides a measure of the concentration of the solution. The units of molarity are the reciprocal of the units of molar volume, yet molarity cannot be defined as inverse molar volume. Why not? By comparing the two definitions, determine a formula for computing the molarity of a solution given its molar volume.

13. In chemistry, the molality of a solution (not to be confused with the similarly spelled term, molarity) is defined as the number of moles of solute per mass (usually, per kilogram) of solvent. Its SI units are mol/kg. Like molarity, the molality thus also provides a measure of the concentration of the solution. Molality has the same SI units as inverse molar mass, yet molality cannot be defined as the inverse of molar mass. Why not? By comparing the two definitions, determine a formula for computing the molality of a solution given its molarity.

14. Consider boiling a pot of water on a stove (with no lid, of course – why?). How would the temperature of the boiling point be affected if the air pressure was higher, or lower, than standard atmospheric pressure? Explain.

15. Why do lakes freeze from the top down, rather than from the bottom up, on a cold night?

16. The equations of thermal physics are expressed in terms of absolute temperature, for which the value must be expressed in degrees Kelvin – not degrees Celsius or Fahrenheit. However, the coefficient of thermal expansion can be expressed in units of per unit Kelvin or per unit Celsius. What makes this possible?

17. Microscopically, what is the distinction between a gas of monatomic oxygen, O, diatomic oxygen, O_2, and ozone, O_3? If you put oxygen atoms in a container at STP, which type of gas will form?

Practice Problems

The selection of practice problems primarily consists of problems that offer practice carrying out the main problem-solving strategies or involve instructive applications of the concepts (or both). The practice problems of this chapter, however, which is dedicated to basic introductory ideas, mostly involve practice with conversions and a few basic relationships – yet may still require you to think your way through the solutions. Many of the concepts of this chapter will not appear in practice problems until later chapters, where related ideas are discussed in much more depth – for example, problems associated with specific heat. You will receive the most benefit if you first try it yourself, then consult the hints, and finally check your answer after working it out again. The hints to all of the problems and the answers to selected problems can be found, separately, toward the back of the book.

Mole number, molar mass, molar volume, mass, density, volume, and surface area

1. How many atoms of each type of element are present in 1.8 g of baking soda ($NaHCO_3$)?

2. A cup contains 423 g of water at STP. What is the molar mass of the water? How many water molecules are in the cup? What is the molar volume of the water? What is the volume of the water?

3. A packet of table sugar ($C_{12}H_{22}O_{11}$) is stirred into a cup of water, which has 20.0 times the mass of the table sugar. What are the mole fractions of the table sugar and the water in the solution?

4. How many grams of diatomic hydrogen (H_2) and of diatomic oxygen (O_2) are needed to make 3.2 L of water at STP?

5. A solid sphere, solid cube, and solid right-circular cylinder with a height equal to its diameter are all made out of the same material and all have the same mass. Find the ratio of their surface areas.

Units and conversions

6. The book *Fahrenheit 451* was named after the temperature at which paper burns. What would the title of this book be in SI units?

7. Is it possible for a temperature to have the same numerical value when it is expressed in degrees Celsius and in degrees Fahrenheit? If so, find this value; if not, explain why it is not possible. Repeat for Kelvin and Fahrenheit. Repeat for Kelvin and Celsius.

8. A temperature has twice its numerical value when it is expressed in degrees Fahrenheit compared to when it is expressed in degrees Celsius. How many times greater is this numerical value, compared to its numerical value in degrees Fahrenheit, when the temperature is expressed in degrees Kelvin?

9. A pressure is expressed as 42 lb./ft.^{2} Express this in psi, N/m^2, atm, mbar, mmHg, and mtorr.

10. Determine the SI units of $\sqrt{P/\rho}$, where P is pressure (not power).

11. Determine the SI units of $k_B T/(d^2 P)$, where d is a distance and P is pressure (not power).

12. Determine the SI units of the constant, a, in the equation $S = a^{1/4} U^{3/4} V^{1/4}$, given that the SI units for entropy are J/K.

13. A system obeys the equation $S^4 = 64nUV^2/a^2$, where a is a constant. The same system also obeys the equation $T^2 = an^b U^c V^d$, where b, c, and d are integers. What must be the values of b, c, and d in order for this equation to be dimensionally viable?

1 Basic Relationships of Thermal Physics

1.1 The Four Laws of Thermodynamics

Zeroth law of thermodynamics: When two objects at different temperature are placed in thermal contact with each other (i.e. they can exchange thermal energy), assuming that the system of two objects is isolated from external influences, heat (thermal energy) spontaneously flows from the higher-temperature object to the lower-temperature object. This exchange of heat warms the cooler object and cools the warmer object until thermal equilibrium is attained. The two objects have the same temperature once thermal equilibrium is reached. This fundamental notion establishes the measurement of temperature as a test of thermal equilibrium.

In order to measure the temperature of the two objects, a third object – a thermometer – is needed. The thermometer can be placed in contact with each object. The thermometer is read when thermal equilibrium is achieved between each object and the thermometer. If both objects are in equilibrium with the thermometer at the same temperature, it can be concluded that the two objects are in thermal equilibrium with each other. This is the essence of the zeroth law of thermodynamics, which serves as the basis for measuring temperature.

According to the zeroth law of thermodynamics, if object A is in thermal equilibrium with object C and object B is also in thermal equilibrium with object C, then objects A and B are in thermal equilibrium with one another. This principle is as fundamental to thermodynamics as the transitive property is to mathematics, which is very similar. The transitive property of mathematics states that if $a = c$ and $b = c$, then $a = b$.

> According to the zeroth law of thermodynamics, if each of two objects, A and B, are in thermal equilibrium with a third object, C, then the two objects, A and B, are also in thermal equilibrium with one another. In this way, a thermometer may represent the third object, C, and can serve as a measure of the temperature of an object as well as the means of establishing thermal equilibrium between multiple objects.

First law of thermodynamics: We will explore the first law mathematically in much depth later in this book (especially, in Sec. 2.5). For now, we will focus on the conceptual significance of this law.

Energy is conserved for a completely isolated system; or, more generally, the total energy of a system plus energy exchanges between the system and its surroundings is constant. Conservation of energy for thermal processes involves internal energy, heat, work, and chemical potential. For a system with constant mole numbers, only internal energy changes, heat, and work need to be considered. In this case, heat added to a system can increase the internal energy of the system and/or allow the system to do work. Heat represents a transfer of thermal energy, and energy is the ability to do work, so it is easy to see that the relationship between internal energy, heat, and work corresponds to conservation of energy.

The first law of thermodynamics expresses conservation of energy in terms of macroscopic measurements of systems of large numbers of particles. For systems with constant mole numbers, the increase in internal energy of a system equals the heat absorbed minus the work done by the system. The change in internal energy is positive if the internal energy increases, heat is positive if heat is absorbed, and work is positive if it is done by the system; instead the change in internal energy is negative if internal energy decreases, heat is negative if it is lost, and work is negative if it is done on the system (in which case the subtraction of negative work effectively makes a plus sign in the equation).

Note that the first law does not state that the energy of a system is constant; this is only true if the system is completely isolated. Rather, the system may gain or lose energy through energy exchanges (namely, heat) between the system and the environment, which is accounted for as part of the first law. Any nonconservative work that may be done (e.g. friction between a piston and the walls of a container) by resistive or dissipative forces is included in such heat exchanges – i.e. it counts as part of the energy that the system loses to (or gains from) its surroundings.

The internal energy change depends upon the temperature change, since temperature is a measure of the average kinetic energy of the molecules, and also depends upon any change in the potential energy of the molecules. For an isochoric or isobaric process, heat may be computed using the equations for specific heat capacity at constant volume or pressure, respectively; however, for other processes, these formulas are not useful (since in those cases neither the volume nor the pressure is constant). If a substance undergoes a phase transition, the heat associated with the phase transition can be related to the latent heat of transformation. The work done depends upon the pressure and the volume change (but generally does not equal the product of the two – that's only the case for work done along an isobar).

> According to the first law of thermodynamics, the heat absorbed by a system with constant mole numbers minus the work done by the system is equal to the change in the system's internal energy. The heat is negative if instead it is released by the system, and the sign of the work changes if work is instead done on the system.[41]

Second law of thermodynamics: Of the many thermodynamic processes that would obey the first law, only some are ever observed to occur in nature. The first law represents the observation that energy is conserved in any thermodynamic process that does occur. The second law represents the observation that many thermodynamic processes that would otherwise conserve energy do not occur in nature. We will explore the second law in more detail, and quantitatively, in Chapter 5.

For example, spontaneous processes occur only one way – i.e. they are irreversible. When ice cubes are placed in beverages at room temperature, heat is observed to flow spontaneously from the liquid to the ice cubes. Although heat could flow from the ice cubes to the liquid, thereby cooling the ice cubes and warming the beverage, without violating the first law of thermodynamics, heat is never observed to flow spontaneously[42] from a colder temperature to a warmer temperature. When a sugar packet is emptied into a cup of tea, the sugar tends to disburse and dissolve in the liquid; if the tea is further stirred with a spoon, the dispersion increases throughout the liquid and more solute is dissolved. But good luck stirring the tea so as to get the sugar back into a single lump!

---

[41] This is sometimes paraphrased as, "You can't get ahead" – i.e. you can't get out more energy than you put in.
[42] However, it can be 'driven' to flow the other way. Yet, this driven flow is not the 'reverse' of the natural flow.

A process need not be spontaneous to be irreversible. After dragging a heavy box along the ground, the ground and box are slightly warmer from the friction. Retracing the path, dragging the box back over the same path, however, produces even more heat from friction, rather than returning the system to its original state. An ice cube may be placed in an empty beaker and the beaker may be heated until the ice melts into water, and although the water may be poured into an ice tray and placed in a freezer to make a new ice cube with the original temperature and pressure, neither process is the reverse of the other. The series of heat changes that occurs over the course of one process does not correspond to the reverse of the other series. This irreversibility is the inherent problem with trying to freeze something to preserve it, then melt it later when it is needed (it won't be quite the same) – e.g. suppose we discovered a frozen caveman and wished to 'un-freeze' him.

There is a cost for trying to achieve the reverse effect of an irreversible process. For example, it is possible to do work for no other purpose than creating heat, and to repeat a thermodynamic cycle over and over while transforming 100% of the work input into heat output. However, the reversal of this, repeating a thermodynamic cycle over and over while transforming heat input into useful work, cannot be done with 100% efficiency; rather, the cost of carrying out this reverse cycle is that there must be some exhaust – and so it is not quite the 'reverse.' It is easy to do work to produce heat (like rubbing your hands together to keep warm on a cold day). Producing heat from work can be done 'for free,' while there is a 'cost' of using heat to do useful work (the inevitable exhaust, which prevents the efficiency from equaling 100%). The main idea behind each case is reversed – i.e. doing work to produce heat, or using heat to do useful work – yet the two cases are not the 'reverse' of one another (as the exhaust does not have a counterpart in the reverse case).

The first law dictates whether or not a process can occur at all, while the second law explains which way the process can occur – i.e. if it will occur forward or backward, and whether or not the forward and backward processes are the thermodynamic reverse of one another. The second law explains why heat can spontaneously flow from an object at higher temperature to an object at lower temperature, but not from lower to higher temperature. It explains why 100% of work can produce heat, but only a fraction of heat can be used to do useful work. It also reveals the distinction between processes that appear to achieve reverse effects, yet are not really the reverse of one another.

The underlying concept behind the second law is entropy. Entropy is a macroscopic property that provides a quantitative measure of the statistical disorder of the molecules of a system. Any macroscopic system inherently consists of a very large number of molecules. The system can be in any one of an even larger number of microstates. Each microstate specifies measureable properties of each molecule, such as kinetic energy or spin angular momentum. Very, very few of such microstates will be very ordered compared to the very, very large number of very disordered states. A microstate where each particle has the same kinetic energy, for example, is extremely ordered, but also extremely unlikely (you have better odds of winning the lottery – twice in a row!). It is much more probable to find the system in one of the multitude of microstates in which there is a random distribution of kinetic energies.

Entropy – like some other thermodynamic quantities, such as temperature and pressure – is well-defined only for a system that is in equilibrium. In equilibrium, every possible microstate is assumed to be equally likely – yet certain groups of microstates are much more likely than others. Given that each microstate is equally likely, the system is much more likely to be in a disordered microstate than in an ordered microstate. The reason is that there are simply many, many more disordered microstates than ordered microstates.

As an example, imagine a rectangular vacuum chamber. If just two molecules enter the chamber, there is a 50% chance of finding both molecules in the same half of the room. With four molecules, the chance of finding all four on the same side is 12.5%. With 8 molecules, it is 0.78%. With 16 molecules, the odds reduce to 100,000 to 3. With 32 molecules, it becomes less than one part in a trillion. Just imagine Avogadro's number of particles! Entropy is thus greater for a system that has more microstates available (in this example, there is more order when all molecules are on the same side).

The entropy of an isolated system has a natural tendency to increase. Put another way, a system tends to approach maximum disorder. If a system is slightly ordered, a natural disturbance or fluctuation is more likely to send the system into a more disordered microstate than a more ordered microstate, as there are many more disordered microstates available.[43] Spontaneous processes therefore result in an increase in entropy. Spontaneous processes are irreversible because they only occur in the direction that results in an increase in entropy. As time increases, the entropy of a spontaneous process increases.[44]

Imagine carrying a stack of papers to your car in a parking lot, when a sudden gust of wind blows the papers. Chances are that the papers will scatter over the parking lot. The papers might not be too disordered initially, but you will probably be in a hurry to pick them up, knowing that as time increases, any sudden disturbance is likely to scatter the papers even more.

All irreversible processes result in an increase in total entropy (that of the system plus that of the surroundings). Heat spontaneously flows from higher to lower temperature, but not vice-versa, because this case results in an increase in entropy: The molecules are more ordered when they can be classified according to one of two possible temperatures than they are if they all have the same equilibrium temperature; thus heat spontaneously flows in the direction that results in a single temperature.[45] Adding sugar to tea, the disorder increases as the sugar disperses and dissolves. Even just stirring the mixture with a spoon, there is more order during the swirling motion than after it settles. The examples with dragging the box and melting an ice cube in a beaker also result in an increase in total entropy. If the process occurs one way and the entropy increases, the reverse process would require a decrease in entropy – and therefore the reverse process is not observed to occur in nature (because it would violate the second law of thermodynamics).

---

[43] All individual microstates are equally probable, unless there is some agent directing otherwise (e.g. spin up and spin down are each 50% likely, unless there is an external magnetic field supplying a preference). Equal likelihood among microstates does not mean that it is equally probable for a system to be in an ordered microstate as a disordered microstate, however, because there are many more disordered states than ordered states.

[44] For this reason, entropy is often called "the arrow of time." As such, measurements of entropy could serve, in principle, as a means of monitoring time.

[45] This should seem counterintuitive at first, and contradictory to the previous example where we stated that there is more order when all of the molecules reside on the same side of the box. Consider the following resolution carefully, and compare it to the previous box example. Two subsystems with N_1 and N_2 particles at temperatures T_1 and T_2, respectively, have fewer microstates available, combined, than when they join to form a single system with $N = N_1 + N_2$ particles at the same temperature T. To see this, imagine that you have four red dice stored in one drawer and six blue dice stored in another drawer. The system with four red dice has $4^6 = 4096$ microstates available – each permutation of four six-sided dice – and the system with six blue dice has $6^6 = 46,656$ microstates available. If you combine the two systems together and the dice remain different color, there are 50,752 combined microstates, as the dice can be distinguished by color. However, if you paint all of the dice green, there are now $10^6 = 1,000,000$ microstates to choose from – much more disorder.

If a thermodynamic process for a macroscopic system begins in one equilibrium state and ends in another equilibrium state, the process will only be observed to occur naturally in the direction(s) (forward or backward) in which the total entropy (of the system plus the surroundings) does not decrease. A process may be driven to occur in a direction that decreases the total entropy, but only at a cost (doing work, whereas the natural process is 'free,' and at the expense of exhaust – i.e. undesired output); and if so, this driven process is not actually the 'reverse' process (e.g. the driven process must result in some exhaust, which is not present in the original process). A gas will quickly disperse to occupy a vacuum, whereas it is hard work to create a vacuum. A gust of wind quickly blows a stack of papers across a parking lot, but you may do some work to decrease the entropy.

This is one form of the second law of thermodynamics: The total entropy (of the system plus the surroundings) can increase or remain the same, but it cannot decrease. The entropy of the system or the surroundings may decrease, but only if the entropy of the other increases by at least the same amount. This form of the second law of thermodynamics explains which processes that obey the first law of thermodynamics may occur in nature.

Reversible processes, on the other hand, occur both forwards and backwards. The total entropy (system and surroundings) remains constant over the course of a reversible process. There is no preferred direction since neither one increases the statistical disorder.[46] All reversible processes are quasistatic; furthermore, all physical processes can only be approximately reversible at best (since it is not possible to completely remove all frictional and dissipative forces).

If the volume of a cylindrical container of gas is suddenly doubled by quickly pulling on a piston, the gas expands. Quickly pushing on the piston, the volume can be halved, returning the system to its initial volume. After quickly changing the volume, the system passes through a series of non-equilibrium states before attaining equilibrium. This series of non-equilibrium states leads to more statistical disorder. The system is no longer in the same thermodynamic state since the entropy has increased. Only a quasistatic process may be reversible, and then only in the absence of frictional or dissipative forces, as in this case the system never deviates more the infinitesimally from equilibrium.

A similar expansion can be achieved via a reversible (and therefore quasistatic) process, in which the piston is pulled only infinitesimally, the system settles back into equilibrium, the piston is pulled infinitesimally again, equilibrium is reached again, and this is repeated numerous times until finally a finite expansion is achieved. Neglecting any friction between the piston and walls, through this series of equilibrium states, the statistical order (combined for the system and surroundings) can be preserved.

It is possible to convert 100% of work input into heat output because this results in an overall increase in entropy (system plus surroundings). The reverse transformation, converting heat input into useful work output, would therefore violate the second law, resulting in a decrease in total entropy (system plus surroundings), if the conversion rate were 100%. Heat can be converted into work through a repeatable thermodynamic cycle – called a heat engine – but only with an efficiency less than 100%, the missing percentage associated with the production of unwanted output (exhaust). The heat engine's cycle may even be reversible: The extra entropy required to make this viable comes at the 'cost' of producing exhaust. It is the conversion of 100% of work input into heat output that is irreversible, as the reverse process would entail a decrease in total entropy (system plus surroundings).

---

[46] Well, the disorder may increase for the system, but if so it will decrease for the surroundings, and vice-versa, if the process is reversible. In either case, the total entropy remains constant for a reversible process.

This is another way of expressing the second law. That is, it is impossible to construct a perfectly efficient heat engine. The Kelvin-Planck form of the second law is similar to this: One cycle of a heat engine cannot achieve no other effect than to absorb energy from a thermal reservoir and produce the same amount of work. The Clausius form of the second law expresses this a little differently, yet equivalently: No cyclical machine can achieve no other effect than to continuously transfer energy from an object at lower temperature to an object at higher temperature without adding energy by doing work. These two forms of the second law are equivalent. Heat spontaneously flows from a higher-temperature object to a lower-temperature object because this results in an increase in total entropy; thus, it would be necessary to do work to achieve the reverse heat flow.

The second law of thermodynamics can be cast in a variety of forms and applies to a broad range of disciplines from cosmology to economics. For example, in biology, plants convert carbohydrates into energy (and water and carbon dioxide) through the oxidation of glucose – a process that increases overall entropy (plant plus soil, air, and any other surroundings). In the reverse reaction, photosynthesis, plants convert water (from the soil) plus carbon dioxide (from the air) into carbohydrates; but this is not for 'free,' as it requires energy from the sun. The sun powers photosynthesis, which would otherwise entail a decrease in total entropy (plant, soil, and air). To see this, consider a seed that grows into a plant that sprouts leaves and blooms flowers – an extremely ordered microstate. The total entropy (plant, soil, air, and sun) actually increases, as the sun increases entropy considerably in the production of the sunlight that powers photosynthesis. We will consider the second law of thermodynamics in much more detail – quantitatively, too – in Chapter 5.

> According to the second law of thermodynamics, the total entropy (system plus surroundings) of a thermodynamic process for a macroscopic system that begins in one equilibrium state and ends in another equilibrium state cannot decrease; for a reversible process, the total entropy (system plus surroundings) will be constant, while for an irreversible process it will increase. The total entropy of the universe cannot decrease. As a result, it is impossible to construct a perfectly efficient heat engine.[47]

Important Distinction. It is important to distinguish between the entropy of the system, the entropy of the system's surroundings, and the total entropy (or overall entropy). It is possible for the entropy of a system to decrease, but only if the entropy of the surroundings increases; similarly, the entropy of the surroundings can decrease only if the entropy of the system increases. The total entropy of the system and surroundings cannot decrease, however.

Conceptual Example. Computers are extremely ordered systems. Try to imagine what your computer looked like before it was made – before the pieces of it were made – before the man-made materials were manufactured. The fact that the matter and energy that makes your computer is extremely ordered compared to how it once existed in nature does not violate the second law of thermodynamics. All of the processes that led to the production of your computer decreased the entropy of the materials and components that was assembled into your computer, but increased the entropy of their surroundings. The total entropy of the universe actually increased with the making of your computer.

---

[47] This is sometimes paraphrased as, "You can't even break even" – i.e. you can't even get back what you put in.

> Several technological breakthroughs had developed historically that led to the development of your computer. That is, it was very challenging for mankind just to come up with the concept for your computer. Also, the parts for your computer were manufactured through a painstaking process, using manufacturing arts that have been perfected over the course of decades. An amazing amount of work was involved in making your computer. Consider, for a moment, how extraordinarily challenging it was to develop and build your computer. Now consider how easy it would be to completely undo all that hard work. You could render your computer utterly unusable in a mere moment. It would be so easy to increase the entropy of your computer system. Nature tends toward an increase in entropy. As a system is disturbed, the system has a tendency to shift toward a more probable microstate. Your computer's current microstate is an extraordinarily improbable microstate. Of course, it was designed and built to withstand minor disturbances. Major disturbances, like hurricanes and earthquakes, would vastly increase its entropy. Even small disturbances can increase its entropy significantly.
>
> Statistically, it's no surprise that computers suffer problems from manufacturing defects, viruses, sudden restarts, overheating, stuck mouseballs, worn mousepads, freezing, internet delays, memory loss... It's much easier to increase the computer's entropy than to decrease it, and even small disturbances cause problems. We take computers for granted, but it's amazing how well they work!

Third law of thermodynamics: Another consequence of the second law of thermodynamics is that it is impossible to reach a temperature of absolute zero. Since absolute temperature is a measure of the average kinetic energy of the molecules of a substance, this makes sense, conceptually, as a lower limit to how statistically ordered an object can get. From a practical point of view, it becomes increasingly more difficult to decrease the temperature of an object as its temperature approaches absolute zero. Experimentally, it seems that it would require an infinite number of operations (the ultimate perfection of a reversible, quasistatic process) to bring an object with finite temperature to absolute zero. Mathematically, the third law can also be derived from the second law macroscopically – i.e. without regarding the object in terms of its molecular composition. As we will learn when we study heat engines quantitatively, one cycle of a reversible Carnot engine has an efficiency equal to $e = 1 - T_c/T_h$,[48] where T_c is the temperature of the lower-temperature reservoir (the exhaust, which receives rejected heat) and T_h is the temperature of the higher-temperature reservoir (the source of heat input). If a reservoir with absolute zero temperature could be found or made, a Carnot heat engine could be constructed with perfect efficiency, in principle (i.e. to the extent that frictional forces and dissipative effects can be minimized, and that the cycle can be carried out quasistatically) – which contradicts the second law of thermodynamics.

> According to the third law of thermodynamics, it is impossible for any object to reach a temperature of absolute zero by performing a finite number of operations, regardless of what idealized procedure may be applied.[49]

---

[48] This formula only applies to one reversible cycle of a Carnot engine. The formula for the efficiency of other types of heat engines is different. It is, unfortunately, much too common for students to attempt to use this formula for heat engines where it does not apply. The Carnot efficiency is significant as it represents the maximum efficiency that can be obtained from any heat engine (the other heat engines resulting in lower efficiency).
[49] This is sometimes paraphrased as, "You can't even back out of the game" – since you can't get to absolute zero.

1.2 Basic Mathematical Requisites of Thermal Physics

Unit vectors: A unit vector is a vector that has a magnitude equal to one mathematical unit.[50] Unit vectors serve a useful directional purpose. The Cartesian unit vectors, $\hat{\imath}$, $\hat{\jmath}$, and $\hat{k}$, point along the positive x-, y-, and z-axes, respectively. Any vector, $\vec{A}$, can be expressed in terms of its Cartesian (or rectangular) components (A_x, A_y, and A_z) and Cartesian unit vectors as $\vec{A} = A_x\hat{\imath} + A_y\hat{\jmath} + A_z\hat{k}$. This representation of a vector shows that the zigzag path that would be followed by traveling A_x units along the x-axis, A_y units along the y-axis, and A_z units along the z-axis is equivalent to the straight-line path from the tail of the vector to its tip – i.e. the components provide information that is equivalent to the magnitude and direction.[51]

Example. Determine the magnitude and direction of the vector $\vec{A} = 4\hat{\imath} - 4\sqrt{3}\hat{\jmath}$.

The components of the vector are the coefficients of the unit vectors: $A_x = 4$ and $A_y = -4\sqrt{3}$. As the components are mutually orthogonal (perpendicular), the magnitude is given by the Pythagorean theorem, $A = \sqrt{A_x^2 + A_y^2} = \sqrt{16 + 48} = 8$,[52] and the direction can be found from trig, $\theta = \tan^{-1}(A_y/A_x) = \tan^{-1}(-\sqrt{3}) = 300°$.[53] This vector lies in the xy plane and is directed $300°$ counterclockwise from the positive x-axis (looking from the positive z-axis).

Position vector: The position vector, $\vec{r}$, is unique because its components are the coordinates of its tip if its tail lies at the origin: $\vec{r} = x\hat{\imath} + y\hat{\jmath} + z\hat{k}$. The position vector of a particle indicates where the particle is relative to the origin of a coordinate system at any given time. The velocity and acceleration vectors are related to the position vector through the usual calculus relationships:

$$\vec{v} = \frac{d\vec{r}}{dt} \quad , \quad \vec{a} = \frac{d\vec{v}}{dt} = \frac{d^2\vec{r}}{dt^2}$$

When applying these derivatives, the Cartesian unit vectors are constant (however, the unit vectors of other coordinate systems – e.g. polar coordinates – are generally not constant).

---

[50] Despite its name, a unit vector is actually unitless, so it's a mathematical unit, not a physical unit. For example, both a force vector and components of force may be expressed in Newtons, which does not permit the unit vectors to have any physical units. The magnitude of a unit vector is one, not one physical unit.

[51] Notice that in three dimensions, two angles are needed in order to specify the direction of a vector, as we will see when we learn about spherical coordinates.

[52] The magnitude of a vector is always positive. The direction comes exclusively from the angle(s), not from the sign of the magnitude. The components, on the other hand, can be positive or negative, as their direction is tied to the vector's direction.

[53] Remember that there is always an alternate angle to consider when using an inverse trig function; most calculators only provide one of the two possibilities. Look at the components to figure out which quadrant the vector lies in. For an inverse tangent, the alternate angle is always $180°$ from its counterpart; inverse cosines and sines do not follow this same rule (instead, you should draw a triangle to figure it out).

Scalar product: Two vectors, $\vec{A}$ and $\vec{B}$, can be multiplied together to result in a scalar. This method of multiplying vectors is called the scalar product (or dot product or, in the context of matrices, the inner product) in order to distinguish it from the cross product. It is used to compute the work done by a force, for example. The scalar product has two useful forms, which are related by the law of cosines:[54]

$$\vec{A} \cdot \vec{B} = AB \cos \theta = A_x B_x + A_y B_y + A_z B_z$$

where A and B are the magnitudes of the vectors and θ is the angle between them.[55] The magnitude of a vector can be found using the scalar product:[56]

$$A = \sqrt{\vec{A} \cdot \vec{A}} = \sqrt{A_x^2 + A_y^2 + A_z^2}$$

Recalling that a unit vector has a magnitude equal to one unit, the magnitude form of the scalar product can be used to work out the scalar product between any two unit vectors:

$$\hat{\imath} \cdot \hat{\imath} = \hat{\jmath} \cdot \hat{\jmath} = \hat{k} \cdot \hat{k} = 1 \; , \; \hat{\imath} \cdot \hat{\jmath} = \hat{\imath} \cdot \hat{k} = \hat{\jmath} \cdot \hat{k} = 0$$

Observe that the scalar product is maximal when two vectors are parallel, zero when they are orthogonal, and most negative when they are antiparallel. That is, the scalar product first takes the projection of one vector onto the axis of the other, and then multiplies this projection with the magnitude of the other.

Example. Find the scalar product of the vectors $\vec{A} = 4\hat{\imath} - 4\sqrt{3}\hat{\jmath}$ and $\vec{B} = 2\sqrt{3}\hat{\imath} + 4\hat{\jmath}$.

The scalar product has two forms. Choose the most efficient form based on whether you know magnitudes and directions or components. In this case, the components were specified: $\vec{A} \cdot \vec{B} = 8\sqrt{3} - 16\sqrt{3} = -8\sqrt{3}$.

Example. Find the angle between the vectors $\vec{A} = \hat{\imath} - \sqrt{3}\hat{k}$ and $\vec{B} = -\sqrt{3}\hat{\imath} + 2\sqrt{2}\hat{\jmath} + \hat{k}$.

The angle between two vectors can be found from the scalar product. First, we need the magnitudes of the given vectors: $A = \sqrt{A_x^2 + A_y^2 + A_z^2} = \sqrt{1 + 0 + 3} = 2$ and $B = \sqrt{B_x^2 + B_y^2 + B_z^2} = \sqrt{3 + 8 + 1} = 2\sqrt{3}$. Next, we use the component formula for the scalar product: $\vec{A} \cdot \vec{B} = -\sqrt{3} + 0 - \sqrt{3} = -2\sqrt{3}$. Finally, the magnitude formula for the scalar product gives $\vec{A} \cdot \vec{B} = -2\sqrt{3} = AB \cos \theta = (2)(2\sqrt{3}) \cos \theta$. Thus, $\cos \theta = -\frac{1}{2}$ or $\theta = 120°$. (Remember, $0 \leq \theta \leq 180°$ for the scalar product.)

---

[54] However, in physics, the law of cosines features a plus sign, $C^2 = A^2 + B^2 + 2AB \cos \theta$, whereas the last term is negative in math. The difference is that in physics we work with the angle between the vectors, whereas in math the law of cosines involves the angle opposite to side C, which is the supplement to the angle used in physics.
[55] Join the two vectors at their tales and take the smaller of the two angles between them – i.e. $0 \leq \theta \leq 180°$.
[56] This is the generalization of the Pythagorean theorem to three dimensions. For example, this formula provides the length of the long diagonal of a 3D rectangular box.

Chain rule: If f is a function of u, which is itself a function of x, the chain rule can be applied to take a derivative of f with respect to x:

$$\frac{df}{dx} = \frac{df}{du}\frac{du}{dx}$$

Example. Given the functions $f(u) = 3u^2$ and $u(x) = 4x^3$, evaluate the derivative of f with respect to x at $x = \frac{1}{2}$.

Applying the chain rule,

$$\frac{df}{dx} = \frac{df}{du}\frac{du}{dx} = (6u)(12x^2) = 72ux^2 = 288x^5$$

Plugging $x = \frac{1}{2}$ into the result, $df/dx(x = \frac{1}{2}) = 9$.

Example. Express acceleration in terms of a derivative with respect to position for 1D motion.

According to the chain rule, $a = dv/dt = (dv/dx)(dx/dt) = v\,dv/dx$. This would be very convenient for integrating the equation of motion for Hooke's law, for example.

Partial derivative: If f is a function of two or more independent variables, a partial derivative of f with respect to one of the variables means to take a derivative while treating the other independent variables as if they were constants. Notationally, the partial derivative looks like a derivative, except that the lower-case del (∂) is used in place of the usual differentiation symbol (d) – e.g. $\partial f/\partial x$. Second-order (and higher-order) partial derivatives can be mixed – e.g. $\frac{\partial^2 f}{\partial x \partial y}$ which means to first compute $\partial f/\partial y$ and then take a partial derivative with respect to x of the result.

Example. Find the partial derivatives of the function $f(x, y) = 3x^2 y - 2y^3$ with respect to each of the independent variables, x and y.

When taking a partial derivative of f with respect to x, hold y constant, and vice-versa:

$$\frac{\partial f}{\partial x} = 6xy \quad , \quad \frac{\partial f}{\partial y} = 3x^2 - 6y^2$$

Total derivative: If f is a function of two independent variables x and y, which themselves are functions of another variable, u, the total derivative of f with respect to u is:

$$\frac{df}{du} = \frac{\partial f}{\partial x}\frac{\partial x}{\partial u} + \frac{\partial f}{\partial y}\frac{\partial y}{\partial u}$$

This is the multivariable generalization of the chain rule. It generalizes to three or more independent variables by simply adding similar terms.

1 Basic Relationships of Thermal Physics

Implicit differentiation: Physical relationships are best expressed on the differential scale; after which relationships on the finite scale can be determined through integration. This is especially important when the integrand(s) would be functions of the integration variables, rather than constants. This can be seen, for example, in the case of work. In general, work only equals the scalar product between force and displacement in the differential form, $dW = \vec{F} \cdot d\vec{s}$. For the special case that the applied force is constant, then work equals the scalar product between force and displacement $W = \vec{F} \cdot \vec{s}$. In the more general case, force is variable and the finite amount of work done equals $W = \int_i^f \vec{F} \cdot d\vec{s}$. This is appropriate, for example, for finding the work in displacing a spring from equilibrium, as the force depends on the displacement from equilibrium ($\vec{F} = -k\vec{s}$ to first order). When developing relationships between differential quantities in physics, it is often necessary to differentiate implicitly. Implicit differentiation is also useful for techniques of integration that involve changing variables.

If f is a function of two independent variables, x and y, the function can be implicitly differentiated according to the product rule as:[57]

$$df = \frac{\partial f}{\partial x} dx + \frac{\partial f}{\partial y} dy$$

If one function $f(x)$ is set equal to another function $g(y)$, both sides of the equation may be implicitly differentiated according to:

$$f(x) = g(y) \rightarrow \frac{df}{dx} dx = \frac{dg}{dy} dy$$

Example. Implicitly differentiate $f = 2A^2 B^3$, where A and B are independent variables.
According to the recipe for implicit differentiation,

$$df = \frac{\partial f}{\partial A} dA + \frac{\partial f}{\partial B} dB = 4AB^3 dA + 6A^2 B^2 dB$$

Example. Express the integral $\int_{x=0}^{2}(4-x^2)^{1/2} dx$ in terms of u where $x = 2 \sin u$.
Implicit differentiation of the substitution equation yields $dx = 2 \cos u \, du$. Substituting this into the integral, $\int_{x=0}^{2}(4-x^2)^{1/2} dx = \int_{u=0}^{\pi/2}(4 - 4\sin^2 u)^{1/2} 2 \cos u \, du = 4 \int_{u=0}^{\pi/2} \cos^2 u \, du$.[58]

Gradient: The gradient, $\vec{\nabla}$, is a differential operator that transforms a multivariable scalar function into a vector quantity:[59]

---

[57] Compare this expression to the total derivative. This is the total differential. This formula serves as the basis for propagation of errors – square each term and set the cross-term to zero if x and y are uncorrelated.

[58] You can improve your math skills and understanding by taking the time to try to fill in the steps. That is, you don't learn by just reading math: Instead, get out a sheet of paper and a pencil, start with the first equation, and see if you agree with the intermediate equation and final equation. Work it out in more steps to check your substitutions and cancellations carefully.

$$\vec{\nabla} = \hat{\imath}\frac{\partial}{\partial x} + \hat{\jmath}\frac{\partial}{\partial y} + \hat{\mathbf{k}}\frac{\partial}{\partial z}$$

It is best to work with the gradient operator applied to a scalar function (since many mistakes are often made working in the pure form):

$$\vec{\nabla} f = \hat{\imath}\frac{\partial f}{\partial x} + \hat{\jmath}\frac{\partial f}{\partial y} + \hat{\mathbf{k}}\frac{\partial f}{\partial z}$$

Conceptually, the gradient represents a topological slope. For example, if the function represents the altitude of earth's surface as a function of east/west and north/south (x and y), there is a steeper slope where the gradient has greater magnitude – since rapid changes in either geographic coordinate produce more positive (or more negative) partial derivatives. If the contours of constant altitude are illustrated on a topological map, the gradient will have a greater magnitude where the contours are more closely spaced. The direction of the gradient is orthogonal to the surface.[60]

Displacement: The initial position vector, $\vec{r}_0$, extends from the origin to the starting position of an object, while the position vector, $\vec{r}$, extends from the origin to the object at time t. The vector subtraction of these position vectors, $\Delta \vec{r} = \vec{r} - \vec{r}_0$, equals the net displacement of the object – a vector that extends from the initial position to the current position. The net displacement can be expressed in Cartesian coordinates as $\Delta \vec{r} = (x - x_0)\hat{\imath} + (y - y_0)\hat{\jmath} + (z - z_0)\hat{\mathbf{k}}$. The differential displacement is $d\vec{s} = \hat{\imath}dx + \hat{\jmath}dy + \hat{\mathbf{k}}dz$.[61]

The total distance traveled, s, on the other hand, is found by integrating the differential arc length: $ds = \sqrt{dx^2 + dy^2 + dz^2}$.[62] The differential arc length is the magnitude of the differential displacement, but the magnitude of the finite displacement vector is always less than or equal to the corresponding finite arc length: $s \geq \|\vec{s}\|$. That is, the arc length is not, in general, the magnitude of the displacement vector.

A derivative of position or displacement with respect to time equals the instantaneous velocity, while a derivative of arc length with respect to time equals the instantaneous speed. The speed does equal the magnitude of the velocity.

Line integral: The integral of the scalar product of a vector with the differential displacement is called a line (or path) integral: $\int \vec{A} \cdot d\vec{s}$. If the initial and final positions are the same – i.e. the path is closed – this is referred to as a closed integral and designated as $\oint \vec{A} \cdot d\vec{s}$. The line integral is very useful both mathematically and conceptually, e.g. for work done by conservative and nonconservative forces.

---

[59] The symbol for the gradient operator is upper-case del, while lower-case del represents a partial derivative.
[60] This is counterintuitive to many students, who are familiar with the conceptual interpretation of a derivative as slope. Indeed, the gradient does represent the slope of a surface, but its direction is perpendicular, rather than tangential, to the surface. (Note that there is not a tangent line, but a tangent plane, in three dimensions.)
[61] The symbol $\vec{r}$ represents position in 3D space. The symbol s represents arc length. The vector analog for arc length is displacement, $\vec{s}$. The vector subtraction $\vec{r} - \vec{r}_0$ is equivalent to the integral of $d\vec{s}$: $\vec{r} - \vec{r}_0 = \int_i^f d\vec{s}$.
[62] In practice, this is applied by factoring – e.g. for a curve in the xy plane, $s = \int \sqrt{1 + (dy/dx)^2}\, dx$.

Integrals of the form $\int_a^b f(x)dx$ are equal to zero if the endpoints are the same – i.e. $b = a$ – for many functions commonly integrated in calculus. However, it is possible to contrive exceptions. For example, consider the integral $\int_{x=0}^{x=0} f(x)dx$ along the path $x=0 \to x=1 \to x=0$ where the function $f(x)$ equals $+1$ if x is increasing and -1 if x is decreasing. In this case, $\int_{x=0}^{x=0} f(x)dx = \int_{x=0}^{x=1} dx - \int_{x=1}^{x=0} dx = 2$. This is not a 'well-behaved' function from calculus class. You might wonder, why consider such a crazy function? Well, it's not crazy: Resistive forces, like friction, act opposite to the direction of motion. If a box is being displaced to the right, friction acts to the left, and vice-versa.

For conservative forces, like gravitational force, the closed line integral equals zero – i.e. $\oint m\vec{g} \cdot d\vec{s} = 0$. For nonconservative forces, like friction, the closed line integral is generally nonzero. In two and three dimensions, forces can seem 'well-behaved' compared to the function defined in the example above, yet still be nonconservative. For example, the force $\vec{F}(x,y) = 3xy^3\hat{j}$, is nonconservative. One way to tell is to compute the line integral for different possible paths. Fortunately, there is a much simpler way: If $\partial F_y/\partial x = \partial F_x/\partial y$, the force is conservative.[63] The line integral is path-independent for a conservative field, but path-dependent for a nonconservative field.

Mean-value theorem: The average value of a function $f(x)$ over an interval $a \leq x \leq b$ can be found from the mean-value theorem of calculus:

$$\overline{f(x)} = \frac{\int_{x=a}^{b} f(x)dx}{\int_{x=a}^{b} dx}$$

The denominator is often written as $b-a$. However, the above form generalizes to account for any weighting factors (distribution functions). The weighted average is

$$\overline{f(x)} = \frac{\int_{x=a}^{b} w(x)f(x)dx}{\int_{x=a}^{b} w(x)dx}$$

Example. Apply the mean-value theorem to find average velocity over a time interval.
For velocity as a function of time, the mean-value theorem gives average velocity as

$$\overline{\vec{v}} = \frac{\int_{t=t_0}^{t} \vec{v}(t)dt}{\int_{t=t_0}^{t} dt} = \frac{\Delta \vec{r}}{\Delta t}$$

The integral of velocity over time is net displacement, so the average velocity of an object equals its net displacement divided by the time interval.

---

[63] This follows in 2D since energy represents capacity to do work. Specifically, work done by a conservative force equals the negative of the change in potential energy; also, for a conservative force, $\vec{F} = -\vec{\nabla}U$. Potential energy is 'well-behaved' if $\frac{\partial^2 U}{\partial x \partial y} = \frac{\partial^2 U}{\partial y \partial x}$, which is equivalent to $\frac{\partial F_y}{\partial x} = \frac{\partial F_x}{\partial y}$. (Nonconservative forces have no potential energy.)

Surface and volume integrals: Differential area and volume elements in Cartesian coordinates are expressed as $dA = \sqrt{(dxdy)^2 + (dydz)^2 + (dzdx)^2}$ and $dV = dxdydz$. For a flat surface lying in the xy plane, for example, the differential area element reduces to $dA = dxdy$.[64] Finite surface area and volume are found from double and triple integrals. When performing a multidimensional integral, one or more of the bounds may have variable limits. If so, that coordinate must be integrated over first. When integrating over one coordinate, treat the others as constants – meaning you can pull a coordinate out of other coordinates' integrals, but you can't pull it out of its own integral.

Example. Integrate $\int x\,dA$ over the triangle bounded by the coordinate axes and the line $y = 1 - x$. If you allow x to enjoy its full freedom – i.e. $0 \leq x \leq 1$ – then y must be limited to $0 \leq y \leq 1 - x$; that's how the math understands the distinction between the triangle and a rectangle. The integral is

$$\int x\,dA = \int\int_{x=0,y=0}^{1,\ 1-x} x\,dy\,dx = \int_{x=0}^{1} x\left(\int_{y=0}^{1-x} dy\right) dx = \int_{y=0}^{1} x(1-x)\,dx = \frac{1}{6}$$

2D polar coordinates: It's much easier to work with 2D polar coordinates than Cartesian coordinates to solve the equations of motion for an object traveling along a circular arc or to find the center of mass of a slice of pie, for example. In these cases, instead of working with horizontal and vertical coordinates, x and y, it is much simpler to work with a radial coordinate r and the angle θ that the radial coordinate makes counterclockwise with the x-axis. The direction comes from θ, such that r is nonnegative.[65]

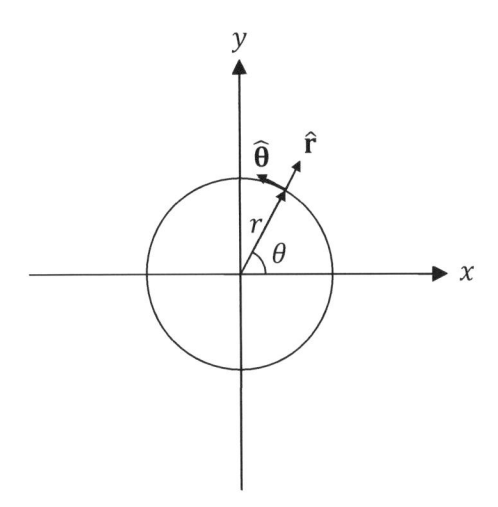

---

[64] If you're just finding area or volume, it is possible to calculate this with a single integral using techniques from calculus – e.g. volume of revolution. However, there are many situations where there is an additional integrand – e.g. electric flux ($\int \vec{E} \cdot d\vec{A}$) for a non-uniform electric field – where those single integral techniques break down.

[65] It would be redundant to allow θ to vary from 0 to 2π and allow r to be positive or negative. The convention is to restrict r to be nonnegative. However, in principle, we could instead restrict θ to lie in the range $0 \leq \theta \leq \pi$ and allow r to be positive or negative. The convention is based on physics: The magnitude of a vector must be nonnegative. In polar coordinates, r, is the magnitude of the position vector, $\vec{r}$.

The 2D polar coordinates are related to the Cartesian coordinates by $x = r\cos\theta$ and $y = r\sin\theta$. The position vector can be expressed in polar coordinates as $\vec{r} = r\hat{r}$, which says to go outward from the origin a distance r at a direction specified by the radial unit vector $\hat{r}$. Comparing the position vector to its 2D Cartesian form, $\vec{r} = x\hat{i} + y\hat{j}$, it follows that $\hat{r} = \hat{i}\cos\theta + \hat{j}\sin\theta$. The tangential unit vector is related to the radial unit vector by adding 90° to θ, $\hat{\theta} = -\hat{i}\sin\theta + \hat{j}\cos\theta$ (or, as you can see, by taking a derivative of $\hat{r}$ with respect to θ).

At the differential level, the radial arc length dr and circular arc length $rd\theta$ give the more general equation for arc length as $ds = \sqrt{dr^2 + (rd\theta)^2}$. The differential area element is $dA = rdrd\theta$.

> **Important Distinction.** The Cartesian unit vectors are constant, but polar, spherical, and cylindrical unit vectors are functions of angular coordinates. When differentiating or integrating vector quantities expressed in polar, spherical, or cylindrical coordinates, it is important to remember that the non-Cartesian unit vectors are not constant – i.e. you may not pull them out of the derivative or integral.

Spherical coordinates: A point within a sphere of radius R_0 can be located by specifying its distance from the origin, r, which may be smaller than R_0, and two angles: The polar angle, θ, measures the angle that the position vector, $\vec{r}$, makes with the z-axis (the pole), and the azimuthal angle, φ, measures the angle of the projection of $\vec{r}$ onto the xy plane counterclockwise from the positive x-axis (as viewed from the positive z-direction).[66] These two angles are equivalent to specifying longitude and latitude.

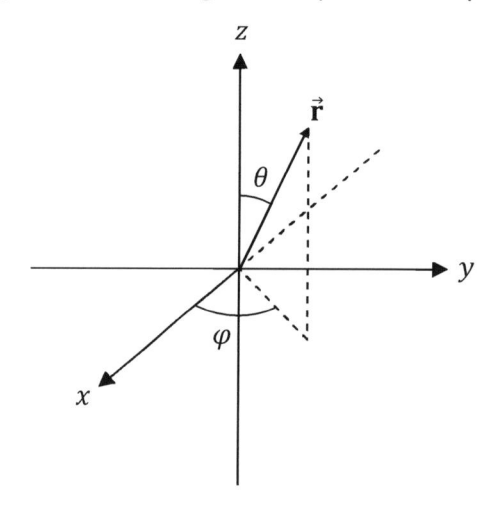

The spherical coordinates are related to the Cartesian coordinates by $x = r\cos\varphi\sin\theta$, $y = r\sin\varphi\sin\theta$, and $z = r\cos\theta$. The fundamental differential arc lengths are dr, $rd\theta$, and $r\sin\theta\, d\varphi$. The surface area of the sphere can be found by integrating $dA = r^2\sin\theta\, d\theta d\varphi$. The differential volume element is $dV = r^2\sin\theta\, drd\theta d\varphi$. The spherical unit vectors are $\hat{r}$, $\hat{\theta}$, and $\hat{\varphi}$. Since $\vec{r} = r\hat{r}$, it follows that $\hat{r} = \hat{i}\cos\varphi\sin\theta + \hat{j}\sin\varphi\sin\theta + \hat{k}\cos\theta$.

---

[66] Unfortunately, it is conventional to call the polar angle θ and the azimuthal angle φ in physics, but to call the polar angle φ and the azimuthal angle θ in math. It's really fun when you're taking physics and math exams on the same day, and both involve problems in spherical coordinates.

Cylindrical coordinates: Adding a third coordinate z to 2D polar coordinates makes (3D) cylindrical coordinates: r_c, θ, and z. The differential area element for the body of the surface of a cylinder is $dA = r_c d\theta dz$ and for the ends is $dA = r_c dr_c d\theta$. The differential volume element is $dV = r_c dr_c d\theta dz$. The cylindrical coordinates are related to the Cartesian coordinates by $x = r_c \cos\theta$ and $y = r_c \sin\theta$. The cylindrical unit vectors are $\hat{\mathbf{r}}_c$, $\hat{\boldsymbol{\theta}}$, and $\hat{\mathbf{k}}$. Since $\vec{\mathbf{r}} = r_c\hat{\mathbf{r}}_c + z\hat{\mathbf{k}}$, it follows that $\hat{\mathbf{r}}_c = \hat{\mathbf{i}}\cos\theta + \hat{\mathbf{j}}\sin\theta$.

1.3 Basic Quantitative Relationships of Thermal Physics

Thermal expansion: The volume of space that a solid or liquid occupies depends upon temperature. For a small fractional change in volume, the change in volume is generally proportional to the change in temperature. In terms of differentials, this is expressed as $dV = \beta V_0 dT$, where β, the 3D coefficient of thermal expansion, depends upon the substance and V_0 is the original volume. Assuming that β is constant, on the finite level the formula becomes $\Delta V = \beta V_0 \Delta T$. For an isotropic solid, the fractional change in length of each of its dimensions is the same, such that the increase in length is greatest along its longest dimension(s). In this case, the solid's linear expansion along its longest dimension(s) is most relevant: $dL = \alpha L_0 dT$. As we will learn later, $\beta = 3\alpha$. The SI units of the coefficients of linear (and 3D) expansion are per degree Kelvin (/K).[67] For most substances, β (and hence α) is positive, meaning that the substance expands as temperature increases. One of the most obvious exceptions is water for combinations of pressure and temperature near its freezing point: Water actually expands when freezing – i.e. the cold water occupies more volume when it changes phase to ice, causing pipes to burst in the winter for unwary homeowners. For this region of pressure and temperature for water, β is negative, and the contraction or expansion of water is more quadratic than it is linear. See Sec. 2.1 for a more in-depth discussion of thermal expansion.

Heat capacity: An object's temperature is often affected by heat exchanged with its surroundings. However, it is possible for an object to absorb or release heat without changing temperature, and for an object to change temperature without exchanging heat with its surroundings.[68] The capacity of heat to affect the temperature of a substance is characterized differentially as đ$Q = CdT$, but this equation is seldom used in practice. Rather, specific heat – heat capacity per unit mass – and molar specific heat – heat capacity per mole – are more apt to be measured. Generally, these are measured either at constant pressure or constant volume. The specific heat at constant pressure or volume is defined by đ$Q = MC_{P,V} dT$ and the molar specific heat at constant pressure or volume by đ$Q = nc_{P,V} dT$. If the specific heat and mass or mole number are constant over the course of the temperature change, these equations become $Q = MC_{P,V} \Delta T$ and $Q = nc_{P,V} \Delta T$ at the finite scale. These equations may be useful for expressing heat in terms of temperature if either the pressure or volume is constant for a process, but not when both pressure and volume change. Heat capacities are discussed again in Sec. 2.2.

---

[67] This can also be expressed as per degree Celsius (/°C), since change in Celsius equals change in Kelvin.
[68] The first law provides a complete description of how all energy transformations – such as work done, phase transitions, or gaining/losing particles – affect the temperature of an object. The first law applies quite generally; heat capacity is most useful for an isobaric or isochoric process – i.e. if pressure or volume are constant – and even then it is usually combined with the first law.

1 Basic Relationships of Thermal Physics

Heat of transformation: When a substance undergoes a phase transition, the amount of heat gained or released equals the product of the mass of substance that changes phase and the latent heat of transformation: $Q = ML$. Heat is released by the substance during freezing, condensation, and deposition, and absorbed during melting, boiling, and sublimation.

First law of thermodynamics: Conservation of energy is expressed by the first law of thermodynamics: The change in internal energy of the system equals the heat gained by the system minus the work done by the system if the mole numbers of the system are constant (otherwise, the exchange of particles also adds or subtracts energy from the system). For constant mole numbers, this is expressed as $dU = đQ - đW$ at the differential scale and $\Delta U = Q - W$ at the finite scale. The change in internal energy is positive ($dU > 0$) if the internal energy of the system increases, heat is positive ($đQ > 0$) if it is added to the system, and work is positive ($đW > 0$) if it is done by (rather than on) the system.[69] The first law of thermodynamics is discussed much more extensively in Sec. 2.5.

Work: A force that contributes toward the displacement of an object does work according to the formula $W = \int_i^f \vec{F} \cdot d\vec{s}$. Only for a constant force does this reduce to $W = \vec{F} \cdot \vec{s}$. At the differential level, $đW = \vec{F} \cdot d\vec{s}$. In thermal physics, it is more convenient to work with pressure and volume than with force and displacement. In the next section, we will see that it is equivalent to express work as $đW = PdV$. At the finite scale, $W = \int_{V=V_0}^{V} PdV$. Here, pressure is nonnegative, so the volume change determines the sign of the work done: Work is positive (done by the system) if the volume increases and is negative (done on the system) if the volume decreases. No work is done for an isochoric (constant volume) process. If the volume changes, work is done (except for a gas undergoing a free expansion – since if the expansion is truly 'free,' there is no pressure). See Sec. 1.4 for a comprehensive discussion of the application of the concept of work in thermodynamics.

Power: The instantaneous rate at which the system is doing work equals the power that the system is delivering: $P = đW/dt$. Power is positive if the system is doing work and negative if work is being done on the system. The same symbol is used for pressure and power, but the distinction should be clear from the context.

Since energy is the capacity to do work, the rate at which energy is being transferred is also a measure of power. The instantaneous rate at which heat is being added to the system is $P = đQ/dt$. In this case, power is positive if heat is being added to the system and negative if it is being removed from the system. The instantaneous rate at which the internal energy of the system is changing is $P = dU/dt$, which is positive if the internal energy is increasing and negative if it is decreasing. In general, the instantaneous rate at which any form of energy is being transformed is $P = dE/dt$.

Pressure: At the differential scale, pressure equals the normal (perpendicular to the surface) component of the applied force divided by the area: $dF_n = PdA$. In general, there is a factor of $\cos\theta$ involved in the normal component of the force. For a force that is distributed evenly and applied normal to the surface, at the finite scale this becomes force per unit area: $P = F/A$.

---

[69] When work is positive (work done by the system), it is being subtracted from heat in the first law; when it is negative (work done on the system), the two minus signs make a positive.

Density: The density of an object may vary from one part to another and so is most appropriately defined for an infinitesimal location as differential mass over differential volume: $dM = \rho dV$. For a substance where the mass is distributed uniformly throughout its volume, this ratio also holds at the finite level – i.e. if density is constant, integration gives $M = \rho V$ – but in the more general case, density is only described correctly in terms of calculus.[70]

Entropy: Defined at the differential scale as differential heat divided by the absolute temperature, $đQ = TdS$, and valid for equilibrium states, the entropy, S, provides a macroscopic measure of the statistical disorder of a system. While entropy is thermodynamically only required to be valid for equilibrium states, the change in entropy can be measured as a system passes from one equilibrium state to another through a series of non-equilibrium states due to the fact that it is path-independent like internal energy: The change in entropy can be computed by integrating over an equivalent reversible path (i.e. having the same initial and final states).

Microscopically, the entropy is proportional to the total number of states available to the system: As the vast majority of available states will be very disordered and only a minute number will be very ordered, more available states provides a greater statistical probability of the system being disordered. To see this, consider a two-state system, analogous to coin-flipping. If you flip two coins simultaneously, there is a 50% chance that the system will be ordered (all heads or all tails), while if you flip ten coins simultaneously, there is a mere 0.2% (or 1 in 500) chance that it will be perfectly ordered (10 heads or 10 tails). Entropy is discussed much more extensively in Chapter 5.

Mole number: The number of moles, n_k, of a particular chemical component present in an object equals the number of molecules, N_k, of that chemical component present in the object divided by Avogadro's number, N_A: $n_k = N_k/N_A$. The mole fraction, x_k, of chemical component k equals its mole number, n_k, divided by the sum of the mole numbers, $\sum_{k=1}^{r} n_k = n$, for all r chemical components: $x_k = n_k/n$. Observe that $\sum_{k=1}^{r} x_k = 1$. The molar volume, v, of the substance equals the volume of the substance, V, per unit mole: $v = V/n$.

Area: At the differential scale, area can be expressed as $dA = \sqrt{(dxdy)^2 + (dydz)^2 + (dzdx)^2}$ in Cartesian coordinates, $dA = \sqrt{(r_c dr_c d\theta)^2 + (dr_c dz)^2 + (r_c dz d\theta)^2}$ in cylindrical coordinates, or $dA = \sqrt{(r^2 \sin\theta \, d\theta d\varphi)^2 + (rdrd\theta)^2 + (r\sin\theta \, drd\varphi)^2}$ in spherical coordinates, for example.[71,72]

---

[70] Students sometimes wonder if the center of mass, moment of inertia, electric field, electric potential, and other integrals in physics involving a non-uniform distribution of mass are just fancy, of if they actually have practical value. Non-uniform densities are actually quite common. For example, celestial objects often have a density that varies radially due to gravitational effects (for which thermodynamics is quite relevant). For a microscopic example, consider a conductor with a net charge in electrostatic equilibrium: A nearby charged object will influence its electrons to be distributed non-uniformly.

[71] The distinction between the cylindrical radial coordinate, r_c, and the spherical radial coordinate, r, is that r_c is the distance from the z-axis, while r is the distance from the origin. The locus of points equidistant from the z-axis, corresponding to r_c equal to a constant, is a right-circular cylinder, whereas the locus of points equidistant from the origin, corresponding to constant r, is a sphere. Observe that $0 \leq \theta \leq 2\pi$ in cylindrical coordinates, but $0 \leq \theta \leq \pi$ in spherical coordinates, while $0 \leq \varphi \leq 2\pi$. Once again, most math textbooks swap θ and φ compared to most physics textbooks.

It is convenient to use Cartesian coordinates for a polygonal area, such as a triangle. For a surface that lies in the xy plane, dA reduces to $dA = dxdy$. Cylindrical coordinates are convenient for finding the area of a circular object or the surface area of the body of a right-circular cylinder. For a surface that lies in the xy plane, dA can also be expressed as $dA = r_c dr_c d\theta$. For the surface area of the body of a right-circular cylinder, $dA = r_c d\theta dz$. Spherical coordinates are convenient for finding the surface area of part or all of the surface of a sphere or the surface area of the body of a right-circular cone. Over the surface of a sphere, $dA = r^2 \sin\theta \, d\theta d\varphi$. For the surface area of the body of a right-circular cone, $dA = r \sin\theta \, dr d\varphi$. Expressed as a vector, $\vec{dA}$ has the same magnitude and its direction is perpendicular to the surface, outward if the surface is curved.

Following are the areas for some common geometric shapes. The area of a triangle with base b and height h is $A = \frac{1}{2}bh$. The area of a circle of radius R is $A = \pi R^2$. The area of an ellipse with semimajor axis a and semiminor axis b is $A = \pi ab$.[73] The surface area of the body of a right-circular cylinder of radius R and height h is $A = 2\pi Rh$ (easily found by unfolding the cylinder into a rectangle), such that its total surface area is $A = 2\pi R^2 + 2\pi Rh$. The surface area of the body of a right-circular cone of base radius R and height h is $A = \pi R\sqrt{R^2 + h^2}$, such that its total surface area is $A = \pi R^2 + \pi R\sqrt{R^2 + h^2}$. The surface area of a sphere of radius R is $A = 4\pi R^2$.[74]

Volume: The differential volume element is expressed as $dV = dxdydz$ in Cartesian coordinates, $dV = r_c dr_c d\theta dz$, and $dV = r^2 \sin\theta \, drd\theta d\varphi$ in spherical coordinates.[75] Following are the volumes for some common geometric shapes. The volume of a right-circular cylinder of radius R and height h is $V = \pi R^2 h$. The volume of a right-circular cone of base radius R and height h is $V = \pi R^2 h/3$. The volume of a sphere of radius R is $V = 4\pi R^3/3$. The molar volume, v, of a substance is its volume per mole: $v = V/n$.

Solid angle: Differential solid angle, $d\Omega$, is defined as the angular part of the differential volume element in spherical coordinates: $d\Omega = \sin\theta \, d\theta d\varphi$. Integrating over a complete sphere, the total solid angle of 3D space is found to be 4π steradians. This two-dimensional angle is called a 'solid' angle. For example, when you look at the sun (which you shouldn't do directly if you're fond of your retinas) and the moon in the sky from the earth, they appear about the same size to your eye – your eyes perceive that their solid angles (or 'apparent' sizes) are roughly the same.

---

[72] These differential area element formulas combine three mutually orthogonal pairs of differential arc lengths together using the 3D generalization of the Pythagorean theorem. In Cartesian coordinates, the three arc lengths are dx, dy, and dz (for which the three mutually orthogonal pairs are $dxdy$, $dydz$, and $dzdx$); in cylindrical coordinates, they are the radial arc length dr_c, circular arc length $r_c d\theta$, and vertical arc length dz; and in spherical coordinates, they are the radial arc length dr, longitudinal arc length $rd\theta$, and latitudinal arc length $r\sin\theta \, d\varphi$ (where the $\sin\theta$ comes from first projecting downward, or upward, onto the xy plane).

[73] The semimajor and semiminor axes of an ellipse are equal to one-half of the symmetric bisectors. You can conceptually understand this area formula by rescaling one of the axes to transform the ellipse into a circle.

[74] It is surprising when students confuse the formula for the circumference of a circle, $C = 2\pi R$, with the formula for the area of a circle, $A = \pi R^2$, or confuse the formula for the surface area of a sphere, $A = 4\pi R^2$, with the formula for the volume of the sphere, $V = 4\pi R^3/3$. Simply inspect the units to avoid such silliness.

[75] Here, all three mutually orthogonal arc lengths – one corresponding to each coordinate – are multiplied together at the differential scale to obtain the differential volume element.

Heat conduction: Thermal energy can be transferred from a high-temperature surface to a low-temperature surface of a substance through thermal conduction. The instantaneous rate, P, at which heat is transferred at a given point in the substance is related to the thermal conductivity, k, of the substance, the cross-sectional area, $\vec{A}$, and the temperature gradient, $\vec{\nabla}T$, by:[76,77]

$$P = -\int k\vec{\nabla}T \cdot d\vec{A}$$

The differential area element, $d\vec{A}$, is perpendicular to the surface of integration, which is the cross-sectional area – i.e. the area of cross section perpendicular to the heat flow. Conceptually, the overall minus sign reflects the fact that heat naturally flows in the direction of decreasing temperature. The thermal conductivity provides a measure of how well the given substance conducts heat; for many materials, it can be looked up in a data table. Heat conduction is introduced thoroughly in Sec. 2.3.

Stefan's law: The instantaneous rate, P, at which all objects emit thermal radiation as a result of thermal vibrations of their molecules is proportional to the fourth power of their absolute temperature, according to Stefan's law: $P_e = \sigma\epsilon A T^4$, where σ is the Stefan-Boltzmann constant, ϵ is the emissivity, and A is the surface area of the object. The instantaneous rate at which all objects absorb thermal radiation from their surroundings is given by a similar equation: $P_a = \sigma\epsilon A T_{env}^4$. The difference is that the emission rate depends upon the temperature of the object, while the absorption rate depends upon the temperature of the surroundings. The net power radiated equals $P_{net} = \sigma\epsilon A(T^4 - T_{env}^4)$. The net power radiated is positive if the object is losing thermal energy (not the same sign convention as heat, but it makes sense in terms of power radiated). The emissivity provides a measure of how well the given substance radiates electromagnetic energy; it can be looked up for many materials in a data table. Stefan's law is considered in more detail in Sec. 2.4.

Ideal gas law: Along an isotherm, the pressure and volume of an ideal gas with constant mole numbers are inversely proportional according to Boyle's law. Along an isobar, the volume and absolute temperature of an ideal gas with constant mole numbers are directly proportional according to Charles's law. Along an isochor, the pressure and temperature of an ideal gas with constant mole numbers are directly proportional according to Gay-Lussac's law. These three gas laws are special cases of the more general ideal gas law: $PV = nRT$. The precise meaning of an ideal gas, the ideal gas law, and the thermal physics of ideal gases is covered extensively in Chapter 3.

---

[76] In this form, it is clear why the rate at which thermal energy is transferred by heat conduction is often called the heat flux: Namely, gravitational flux, electric flux, and magnetic flux are similarly defined as integrals of the scalar product of the corresponding field and the differential area element vector. These fields can similarly be expressed in terms of gradients of scalar potentials (except in the case of magnetic field, which is a curl of the vector potential, $\vec{\nabla} \times \vec{A}$). The flux of thermal conduction field lines is perfectly analogous to gravitational and electric field lines, which flow from higher to lower temperature. Regions where thermal field lines are more concentrated correspond to regions of greater heat flux.

[77] We will learn how to treat this formal integral in the next chapter.

Equipartition of energy: According to the law of equipartion of energy, the total kinetic energy of a system at thermal equilibrium is divided equally, on average, among its independent degrees of freedom. For a degree of freedom for which kinetic energy varies quadratically (such as a component of velocity makes to translational kinetic energy, since $K = \frac{1}{2}mv_x^2$, or as a component of angular velocity makes to rotational kinetic energy, since $K = \frac{1}{2}I_x\omega_x^2$), its contribution to the total kinetic energy is $\frac{1}{2}k_B T$. For example, a monatomic ideal gas has 3 degrees of freedom ($K = 3k_B T/2$), corresponding to translation along each of the three dimensions of space, a diatomic ideal gas has 5 independent components ($K = 5k_B T/2$), where the two additional degrees of freedom come from angular momentum about the two axes perpendicular to the diatomic bond (the angular momentum about the axis of the bond itself being negligible in comparison).[78]

Law of Dulong and Petit: The molar specific heat at constant volume, c_V, for a solid substance is observed to approach the value $3R$ with increasing temperature, independent of the properties of the solid. At low temperatures, the value for c_V departs from the law of Dulong and Petit, vanishing in the limit that the absolute temperature approaches zero. The value $3R$ is associated with the law of equipartition of energy, and the low-temperature effect is attributed to quantum effects.

Efficiency of a heat engine: The efficiency of a heat engine equals ratio of the useful work produced as output to the heat intake from a heat reservoir: $e = W_{out}/Q_{in}$. All heat engines expel some heat to a cold reservoir as exhaust. The net work done during one thermodynamic cycle is, according to the first law of thermodynamics, $W_{out} = Q_{in} - |Q_{out}|$. It is therefore inherently impossible to construct a perfectly efficient heat engine – i.e. the efficiency must be less than unity: $e = 1 - |Q_{out}|/Q_{in}$. We will apply this definition to a variety of heat engines in Chapter 5.

1.4 Work: Interpreting Pressure-Volume (P-V) Diagrams

P-V diagram: A P-V diagram illustrates the relationship between pressure and volume as a system undergoes one or more thermodynamic processes. It does not provide a complete description of the thermodynamic behavior of the system. Rather, a P-V diagram represents the mechanical behavior of the system; it provides information about the work done by (or on) the system.

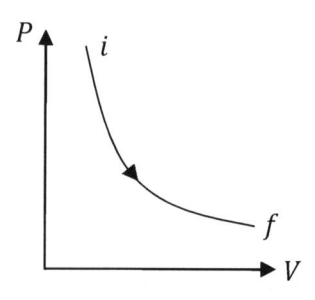

---

[78] This is neglecting vibrational energies, which are more subtle as they involve quantum mechanical effects.

***P-V* curves**: The *P-V* graphs for some basic thermodynamic processes include:[79]
- Compression – a curve that heads upward (pressure increases).
- Decompression – a curve that heads downward (pressure decreases).
- Expansion – a curve that heads to the right (volume increases).
- Contraction – a curve that heads to the left (volume decreases).
- Isochor – a straight vertical line (constant volume).
- Isobar – a straight horizontal line (constant pressure).
- Free expansion – a straight horizontal line along the *V*-axis (no pressure).
- Isotherm – often, a curve that resembles the hyperbola $P = a/V$ (constant temperature).[80]
- Adiabat – often, a curve that has the form $P^a V^b = c$ (no heat exchanged).[81]
- Thermodynamic cycle – a closed curve (same initial and final states).

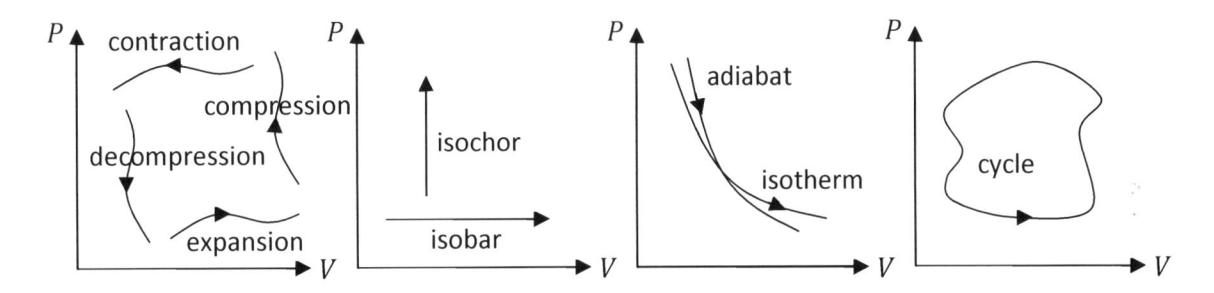

Thermodynamic work: Mechanical work is done by a force, $\vec{F}$, that contributes toward the displacement, $\vec{s}$, of an object as $W = \int_i^f \vec{F} \cdot d\vec{s}$. In thermodynamics, however, it is more convenient to work with pressure and volume than force and displacement. A gas that expands or contracts is displaced three-dimensionally along a volume, and the force that a fluid exerts on the walls of its container results in pressure. The pressure involves the normal component of the force, which is along the outward displacement of the fluid, such that the scalar product is simply $\vec{F} \cdot d\vec{s} = \pm F_n ds$ (where the positive sign corresponds to an expansion). The differential volume change can be separated into an outward displacement and cross-sectional area: $dV = A ds$. Therefore, $F_n ds = \left(\frac{F_n}{A}\right) A ds = P dV$.

---

[79] Strictly speaking, *P-V* curves correspond to quasistatic processes, as pressure and volume are only well-defined variables throughout the system for equilibrium states. Irreversible processes carried out between two equilibrium states are represented simply as two points – the initial and final equilibrium states.

[80] Two variables that have a reciprocal relationship – i.e. $y = 1/x$ – result in a hyperbola that is symmetric about the line $y = x$; as either variable increases, the other decreases. Adding a constant a, as in $y = a/x$, distorts the symmetry. Pressure and volume share such a relationship for an ideal gas with constant mole numbers when temperature is constant: $P = nRT/V$. Not all systems behave as ideal gases, but for fixed temperature, a compression (increase in pressure) generally decreases the volume, and increasing volume generally decreases the pressure – assuming both pressure and volume are free (e.g. a gas enclosed in a rigid container can't expand).

[81] It is generally expected that $b > a$, as in the case of an ideal gas (for which $a = 1$ and $b = \gamma = c_p/c_V > 1$). If so, compared to an isotherm, an adiabat will have a little higher pressure at lower volume and a little lower pressure at higher volume – which is important because isotherms and adiabats often look very similar. However, not all systems follow a relationship of this form for an adiabatic process.

Thus, in thermodynamics, work is done when a system changes volume: $W = \int_{V=V_0}^{V} PdV$.[82] Observe that work is positive when volume increases and negative when volume decreases. Positive work represents work done by the system, while negative work represents work done on the system.

Area under a *P-V* curve: In calculus, a definite integral conceptually represents the 'area' under the curve. Thus, the 'area' under a *P-V* curve represents the work done by the system, where the 'area' is positive for an expansion and negative for a contraction.[83] When the shape of the curve is algebraically known, in principle the area can be found algebraically by integrating the expression. However, for some expressions, such as e^{-x^2}, no algebraic antiderivative exists; also, sometimes you do not have a smooth curve, but a set of experimental data points. Thus, sometimes the area must be computed via numerical integration techniques.

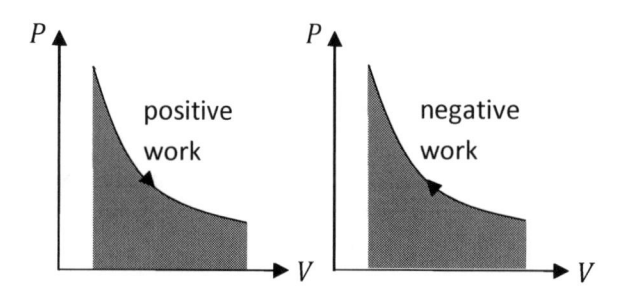

Area for a *P-V* cycle: A closed curve on a *P-V* diagram represents one complete thermodynamic cycle. Part of the cycle involves an expansion and part involves a contraction. Thus, work is positive for part of the cycle and negative for the rest. The work under the lower part of the *P-V* cycle is common to both and cancels out from the net work. Thus, the 'area' inside the closed curve is the net work done. The net work done during a cycle is positive for a clockwise path and negative for a counterclockwise path.

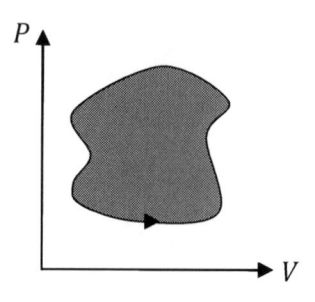

---

[82] An exception is when a gas expands freely (i.e. the gas is uncontained), in which case no work is done because the pressure is zero.
[83] The integral is like 'area' geometrically in that the area under the curve can be divided up into geometric shapes, like rectangles and triangles, and can be computed from geometric formulas like one-half base times height. However, in this case, the dimensions are those of pressure and volume, not distance, and so this 'area' does not have units of m^2.

Work done for common processes: The area under some common P-V curves include:
- Isochor – zero work is done because the volume remains constant ($\Delta V = 0$).
- Isobar – pressure is constant: $W = \int_{V=V_0}^{V} P dV = P \int_{V=V_0}^{V} dV = P(V - V_0) = P\Delta V$.[84]
- Free expansion – zero work is done because there is no pressure ($P = 0$).
- Isotherm – if the relationship between pressure and volume is known for fixed temperature, the integral can be performed by expressing the pressure in terms of the volume. For example, if $P = a/V$, then the work done is: $W = \int_{V=V_0}^{V} P dV = \int_{V=V_0}^{V}(a/V) dV = a \int_{V=V_0}^{V} V^{-1} dV = a \ln(V/V_0)$.[85]
- Adiabat – if the relationship between pressure and volume is known for the adiabat, the integral can be performed by expressing the pressure in terms of the volume. For example, if $P^a V^b = c$, where $a \neq b$, then the work done is:[86]

$$W = \int_{V=V_0}^{V} P dV = \int_{V=V_0}^{V} \left(\frac{c}{V^b}\right)^{1/a} dV = c^{1/a} \int_{V=V_0}^{V} V^{-b/a} dV = \frac{c^{1/a}}{1 - b/a}\left(V^{1-b/a} - V_0^{1-b/a}\right)$$

Path dependence: The work done by a system is generally path-dependent as the 'area' under the curve connecting an initial state to a final state depends upon the shape of the curve. Heat is similarly path-dependent. However, internal energy, which is a perfect differential (as notationally distinguished: dU, $đW$, and $đQ$), is path-independent. Thus, for a thermodynamic cycle, the net change in internal energy is zero (though it may increase or decrease during the cycle, the sum of such increases and decreases is zero), while work is generally done and heat is generally exchanged.

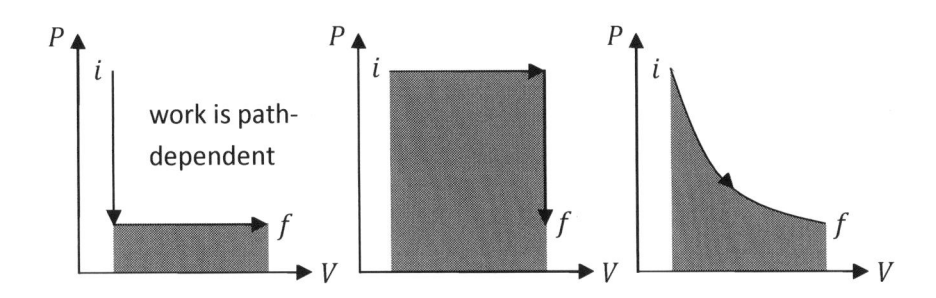

---

[84] This equation is not true in general, but only applies to work done along an isobar. If pressure is not constant, you may not pull it out of the integral. When pressure may be changing, you need to rewrite the pressure in terms of the volume before you can integrate.

[85] The relationship $P = a/V$ does not hold for all isothermal processes. It does hold, for example, for an ideal gas. In the last step, the logarithmic identity for the difference of logs is employed: $\ln(x/y) = \ln(x) - \ln(y)$.

[86] The relationship $P^a V^b = c$ applies to many, but not all, adiabatic processes. For an ideal gas, for example, $a = 1$ and $b = \gamma = c_p/c_V$. Note that temperature is generally not constant for an adiabat – i.e. adiabats are generally not isothermal (and isotherms are generally not adiabatic). For an isotherm, you may rewrite pressure in terms of both volume and temperature and compute the work done knowing that temperature is constant, but for an adiabat temperature is not constant and so you need pressure in terms of volume only. (For an adiabat, generally $a \neq b$; however, for the special case, $a = b$ the work done is logarithmic, like the $P = a/V$ isotherm).

Example. A system completes the thermodynamic cycle illustrated below, where $V_A = 10 \text{ m}^3$, $V_B = 2 \text{ m}^3$, $P_B = 5$ kPa, and $P_C = 20$ kPa. Determine the work done by the system for each process and for the complete cycle. Also determine the net heat exchange and internal energy change for the complete cycle.

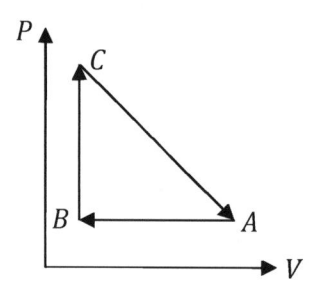

The work done for each process equals the area under each curve. The work done during the isobaric contraction, $A \to B$, equals the area of the rectangle below the line $\overline{AB}$: $W_{AB} = $ base × height $= (V_B - V_A)P_B = (2 \text{ m}^3 - 10 \text{ m}^3)\, 5 \text{ kPa} = -40$ kJ. It is negative since the volume decreased. No work is done during the isochoric pressurization, $B \to C$, as the volume did not change: $W_{BC} = 0$. The work done along the last process, $C \to A$, equals the area of the trapezoid below the line $\overline{CA}$: $W_{CA} = $ base × ave. height $= (V_A - V_C)(P_A + P_C)/2 = (10 \text{ m}^3 - 2 \text{ m}^3)(5 \text{ kPa} + 20 \text{ kPa})/2 = 100$ kJ. It is positive since the volume increased. The net work equals the area of the triangle: $W_{net} = $ base × height$/2 = (V_A - V_C)(P_C - P_B)/2 = (10 \text{ m}^3 - 2 \text{ m}^3)(5 \text{ kPa} - 20 \text{ kPa})/2 = 60$ kJ. Note that $W_{net} = W_{AB} + W_{BC} + W_{CA} = 60$ kJ.

The net internal energy change must be zero for a complete cycle because, unlike work and heat, it is path-independent: $\Delta U_{net} = 0$. The net heat exchange can then be found from the first law: $\Delta U_{net} = Q_{net} - W_{net}$. It is $Q_{net} = \Delta U_{net} + W_{net} = 0 + 60$ kJ $= 60$ kJ.

Example. Determine the work done along the isotherm illustrated below, where $V_A = 4 \text{ m}^3$ and $V_B = 20 \text{ m}^3$. The equation of the isotherm is $PV = 80$ kJ.

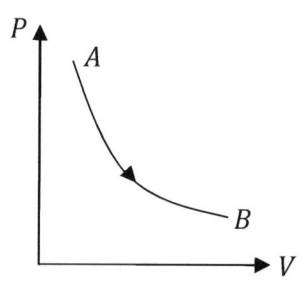

The area under the curve is given by the work integral:

$$W = \int_{V=V_A}^{V_B} P dV = \int_{V=V_A}^{V_B} \frac{80 \text{ kJ}}{V} dV = 80 \text{ kJ} \int_{V=V_A}^{V_B} \frac{dV}{V} = 80 \text{ kJ} \ln\left(\frac{V_B}{V_A}\right) = 80 \text{ kJ} \ln(5) = 1.3 \times 10^2 \text{ kJ}$$

1.5 Phase Transitions: Interpreting Pressure-Temperature (P-T) Diagrams

***P-T* diagram**: A P-T diagram for a substance is called a phase diagram as it illustrates what phase (i.e. solid, liquid, or gaseous) the substance will exist in for a given combination of temperature and pressure. It does not provide a complete description of the thermodynamic behavior of the substance, nor does it provide information about work done. It does provide useful information about the phase behavior and phase transitions of a substance.[87]

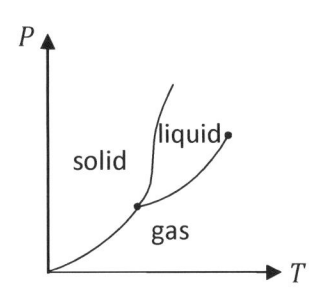

Phase curves: The phase curves show the boundary between two phases – solid-liquid, liquid-vapor, or solid-vapor – of a substance. The phase curve is also referred to as the coexistence curve.

Phase transitions: If the pressure or temperature (or both) of a substance is changed such that the phase curve is crossed, the substance generally undergoes a phase transition.

> **Conceptual Example**. Consider the three phases in the illustration above. The liquid can be cooled along an isobar (i.e. at constant pressure), decreasing its temperature until it freezes into solid. The liquid can also be decompressed along an isotherm (i.e. at constant temperature) until it vaporizes into gas. If the solid increases temperature along an isobar, it can melt into liquid or sublimate into gas, depending upon whether or not the pressure is above or below the triple point (where the three phase curves intersect). Both phase transitions are also possible if the solid is instead decompressed along an isotherm.[88]

Critical point: The liquid-vapor phase curve ends abruptly at the critical point. The properties of the two phases approach one another as the pressure and temperature are increased toward the critical point. Beyond this critical pressure and temperature, there is no clear distinction between the liquid and gaseous phases and the heat of vaporization is zero. The liquid and gaseous phases merge into a single phase, called a supercritical fluid, beyond the critical pressure and temperature.

---

[87] The phase diagram can vary considerably from the example illustrated here. This example serves to illustrate the key features, such as phase curves and critical points.
[88] The same is not true for water, which has an anomalous solid-liquid phase curve with negative slope. In the case of water, melting along an isotherm requires compression, rather than decompression. This is because water anomalously expands when freezing into ice.

1 Basic Relationships of Thermal Physics

Triple point: For a substance that exists in the solid, liquid, and gaseous phases, if the three phase curves intersect, this point of intersection is known as the triple point. The substance coexists in all three phases at the triple-point pressure and temperature.

Clausius-Clapeyron equation: The slope of the phase curve, which according to calculus equals the derivative of the pressure with respect to the temperature, is given by the Clausius-Clapeyron equation:

$$\frac{dP}{dT} = \frac{L}{T\Delta V}$$

where L is the latent heat and ΔV equals the amount by which the volume of the substance would change as a result of the phase transition at that point on the phase curve.[89]

Example. The Clausius-Clapeyron equation has the form, $dP/dT = aP/T^2$, for a system, where a is a constant. Determine the equation of the phase curve for this system.

Since a is constant, the Clausius-Clapeyron equation can be integrated after separating the two variables, P and T:

$$\frac{dP}{dT} = \frac{aP}{T^2}$$

$$\int_{P=P_0}^{P} \frac{dP}{P} = \int_{T=T_0}^{T} \frac{adT}{T^2}$$

$$\ln\left(\frac{P}{P_0}\right) = -a\left(\frac{1}{T} - \frac{1}{T_0}\right)$$

Metastable states: If the pressure and/or temperature of a substance are slowly changed such that the phase curve is crossed, yet a phase transition has not occurred, the substance is said to be in a metastable state. A sudden perturbation would likely result in a phase transition. Metastable states form as a result of supercooling, superheating, and supersaturation.

Conceptual Example. Just enough sugar is added to a cup of iced tea such that the solution is saturated – i.e. any more sugar added would be left undissolved. If the iced tea is slowly cooled, it can become supersaturated – i.e. if you had first cooled the iced tea and then added the sugar, some of the sugar would not have dissolved. If you now stir the solution, it will likely no longer be supersaturated.

---

[89] Note that the latent heat of vaporization and density (and therefore the change in volume) are, in general, a function of pressure and temperature. Hence, you can't simply treat L and ΔV as constants (since they aren't) and proceed to integrate to find the equation of the phase curve. However, if you know the equation of the phase curve, you can take a derivative of pressure with respect to temperature to determine its slope; and if you want to integrate the Clausius-Clapeyron equation, you need to express L and ΔV in terms of P, T, and constants. Additionally, the Clausius-Clapeyron equation may serve as a useful approximation for small changes in pressure and temperature – i.e. $\Delta P/\Delta T \approx L/T\Delta V$ if ΔP and ΔT are small (while ΔV need not be small – and generally isn't).

Conceptual Questions

The selection of conceptual questions is intended to enhance the conceptual understanding of students who spend time reasoning through them. You will receive the most benefit if you first try it yourself, then consult the hints, and finally check your answer after reasoning through it again. The hints and answers can be found, separately, toward the back of the book.

1. Each of the four vectors illustrated below has a magnitude of 5 m and a tip that lies at one corner of a regular pentagon. What are the magnitude and direction of the resultant?

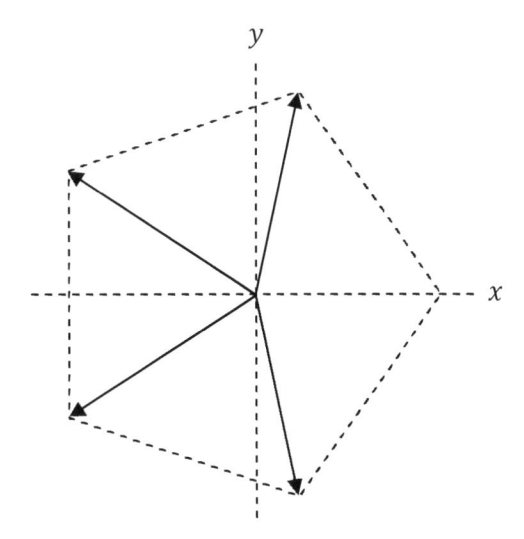

2. When a thermometer is placed in a pot of boiling water, it indicates that the temperature of the water is 100° C. How does the temperature of the thermometer compare to that of the water? Explain.

3. How is a thermos bottle effective at keeping coffee warm, yet also effective at keeping soda cool? What is different when the first law of thermodynamics is applied, conceptually, to the thermos bottle compared to a coffee mug that explains this?

4. A liquid is poured into a thermos bottle. The thermos bottle is then shaken vigorously. Is heat added to the liquid? If so, where did it come from; if not, how do you know this? Is work done on the liquid? Explain. Does the internal energy of the liquid change? Explain. Does the temperature of the liquid change? Explain. Describe the entropy changes of the liquid, the liquid's surroundings, and the overall entropy. Be sure to use the first and/or second laws of thermodynamics in your answers.

5. Absolute zero can be approached using a series of experimental techniques that effectively makes the molecules of the system move more slowly on average. The temperature of the system can also be lowered so that the molecules' individual average speeds are not reduced. How is this possible?

6. A box is pushed 20 m to the east along horizontal ground. Is this process thermodynamically reversible? Explain. Suppose that the box is now pushed 20 m to the west, returning the box to its original position (at rest). Describe any energy (including mechanical energy, internal energy, and heat exchanges) and/or entropy changes that occur for the box, its surroundings, and overall. Be sure to use the first and/or second laws of thermodynamics in your answers.

7. How do each of the first and second laws of thermodynamics relate to the possibility of designing a machine that remains in perpetual motion (i.e. without any external energy input once it is set in motion)? Now consider whether or not the universe as a whole is in perpetual motion. How does this relate to your response regarding a possible perpetual motion machine?

8. The second law of thermodynamics applies to the physical universe. Does it also apply to thought?[90] Specifically, what tends to happen to the entropy of human thought over the course of time – both for individuals and for the human race as a whole? Now consider that these thoughts are related to the physical brains of the humans who think them. What happens to the entropy of the human brain over the course of time? Discuss this in the context of the second law of thermodynamics.

9. How is it possible for the net work to be zero for a complete cycle that does not consist of a reversible process plus its reverse process (i.e. the P-V curve does not retrace its footsteps in the opposite direction, but actually makes a loop)? What can you say about the heat exchanged between the system and surroundings for the cycle in this case – and under what assumption?

10. For a complete cycle, is the work really path-dependent, or is it just area-dependent? That is, does the net work depend upon the shape of the path, or just the area enclosed by the shape? For example, would it make a difference if the path were circular, triangular, or square, so long as the path encloses a given amount of area?

11. For a complete cycle, do heat and work depend upon the choice of the initial point (which also serves as the final point)? Explain.

Practice Problems

The selection of practice problems primarily consists of problems that offer practice carrying out the main problem-solving strategies or involve instructive applications of the concepts (or both). However, this particular chapter mostly provides practice in the mathematical requisites. Techniques that will be described in more depth in future chapters will not appear in practice problems in this chapter – for example, this includes problems associated with the first law of thermodynamics. You will receive the most benefit if you first try it yourself, then consult the hints, and finally check your answer after working it out again. The hints to all of the problems and the answers to selected problems can be found, separately, toward the back of the book.

---

[90] This subject was considered by Ralph Waldo Emerson.

Vectors, unit vectors, position vectors, and the scalar product

1. (A) In 2D space, there are two possible vectors with a magnitude equal to -2 times its x-component. Explain. Express each in terms of Cartesian unit vectors. (B) In 3D space, how many vectors have a magnitude equal to -2 times its x-component and 4 times its z-component? Express each in terms of Cartesian unit vectors.

2. Three vectors lying in the xy plane have equal magnitudes, and one of the vectors has a direction of $90°$ (measured counterclockwise from the positive x-axis). Their resultant is zero. What are the directions of the other two vectors? Prove your answer.

3. Two vectors are defined as $\vec{A} = 2\hat{\imath} - \hat{\jmath}$ and $\vec{B} = 3\hat{\imath} + 2\hat{\jmath}$. Determine (A) the magnitude of $\vec{A}$, (B) the direction of $\vec{B}$, (C) the value of $\vec{A} \cdot \vec{B}$, (D) the angle between $\vec{A}$ and $\vec{B}$, (E) the magnitude and direction of $\vec{A} + \vec{B}$, (F) the magnitude and direction of $\vec{A} - \vec{B}$, (G) and the magnitude and direction of $2\vec{B} - 3\vec{A}$.

4. The position of a particle is given by $\vec{r}(t) = (t-2)^2\hat{\imath} - 4t\hat{\jmath}$, where position comes out in meters when time is input in seconds. Determine (A) the initial velocity of the particle, (B) the acceleration of the particle at $t = 4$ s, and (C) the speed of the particle at $t = 5$ s.

5. The velocity of a particle is given by $\vec{v}(t) = t^2\hat{\imath} - (2t+1)\hat{\jmath}$, where velocity comes out in m/s when time is input in seconds. Determine (A) the magnitude of the initial velocity of the particle, (B) the magnitude of the acceleration of the particle at $t = 3$ s, (C) the speed of the particle at $t = 2$ s, and (D) the net displacement of the particle from $t = 0$ to $t = 4$ s.

6. Three vectors are defined as $\vec{A} = 2\hat{\imath} + \hat{\jmath} - \hat{k}$, $\vec{B} = 3\hat{k} - \hat{\imath}$, and $\vec{C} = 2\hat{\jmath} - \hat{k} + \hat{\imath}$. Determine (A) $(\vec{A} + \vec{B}) \cdot \vec{C}$ and (B) $(\vec{C} - \vec{A}) \cdot (3\vec{A} + 2\vec{B})$.

Chain rule, partial derivatives, implicit differentiation, and the gradient

7. Given that $f(\theta) = 2\theta \sin 3\theta$ and $\omega = d\theta/dt$, express df/dt in terms of ω and θ.

8. The velocity of a particle moving along the x-axis is given by $v(t) = 3t^3 - 6t$, where velocity comes out in m/s when time is input in seconds. Evaluate dv/dx at $t = 2$ s.

9. Find the partial derivatives of the function $f(x,y) = 2x^4 - x^2y^2 + 3xy^3$ with respect to each of the independent variables, x and y.

10. Given that $f(u,v) = u^2/v$, $u = x \sin y$, and $v = \cos xy$, find the partial derivatives of f with respect to each of the independent variables, x and y.

11. Find the gradient of the function $f(x,y,z) = \dfrac{2x^2 + 3xy - y^2}{z^2}$.

1 Basic Relationships of Thermal Physics

12. Evaluate the gradient of the function $f(x,y,z) = 30[(x-3)^2 + (z-4)^2]^{-1}$ at the origin.

13. Implicitly differentiate $z = \frac{x^2}{y+1} - \frac{xy}{x-1}$, where x and y are independent variables.

14. Implicitly differentiate both sides of $x \sin x = y^4 - 4y$.

Displacement, arc length, line integrals, and the mean-value theorem

15. A particle travels halfway around a circle in the xy plane that is centered about the origin, starting at the point $(\sqrt{3}\text{ m}, -1.0\text{ m})$. The particle travels with a constant speed of 4.0 m/s. Find (A) the net displacement of the particle, (B) the total distance traveled by the particle, (C) the average speed of the particle, and (D) the average velocity of the particle.

16. A particle travels along the parabola $y = 9 - 2x^2$, where y comes out in meters when x is input in meters, starting at $x = -1.0$ m and finishing at $x = 2.0$ m. Find (A) the net displacement of the particle and (B) the total distance traveled by the particle.

17. Evaluate the integral $\int \vec{A} \cdot d\vec{s}$, where $\vec{A}(x,y) = 2x^3\hat{i} - xy^2\hat{j}$, over each of the following paths: (A) the straight line connecting the origin to (2,3) and (B) along the route $(0,0) \to (2,0) \to (2,3)$. (C) Is the field $\vec{A}$ a conservative field? Check this directly and also using the partial derivative method.

18. Determine the average value of the function $\sin^2 \theta$ over one complete cycle.

Polar coordinates, spherical coordinates, cylindrical coordinates, and multiple integrals

19. A particle travels with uniform circular motion in a circle of radius R_0 lying in the xy plane and centered about the origin. Its position vector is given by $\vec{r} = R_0\hat{r}$, where the direction of the radial unit vector $\hat{r}$ is specified by the angle θ, which is measured counterclockwise from the positive x-axis. Since the angular speed, ω_0, of the particle is constant in the case of uniform circular motion, $\theta = \omega_0 t$. Apply successive derivatives with respect to time in order to derive equations for the velocity and acceleration of the particle. Express your results in terms of the polar unit vectors.

20. Water is poured into a trough, which is shaped like one-half of a right-circular cylinder, until the water level reaches half the height of the trough. What fraction of the volume of the trough is filled with water?

21. Water is poured into a hemispherical bowl until the water level reaches half the height of the bowl. What fraction of the volume of the bowl is filled with water?

22. A glass sphere is painted in two colors as follows: The top third of its height is black and the bottom two-thirds of its height is white. (A) What fraction of the total paint used is black? (B) Viewed from the center, what is the solid angle corresponding to the black paint?

Density and pressure

23. A rod lying along the x-axis with one end at the origin has non-uniform density $\rho(x) = \beta x$, where β is a constant. The rod is cut in half. How much does each end weigh in terms of its total weight?

24. A solid disc with radius R_0 lying in the xy plane, centered about the origin, has non-uniform density $\rho(r) = \beta r$, where β is a constant. What fraction of its mass lies in the range $0 \leq r \leq R_0/2$?

25. An umbrella has the shape of a hemisphere with a radius of 60 cm. The rain falls straight down on the umbrella, which is held vertically. The raindrops exert an approximately constant force of 3.0 N downward on the umbrella. What average pressure does the rain exert on the umbrella?

Work and P-V diagrams

26. A system completes the thermodynamic cycle illustrated below on the left, where $V_A = 5 \text{ m}^3$, $V_B = 20 \text{ m}^3$, $V_C = 25 \text{ m}^3$, $P_A = 25 \text{ kPa}$, and $P_C = 10 \text{ kPa}$. Determine the work done by the system for each process and for the complete cycle.

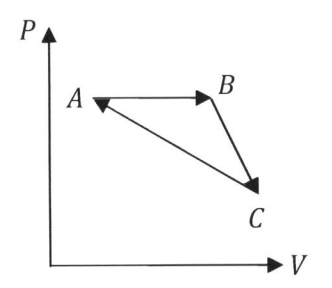

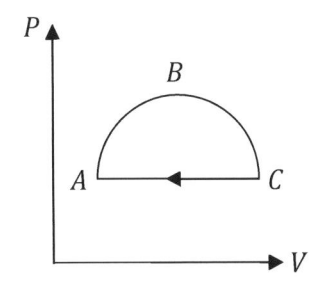

27. A system completes the thermodynamic cycle illustrated above on the right, where $V_A = 5 \text{ m}^3$, $V_C = 25 \text{ m}^3$, $P_A = 20 \text{ kPa}$, and $P_B = 60 \text{ kPa}$. The curved path has the shape of a semicircle.[91] Determine the work done by the system for each process and for the complete cycle.

28. A fluid is governed by the equation $PV^2 = aT^2$, where a is a constant. Express the work done by the fluid along each of the following paths in terms of the initial pressure P_0 and initial volume V_0. (A) The fluid expands from V_0 to $4V_0$ along an isotherm. (B) The fluid is heated from T_0 to $4T_0$ along an isobar. (C) The fluid is cooled from T_0 to $T_0/4$ along an isochor.

29. A fluid doubles its volume according to the equation $P^2V^3 = const$. Determine the work done by the fluid in terms of its initial pressure P_0 and its initial volume V_0.

---

[91] Technically, the shape is a semi-ellipse since the pressure and volume axes have different dimensions – therefore the major and minor axes of the ellipse inherently cannot be equal. However, an ellipse can be transformed into a circle by rescaling the axes. For example, an ellipse centered about the origin with semimajor axis a along x and semiminor axis b along y can be rescaled into a circle via the transformation $x \to x/a$ and $y \to y/b$. Rescaling the axes thus, it is very easy to see that the area of an ellipse is equal to πab.

2 Basic States and Processes in Thermal Physics

2.1 Thermal Expansion of Liquids and Solids

Thermal expansion: As might be expected intuitively from the conceptual interpretation of absolute temperature, the average interatomic distance tends to increase as the temperature of a liquid or solid[92] substance increases. This increase in the average distance between atoms causes the volume of the substance to expand. For small relative expansions, the fractional change in volume is observed to be approximately linear with the temperature change. Many solids expand isotropically (uniformly in all directions), in which case the fractional change in length measured along a linear dimension is also directly proportional to the temperature change.

However, there are some notable exceptions to this rule. In particular, water tends to expand when freezing, which is why pipes sometimes burst during cold weather. Above 4° C near atmospheric pressure, water expands with decreasing temperature and contracts with increasing temperature, and the expansion or contraction of water around this temperature is more quadratic than it is linear.

Linear expansion thermometer: The common mercury-in-glass thermometer is based on thermal expansion. Mercury in a glass tube expands and contracts with temperature changes. The fractional length of the expanded or contracted mercury is proportional to the temperature change. Thus, the position of the mercury in the glass tube provides a measurement of temperature.

Three-dimensional expansion: A solid or liquid substance that increases in temperature by a differential amount dT expands (or contracts) by a differential amount $dV = \beta V_0 dT$, where the coefficient of thermal (volume) expansion, β, is characteristic of the substance and V_0 is the volume prior to the expansion (or contraction). For a small fractional change in volume, $\Delta V/V_0$, β is approximately constant and integration yields $\Delta V = \beta V_0 \Delta T$. The coefficient of thermal (volume) expansion, β, is generally positive; for water near its freezing point, however, β is negative. The SI units of β are per degree Kelvin (K^{-1}).

Example. Derive an equation that relates the initial and final radii for a metal sphere that expands isotropically when it is heated, assuming that its fractional volume change is small.

The volume of the sphere is related to its radius by $V = 4\pi R^3/3$:

$$R = \sqrt[3]{3V/4\pi} = \sqrt[3]{3(V_0 + \Delta V)/4\pi} = \sqrt[3]{3(V_0 + \beta V_0 \Delta T)/4\pi} = \sqrt[3]{(R_0^3 + \beta R_0^3 \Delta T)} = R_0 \sqrt[3]{(1 + \beta \Delta T)}$$

---

[92] A gas is very often confined in a container from all sides, in which case the gas itself has no room to expand unless the container expands with it. Also, a gas naturally tends to expand – i.e. no increase in temperature is needed. Just increase the volume of its container and the gas will naturally fill up the space.

Isotropic expansion: When a solid expands isotropically, the fractional increase in length is the same in all directions. Consider a rectangular solid with original length L_0, width W_0, and height H_0. Its increased volume V can be expressed as:[93]

$$V = LWH = (L_0 + \Delta L)(W_0 + \Delta W)(H_0 + \Delta H)$$
$$LWH = L_0 W_0 H_0 + \Delta L W_0 H_0 + \Delta W L_0 H_0 + \Delta H L_0 W_0 + \mathcal{O}(\Delta^2)$$
$$\Delta V = \Delta L W_0 H_0 + \Delta W L_0 H_0 + \Delta H L_0 W_0 + \mathcal{O}(\Delta^2)$$
$$\frac{\Delta V}{V_0} = \frac{\Delta L W_0 H_0 + \Delta W L_0 H_0 + \Delta H L_0 W_0}{L_0 W_0 H_0} + \mathcal{O}(\Delta^2)$$
$$\frac{\Delta V}{V_0} = \frac{\Delta L}{L_0} + \frac{\Delta W}{W_0} + \frac{\Delta H}{H_0} + \mathcal{O}(\Delta^2)$$
$$\frac{\Delta V}{V_0} = \beta \Delta T = 3 \frac{\Delta L}{L_0} + \mathcal{O}(\Delta^2)$$
$$\Delta L = \frac{\beta}{3} L_0 \Delta T - \mathcal{O}(\Delta^2)$$

This result exemplifies the following principle: For a solid that expands isotropically, the increase in length along any of its dimensions may be thought of as a linear expansion of the form $\Delta L = \alpha L_0 \Delta T$, for a small fractional increase in length, where the coefficient of linear expansion is related to the coefficient of thermal (volume) expansion by $\beta = 3\alpha$.

Linear expansion: For a solid that expands isotropically, since the fractional increase in length is the same for each of its dimensions, the increase in length is largest for its longest dimension(s).[94] Thus, the linear expansion along the solid's longest dimension(s) is most relevant: $dL = \alpha L_0 dT$.

Example. A metal tape measure is designed for use at room temperature. On a hot day, for which the metal expands by 2%, the tape measure is used to measure the length of a wooden pole. One end of the pole lies at the 0-cm mark and the other end lies at the 200-cm mark. How long is the pole?

The tape measure is expanding by 2%, but the wooden pole's expansion will be negligible compared to that of the metal. Conceptually, it is clear that the wooden pole is longer than 200 cm. That is, if we cool the system (metal tape measure plus wooden pole), the wooden pole will remain approximately the same length while the metal tape measure will contract – i.e. the markings will get closer together and the wooden pole will extend beyond the 200-cm mark. Since the tape measure expanded by 2%, the wooden pole is 2% longer than 200 cm: $200 \text{ cm} (1 + 2\%) = 204 \text{ cm}$. Thus, the wooden pole is 204 cm long.[95]

---

[93] The $\mathcal{O}(\Delta^2)$ represents all terms of order Δ^2 or higher (such as $L_0 \Delta W \Delta H$). Since the fractional change is assumed to be small, squaring the change is even smaller – such that these higher-order terms represent a negligible contribution. In the next-to-last step, the three fractional changes are set equal because the expansion is isotropic.

[94] For example, a 10% increase enhances a 1-cm dimension by 1 mm, but enhances a 0.2-cm dimension by only 0.2 mm. The fractional increase is the same, but the increased amount is different.

[95] If instead a wooden meterstick is used to measure a metal pole, then the meterstick reading will be greater at the higher temperature. In both cases, the metal expands, but they differ in which measured reading is larger.

Two-dimensional expansion: Consider a solid that has constant cross section, such as a right cylinder (not necessarily circular) or a cuboid (a rectangular box). If it expands isotropically, the cross-sectional area expands according to $dA = 2\alpha A_0 dT$. The coefficient of areal expansion equals 2α (or $2\beta/3$).[96]

Example. A right-circular cylinder made of metal has an initial radius of 4 cm and length of 20 cm. The metal is heated until its length becomes 21 cm. By what factor does its cross-sectional area increase?
  The length increased by 5%. Therefore the fractional area increase will be 10% (since the coefficient of area expansion equals 2α) – i.e. the cross-sectional area increases by a factor of 1.1.

2.2 Thermal Equilibrium: Calorimetry

Calorimeter: A device that quantitatively measures how much heat is exchanged is called a calorimeter. A simple calorimeter consists of a container of liquid with a thermometer.

Calorimetry: The scientific technique of measuring heat exchange with a calorimeter is referred to as calorimetry. If an object is placed in the liquid of a simple calorimeter, if the solid and liquid have different initial temperatures, heat is exchanged until thermal equilibrium is attained. The heat exchanges between the sample, liquid, container, and the container's surroundings are related by conservation of energy. The heat gained or lost by any single material is related to its specific heat, mass, and temperature change.

Thermal equilibrium: The temperature of the thermometer immersed in the liquid of a simple calorimeter changes as heat exchanges occur. Thermal equilibrium is reached when the heat exchanges cease – i.e. when the temperature stabilizes.[97] At thermal equilibrium, the sample and liquid have the same temperature.

Specific heat capacity: At the differential scale, the amount of heat gained or lost by a substance is related to its temperature change by its mass and specific heat capacity: $đQ = MC_{P,V} dT$, where C_P is appropriate for an isobar (constant pressure) and C_V is appropriate for an isochor (constant volume). If the mass and specific heat capacity are constant, integration yields $Q = MC_{P,V} \Delta T$.

Conservation of energy: The first law of thermodynamics expresses conservation of energy in a way that accounts for internal energy changes, heat exchanges, and work done. For a simple calorimeter, the work is negligible and conservation of energy simplifies to state that the total heat gained plus the total heat lost equals zero (since heat loss is negative): $\sum Q_g + \sum Q_\ell = 0$.[98]

---

[96] Linear, areal, and volume expansions come in the ratio $\alpha: 2\alpha: 3\alpha$.
[97] If any substance changes phase, the heat it releases or absorbs first goes toward supplying the energy needed for its phase transition before changing its temperature. That is, the temperature of a material may be temporarily constant while it changes phase, and then begin to change once the phase transition is complete.
[98] Alternatively, you may set the net heat gain equal to the absolute value of the net heat loss: $\sum Q_g = |\sum Q_\ell|$.

When applying conservation of energy to a calorimeter, if any substance changes phase, it is necessary also to include the energy that supplies the phase transition: $Q = ML$, where L is the latent heat of transformation. For example, if an ice cube is placed in a thermally insulated cup of water where the only heat exchanges occur between the ice and water, the heat lost by the water partly goes into melting the ice cube and partly goes into raising the temperature of the resulting ice water until all of the water reaches the equilibrium temperature.[99]

Example. A solid sample initially at temperature T_s is placed in a thermally insulated cup of liquid that is initially at temperature $T_\ell < T_s$. Treating the system as the solid sample and the liquid, and neglecting any heat exchanges between the system and surroundings, derive an equation for the equilibrium temperature in terms of the initial temperatures assuming that no phase changes occur.

First, apply the first law of thermodynamics to express conservation of energy assuming that a negligible amount of work is done (since the volume changes will be negligible). Specifically, the heat gained by the liquid plus the heat lost by the solid equals zero. Conceptually, you may find it simpler to set the heat gain equal to the absolute value of the heat loss, in which case you don't have to keep track of the minus signs:[100,101]

$$Q_g^\ell = |Q_\ell^s|$$
$$M_\ell C_\ell (T_e - T_\ell) = M_s C_s (T_s - T_e)$$
$$T_e = \frac{M_\ell C_\ell T_\ell + M_s C_s T_s}{M_\ell C_\ell + M_s C_s}$$

Example. A block of ice with initial temperature T_i is placed in a thermally insulated cup of water with initial temperature T_w. All of the ice melts. Assuming that the only heat exchanges occur between the ice and water, derive an equation for the equilibrium temperature in terms of the initial temperatures.

The heat lost by the water first raises the temperature of the ice to its melting point, $T_m = 0°C$, then releases energy as the ice melts, and then raises the temperature of the ice water to equilibrium:

$$Q_g^{ice} + M_i L + Q_g^{ice\,water} = |Q_\ell^{water}|$$
$$M_i C_i (T_m - T_i) + M_i L + M_i C_w (T_e - T_m) = M_w C_w (T_w - T_e)$$
$$T_e = \frac{M_i C_i T_i - M_i L + M_w C_w T_w}{(M_i + M_w) C_w}$$

---

[99] In the case of a phase change, work is done because the volume changes (e.g. the same mass of water occupies much more volume in the form of steam).

[100] While this is fine for calorimetry, you will have to pay attention to the signs of heat and work in the more general applications of the first law of thermodynamics (Sec. 2.5). When you write down the temperature changes, make sure that the signs are consistent on both sides of the equation. In this case, since $T_\ell < T_s$ initially, the temperature of the solid will decrease and the temperature of the liquid will increase.

[101] Focus on understanding the strategy and concepts. Don't memorize the equations derived in examples; rather, learn how to derive the equations. It is best to know the small set of fundamental equations and how to use them. Most derived equations are special cases with limited application. For example, in this case, if you only know the derived equation, what would you do if there are heat exchanges between three substances, rather than just two? If you understand the strategy and concepts, you will be able to apply it to more general situations.

2.3 A Steady State: Heat Conduction

Methods of heat transfer: Thermal energy is often transferred by one or more of three common methods – conduction, convection, and radiation. Heat is transferred through a solid through thermal conduction, in which the conduction occurs continuously through the volume of the object. As one portion of a solid is heated, atoms in that region begin to vibrate with greater amplitudes about their equilibrium positions. As they do so, they can transfer kinetic energy to their neighbors during collisions. These collisions ultimately transfer thermal energy throughout the volume of the object – conducting heat. Heat conduction occurs when one end of a fireplace poker is placed in a fire and the other end feels warm, and also through the thickness of a window pane that separates a warm room from cold weather outdoors. Thermal conduction is particularly efficient for metals, which tend to be good conductors of heat. Thermal conduction also occurs through fluids. However, convection is generally more efficient than thermal conduction for fluids.[102]

Whereas conduction transfers heat through interatomic collisions, convection transfers heat through the actual movement of the fluid. As one portion of a fluid is heated, that portion expands; hence, it becomes less dense than the rest of the fluid and rises upward according to Archimedes' principle.[103] This is the origin of the phrase, "Heat rises." A fire or a room heater thus produces a density gradient in the air around it, causing thermal energy to be transferred throughout the air around it by convection. Another example of convection is in the convection currents that form as a pot of water is heated: Again a density gradient is established with less dense layers at the bottom where the stove's heat is directly applied, allowing thermal energy to be transferred upward through convection currents.

All objects exchange thermal radiation with their surroundings. Macroscopic[104] objects both continuously radiate electromagnetic waves to their surroundings and absorb electromagnetic waves from their surroundings. Macroscopic objects consist of large numbers of molecules, which have kinetic energy. At any instant many of these molecules are engaged in thermal vibrations that emit electromagnetic waves. Thus, all macroscopic objects are always emitting thermal radiation. The universe is also filled with an abundance of electromagnetic radiation, which all macroscopic objects continuously absorb.[105] The sun provides the vast majority of electromagnetic energy that heats the earth (and therefore objects on earth). By comparison, thermal conduction and convection from the sun to the earth are virtually impossible, since the region between the sun and earth is a virtual vacuum.

---

[102] Fluids include both liquids and gases. It's easy to remember that fluids include the phases of matter that flow.

[103] According to Archimedes' principle, an object submerged in a fluid experiences a buoyant force equal to the weight of the displaced fluid (not, in general, equal to the weight of the displaced object – that's only true in the case of flotation, where only a fraction of the object is actually submerged in liquid). As a result, an object rises upward if it has less density than the fluid (like a hot air balloon) and falls downward if its density exceeds that of the fluid.

[104] Macroscopic could actually mean microscopic. In thermodynamics, macroscopic is used in reference to any objects or system that has a statistically very large sample of molecules, which includes objects on the order of a micron (micrometer) in size – i.e. 'microscopic' – or smaller.

[105] Indeed, an object would have to be extremely well insulated from the environment in order to minimize this. "You can run, but you can't hide!"

Thermal conduction: When two surfaces of a solid object are at different temperatures, thermal energy may be transferred from the higher-temperature surface to the lower-temperature surface via conduction. In this case, the instantaneous rate at which thermal energy is transferred is proportional to the temperature gradient and the cross-sectional area by a constant, called the thermal conductivity, which depends upon how well the material conducts heat. The proportionality with cross-sectional area can be understood through analogy with electrical conduction: In the case of electric current, the electrons have more collisions, on average, passing through a thin conductor than a thick conductor, all else being equal, just as opening lanes on a freeway tends to relieve traffic, while closing lanes tends to impede traffic.[106] A larger temperature difference between the two surfaces or less spacing between the two surfaces yields a greater temperature gradient, explaining the proportionality with the gradient.

At a given point in the substance, the instantaneous rate, P, at which heat is transferred is related to the thermal conductivity, k, of the substance, the cross-sectional area, $\vec{A}$, and the temperature gradient, $\vec{\nabla}T$, by:

$$P = -\int k\vec{\nabla}T \cdot d\vec{A}$$

The differential area element, $d\vec{A}$, is cross-sectional area – i.e. the integral is over the surface area that is perpendicular to the direction of the heat flow; the direction of $d\vec{A}$, however, is along the heat flow. The overall factor, $-k$, provides a measure of how well the given substance conducts heat and includes a minus sign because heat flows naturally in the direction of decreasing temperature.

A natural heat flow is along the negative of the temperature gradient, such that $\vec{\nabla}T \cdot d\vec{A}$ reduces to $-\|\vec{\nabla}T\|dA$. The temperature gradient is often a function of the coordinate perpendicular to $d\vec{A}$, while constant over the cross-sectional area (the surface of integration). If so, it is convenient to work in an orthogonal coordinate system where two coordinates are tangent to the cross-sectional area A and the third coordinate, call it s, is along the heat flow, such that the magnitude of the gradient becomes $\|\vec{\nabla}T\| = |dT/ds|$. If the thermal conductivity is uniform over the surface of integration, it can come out of the integral. In these cases, the instantaneous rate at which heat is transferred simplifies to:[107]

$$P = k\left|\frac{dT}{ds}\right|A$$

This instantaneous rate of energy transfer P is a function of s, which is a coordinate along the heat flow, and in this case is uniform over the cross-sectional area (which may also be a function of s).

---

[106] For a wire with uniform resistivity ρ, uniform cross-sectional area A, and length L, the resistance equals $R = \rho L/A$. The conductivity σ is the reciprocal of resistivity: $\sigma = 1/\rho$. Thus, $R = L/\sigma A$. According to Ohm's law, the voltage ΔV equals the current I times the resistance R: $\Delta V = IR$. Therefore, $\Delta V = IL/\sigma A$, or for a given voltage, a greater cross-sectional area results in more current, for the same material and length of wire.

[107] This equation does not apply (hence direct integration is instead needed) if (a) the heat flow is not spontaneous (i.e. forced to differ from its natural route), if (b) the thermal conductivity is not constant along the cross-sectional area (i.e. if the thermal conductivity is non-uniform for the substance), or if (c) you choose a coordinate system other than the one described here (such a coordinate system is natural for a very symmetric cross-sectional area, like a sphere, cylinder, or plane, but can be awkward in general).

This formula also assumes that a steady state has been reached. The direction of the heat flow is settled and temperature is well-defined for the steady state. The instantaneous rate of heat flow, P, is also the same at all cross sections in the above formula for the steady state, since energy must be conserved: However much energy passes through the cross-sectional area defined by one value of s, the same amount of energy must pass through the cross-sectional areas defined by any other value of s (otherwise, some energy was gained or lost).

This first-order differential equation is separable as long as the thermal conductivity k is not a function of both T and s:[108] That is, the temperature, T, and heat flow coordinate, s, can be separated on different sides of the equation. Note that the cross-sectional area A is generally a function of s. If indeed the equation is separable, both sides can then be integrated:

$$P \int_{s=s_0}^{s} \frac{ds}{kA} = \int_{T=T_l}^{T_h} dT = T_h - T_l = \Delta T$$

where the heat flows spontaneously from the surface with higher temperature, T_h, to the surface of lower temperature, T_l. This is generally the most practical form of the equation for deriving an expression for instantaneous rate of heat flow (provided that the various assumptions apply to the specific problem at hand). Here we have assumed that k is not a function of temperature.

Example. Derive an equation for the steady-state rate of heat transfer by thermal conduction across the length L of a thermal conductor with uniform cross-sectional area A and uniform thermal conductivity k, where the two ends have different temperature and the body is thermally insulated.

In this case, we may apply the equation above. The cross-sectional area is constant along the length of the conductor and therefore can come out of the integral along with k, which is also constant. As heat flows along the length of the conductor, we can use the cylindrical or rectangular coordinate z for the heat flow coordinate s.

$$P \int_{s=s_0}^{s} \frac{ds}{kA} = \frac{P}{kA} \int_{z=0}^{L} dz = \frac{PL}{kA} = \Delta T$$

$$P = kA \frac{\Delta T}{L}$$

Conceptually, it should make sense that P is greater for a material that conducts heat better (one with a higher value of k), a shorter conductor (one with a shorter L), or a greater temperature difference (larger ΔT). For the effect of cross-sectional area, compare with electrical conduction: In either case, the transfer of energy occurs at a greater rate if the conductor has greater cross-sectional area because this is like opening more lanes on a freeway during rush-hour traffic. Observe that the result for P is uniform throughout the conductor for this particular geometry and direction of heat flow.

---

[108] In obtaining the previous equation, we only required k to be uniform over the cross-sectional area; hence, it may still be a function of the heat flow coordinate, s.

***R* value**: The ratio of the length of the conductor separating the two surfaces of different temperature to the thermal conductivity is called the *R* value of the conductor: $R = L/k$. The result of the previous example, involving a thermal conductor with uniform cross-sectional area and uniform thermal conductivity, can be expressed in terms of the *R* value as:

$$P = A\frac{\Delta T}{R}$$

Example. A group of N rectangular conducting slabs are placed in series, as illustrated below. The slabs make thermal contact at the shared walls. The far ends are at different temperatures T_h and T_l, where $T_h > T_l$. The remaining walls are thermally insulated. Each slab has its own thermal conductivity, k_i, and length, L_i; and the thermal conductivity of each slab is uniform throughout. Derive a steady-state formula for the effective R value for the series in terms of their individual R_i values.

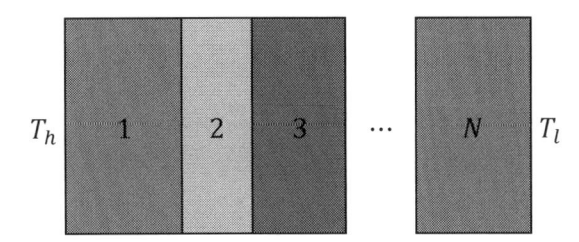

In series, once a steady state is reached, conservation of energy requires that $P_1 = P_2 = \cdots = P_N$ – because if the instantaneous rates of energy transfer for two slabs were different, energy would either be gained or lost. The overall temperature difference, $\Delta T = T_h - T_l$, must equal the sum of the temperature differences across each of the slabs: $\Delta T = \Delta T_1 + \Delta T_2 + \cdots + \Delta T_N$. These temperature differences can be expressed in terms of the instantaneous energy rates as:

$$\frac{PR_s}{A} = \frac{P_1 R_1}{A_1} + \frac{P_2 R_2}{A_2} + \cdots + \frac{P_N R_N}{A_N}$$

The cross-sectional area and instantaneous rate of energy transfer cancel, since these are the same for each slab. Therefore, R values add for thermally conducting slabs (with equal cross-sectional areas) connected in series:

$$R_s = R_1 + R_2 + \cdots + R_N$$

In terms of the thermal conductivities,

$$\frac{L}{k_s} = \frac{L_1}{k_1} + \frac{L_2}{k_2} + \cdots + \frac{L_N}{k_N}$$

The same instantaneous rate of heat transfer would result if the series were replaced by a single conducting slab of the same total length, $L = L_1 + L_2 + \cdots + L_N$, and equivalent conductivity k_s.

Example. A spherical conducting shell with inner radius a and outer radius b has uniform thermal conductivity k. Its inner and outer walls have different temperatures. Derive an equation for the steady-state rate at which heat is transferred by thermal conduction across the conducting shell.

In this case, the cross-sectional area is not constant, but varies radially. Conservation of energy again requires that P be uniform throughout the conductor. The heat flows radially, such that the radial coordinate r of spherical coordinates serves as a convenient heat flow coordinate:

$$P \int_{s=s_0}^{s} \frac{ds}{kA} = \frac{P}{k} \int_{s=a}^{b} \frac{dr}{A} = \Delta T$$

The cross-sectional area is $A = 4\pi r^2$. Substituting this into the denominator,

$$\frac{P}{k} \int_{s=a}^{b} \frac{dr}{4\pi r^2} = \frac{P}{4\pi k} \frac{b-a}{ab} = \Delta T$$

$$P = 4\pi k \Delta T \frac{ab}{b-a}$$

Electrical conduction: Thermal conduction has several analogies with electrical conduction, since the properties of metals that tend to make them good conductors of electricity also tend to make them good conductors of heat. These analogies are particularly instructive for students who have already studied calculus-based electromagnetism.

Temperature, T, is analogous to electric potential, V. An electric potential difference, ΔV, between two surfaces of a conductor creates an electric field, $\vec{E}$, in the conductor, which accelerates electrons, creating electric current, I. Similarly, a temperature difference, ΔT, between two surfaces of a conductor creates a flow of heat through the conductor. The negative temperature gradient, $-\vec{\nabla}T$, is analogous to electric field, $\vec{E}$. This follows as electric field equals the negative of electric potential: $\vec{E} = -\vec{\nabla}V$. Emf (which only differs from terminal potential difference, ΔV_t, due to internal resistance of a power supply) equals the negative line integral of electric field: $\Delta V = -\int \vec{E} \cdot d\vec{s}$. Temperature difference can similarly be expressed as a line integral of the temperature gradient: $\Delta T = \int \vec{\nabla}T \cdot d\vec{s}$. Thus, temperature difference, ΔT, can be thought of as sort of a "thermal emf," and negative temperature gradient, $-\vec{\nabla}T$, can be thought of as a sort of "heat field."

The electric flux, Φ_e, through a surface equals the integral of the scalar product of electric field, $\vec{E}$, and the differential area element, $d\vec{A}$, over that surface: $\Phi_e = \int \vec{E} \cdot d\vec{A}$. Conceptually, electric flux, Φ_e, provides a measure of the relative number of electric field lines entering versus exiting the surface.[109] A similar integral over the temperature gradient, i.e. $-\int \vec{\nabla}T \cdot d\vec{A}$, therefore provides a measure of heat flux – the number of thermal field lines passing through a given surface.

---

[109] Technically, there are field lines everywhere. We're not comparing infinities, though. When we make a map of electric field lines, we draw a diagram to scale, where the scale is how many field lines to draw per Coulomb of source. In this way, there are a finite number of field lines passing through any surface.

On such topological maps – namely, electric field maps and temperature gradient maps[110] – field lines exit and enter sources. For an electrostatic electric field map, the sources are electric charges; electric field lines (aka lines of force) exit positive charges and enter negative charges. On a dynamic electric field map – e.g. where a DC power supply provides a potential difference between two surfaces of a conductor – electric field lines flow from the higher-potential surface to the lower-potential surface. Thermal field lines similarly run from the higher-temperature surface to the lower-temperature surface. Relative highs and lows are analogous to positive and negative electric charges, but there is an important difference: All absolute temperatures are positive, while charges can be positive or negative.[111] Just as Gauss's law states that the net electric flux passing through any closed surface is proportional to the net charge enclosed by the surface, $\oint \vec{E} \cdot d\vec{A} = q_{enc}/\epsilon_0$, Gauss's law for thermal conduction similarly states that the net heat flux passing through any closed surface is proportional to the net temperature source enclosed (which requires more thought, and drawing out a few maps, to understand compared to the more obvious case of counting up net electric charge).

Electric current, I, is the instantaneous rate of flow of electric charge, q: $I = dq/dt$. The analogous thermal current is the instantaneous rate of flow of heat, which we have denoted as P. Electric current, I, is most formally related to the current density, $\vec{J}$, by: $I = \int \vec{J} \cdot d\vec{A}$. Current density is proportional to the electric field inside the conductor: $\vec{J} = \sigma \vec{E}$. Since electric field, $\vec{E}$, is analogous to the negative temperature gradient, $-\vec{\nabla}T$, current density, $\vec{J}$, is analogous to $-k\vec{\nabla}T$. Compare the formula for the instantaneous rate of heat flow, $P = - \int k\vec{\nabla}T \cdot d\vec{A}$, to the relationship between electric current and current density, $I = \int \vec{J} \cdot d\vec{A}$. Now it is easy to see that P is indeed a sort of thermal current. Just as Ohm's law ($\Delta V = IR$) relates electric potential difference, ΔV, to electric current, I, by a factor of electric resistance, R, one could similarly define a thermal resistance, R_{th}, as the ratio of temperature difference, ΔT, to thermal current, P: $\Delta T = PR_{th}$.

Thermal conductivity, k, is analogous to electrical conductivity, σ, which is the inverse of resistivity, $\rho = 1/\sigma$. Thus, thermal conductivity, k, is analogous to inverse resistivity, $1/\rho$. For an electric conductor of length L with constant cross-sectional area A, resistance, R, is related to resistivity, ρ, by $R = \rho L/A$. For a similar thermal conductor, L/kA provides a measure of thermal resistance. This can also be written as R number divided by cross-sectional area (so R number is proportional to, but not quite the same as, thermal resistance). The equivalent resistance of a series of N electric resistors equals the sum of the resistances: $R_s = R_1 + R_2 + \cdots + R_N$. In an example, we found a similar formula for N thermal conductors connected in series. The equivalent resistance of N parallel electric resistors is found by adding the resistors in reciprocal: $1/R_s = 1/R_1 + 1/R_2 + \cdots + 1/R_N$. A similar relationship holds for thermal resistance. If all of the thermal conductors have the same length and cross-sectional area, note that whereas thermal resistance adds in series and adds in reciprocal in parallel, thermal conductivity in this case adds in parallel and adds in reciprocal in series (just the opposite).

---

[110] The temperature gradient map discussed here is much simpler than a weather map, which also involves other thermal physics quantities, such as pressure.

[111] It is interesting to compare electricity, gravity, and thermal conduction. The source of electric field – electric charge – can be positive or negative, and opposites attract while likes repel. The source of gravitational field – mass – can evidently only be positive, and like masses (for there are no unlike masses) attract. The source of a thermal field (i.e. temperature gradient) is a relative high or low temperature, which can only be positive, and heat flows from high to low temperatures just as electric field runs from positive to negative charge.

Kirchhoff's rules[112] also apply to thermal circuits. In this case, the junction rule applies to thermal current (P) and the loop rule applies to temperature differences (ΔT). Whereas the junction rule corresponds to conservation of electric charge in electric circuits, it corresponds to conservation of thermal energy in thermal conduction circuits.[113] The loop rule states that the sum of the temperature changes will add to zero for a closed loop in a thermal conduction circuit. Observe that thermal current (P) is the same for thermal conductors in series, whereas the temperature difference (ΔT) is the same for thermal conductors in parallel.

2.4 Radiative Equilibrium: Electromagnetic Radiation

Stefan's law: The thermal energy that an object radiates in the form of electromagnetic waves tends to cause its temperature to decrease.[114] Similarly, the thermal energy that an object absorbs from its surroundings tends to cause its temperature to increase. This electromagnetic radiation is readily visible for metals heated to high temperatures, as they glow (like the heating element in an electric grill). For most objects, there is not enough visible thermal radiation to detect with the eye, which is why it is very difficult to see on a moonless night far away from any light sources – unless you happen to have infrared goggles to pick up electromagnetic frequencies that would otherwise be unseen. As all macroscopic objects continuously radiate and absorb thermal radiation, these two effects are always competing.

Consider an object that has constant volume and constant mole numbers and does not approach a combination of pressure and temperature that would result in a phase transition. If the instantaneous rate at which the object emits thermal radiation exceeds the instantaneous rate at which it absorbs thermal radiation from its surroundings, its temperature decreases; if the absorption rate is greater than the emission rate, its temperature increases; and if the rates are balanced, the system is in radiative equilibrium with its surroundings and its temperature remains constant. Heat tends to flow from higher temperature to lower temperature, which implies that the emission and absorption rates depend upon the temperature of the object and surroundings: For example, if the object is at higher temperature than its surroundings, its emission rate will exceed its absorption rate, which causes the object to cool and the surroundings to warm until equilibrium is achieved. According to Stefan's law, the instantaneous rate at which an object emits thermal radiation is proportional to the fourth power of the object's temperature ($P_e = \sigma \epsilon A T^4$), whereas the corresponding absorption rate is proportional to the fourth power of the temperature of the object's surroundings ($P_a = \sigma \epsilon A T_{env}^4$). The formulas are identical except for which temperature to use, such that the rates are exactly equal when radiative equilibrium is attained. The net power radiated is $P_{net} = \sigma \epsilon A (T^4 - T_{env}^4)$, which is positive if the object is losing thermal energy through radiation.

---

[112] Observe that Kirchhoff has two h's, two f's, but no 'ch' sound (rather, it's a hard 'c' as in 'cough').

[113] Thermal energy is conserved for thermal conduction circuits, but not in general – e.g., if work is done as the system changes volume. The first law expresses conservation of energy in thermal physics more generally.

[114] For an object that has fixed volume, no work is done. According to the first law, the heat it loses results in a decrease in internal energy, which generally means less average kinetic energy of its molecules and hence lower temperature (unless some or all of the heat goes toward a phase transition; and assuming no molecules are gained or lost). Similarly, a heat gain in the absence of work done tends to result in an increase in temperature.

The emissivity, ϵ, provides a measure of how well the substance radiates electromagnetic energy; ϵ approaches 1 for a perfect absorber/emitter (corresponding to a blackbody radiator) and ϵ approaches 0 for a perfect reflector. The area, A, is the surface area of the object. The proportionality constant, σ, is the Stefan-Boltzmann constant ($\sigma = 5.670 \times 10^{-8}$ W/m^2/K^4).

Radiative equilibrium: An object is in radiative equilibrium with its surroundings if the net power radiated equals zero – i.e. the instantaneous rate at which it emits thermal radiation equals the instantaneous rate at which it absorbs thermal radiation from its surroundings. This occurs when the object and its surroundings have the same temperature.

Example. Assuming that its emissivity remains unchanged, if a star suddenly expands to twice its radius and its temperature halves, what would happen to the amount of energy it radiates each second?
The ratio of its final emission rate to its initial emission rate is:

$$\frac{P_e}{P_{e0}} = \frac{\sigma \epsilon A T^4}{\sigma \epsilon A_0 T_0^4} = \frac{\pi R^2}{\pi R_0^2}\left(\frac{T}{T_0}\right)^4 = \left(\frac{R}{R_0}\right)^2 \left(\frac{T}{T_0}\right)^4 = 2^2 \left(\frac{1}{2}\right)^4 = \frac{1}{4}$$

The amount of energy that it emits each instant would decrease by a factor of 4.

Intensity: Intensity, I, is defined as power per unit area: $I = P/A$. In the context of thermal radiation, where P designates the instantaneous rate that electromagnetic energy is emitted or absorbed, the quantity P/A is sometimes termed the radiancy – which is the instantaneous rate that energy is emitted or absorbed per unit area. For radiation that is emitted spherically – such as that emitted by a point-source or sphere – the area is that of a sphere, $A = 4\pi r^2$, where r is the distance from the source. That is, as you get further from the source, the intensity decreases by a factor of r^2. (The total energy passing through a sphere each instant is the same for any size sphere since energy is conserved.)

Example. One planet is three times further from its star than another planet. By what factor is the intensity of the thermal radiation that it receives from its star less than it is for the other planet?
The ratio of the two intensities is:

$$\frac{I_2}{I_1} = \frac{P/A_2}{P/A_1} = \frac{A_1}{A_2} = \frac{\pi R_1^2}{\pi R_2^2} = \left(\frac{R_1}{R_2}\right)^2 = \left(\frac{1}{3}\right)^2 = \frac{1}{9}$$

The more distant planet receives thermal radiation with an intensity that is 9 times less.

2.5 Conservation of Energy: The First Law

The first law of thermodynamics: The first law of thermodynamics expresses the law of conservation of energy as it applies to thermal physics. For a system with constant mole numbers, internal energy change, heat added to the system, and work done by the system are related by $dU = đQ - đW$ at the differential scale and $\Delta U = Q - W$ at the finite scale. If mole numbers are not constant, additional term(s) must be added, since all of the particles that make up the system carry energy. A positive $đQ$ is interpreted as heat added to the system, while a negative $đQ$ is interpreted as heat lost by the system. A positive $đW$ is interpreted as work done by the system, while a negative $đW$ is interpreted as work done on the system (in this case, a negative number is subtracted from $đQ$ in the equation, in which case the two minus signs effectively make a plus sign). A positive dU corresponds to an increase in the system's internal energy, while a negative dU corresponds to a decrease in the system's internal energy.

Heat, Q, includes exchanges of thermal energy between the system and environment and also accounts for any heat absorbed or emitted during a phase transition. If pressure or volume are constant – i.e. if the process is isobaric or isochoric – then Q can be calculated using the equations for specific heat capacity ($đQ = MC_{P,V}dT$, or the molar form $đQ = nc_{P,V}dT$). In these two cases, heat is associated with a temperature change. It is important to realize that these equations do not hold in general: They do not hold when pressure and volume are both changing. It is also important to realize that heat can be exchanged without causing a temperature change. Indeed, Q is often nonzero for isothermal processes. An adiabatic process implies that Q is zero, but an isothermal process does not. If the process is not adiabatic, isobaric, or isochoric – e.g. it may be isothermal – then you may be able to use the first law to solve for heat (if you can compute the change in internal energy and work done by other means); otherwise you need a different formula for heat (i.e. don't use the specific heat equations when neither pressure nor volume is constant, and don't set heat to zero unless the process is adiabatic).

Work is done when volume changes (and when pressure is nonzero). For an isochoric process (constant volume), W is zero; and for a free expansion (since pressure is zero), $W = 0$. Otherwise, you can compute work through integration, $W = \int_{V=V_0}^{V} PdV$, or if a graph is given, from the area under a P-V curve. Pressure can only be pulled out of the integral for an isobaric process (constant pressure). Otherwise, you need to use formulas that apply to the problem to express pressure in terms of volume before you can integrate. Work will be positive if the volume increases.

Since temperature is a measure of the average kinetic energy of the molecules, internal energy generally depends upon the temperature change. For this reason, the change in internal energy may be zero for an isothermal process. However, internal energy also depends upon the potential energy of the molecules, so both temperature and the potential energy of the molecules must be accounted for to determine whether or not the internal energy changes and, if so, how much. There are formulas for the internal energy of specific systems, such as for an ideal gas, which we shall learn in later chapters. For a specific system, if you know the formula for the internal energy, you may use it; it may also be useful to use the first law to solve for the change in internal energy.

Internal energy is path-independent, whereas heat and work depend very much on the path. Once you are able to determine the change in internal energy for any path connecting two states, you know this internal energy change for any other path. For a cycle (i.e. a closed path), $\Delta U = 0$.

The 1ˢᵗ law for common processes: The relationship between heat, work, and internal energy for some common P-V curves, for a system with constant mole numbers, include:

- Isochor – volume is constant ($\Delta V = 0$). No work is done ($W = 0$). Heat can be expressed in terms of the heat capacity at constant volume ($đQ = MC_V dT$ or $đQ = nc_V dT$). The change in internal energy equals the heat exchanged ($\Delta U = Q$).
- Isobar – pressure is constant and so can be pulled out of the work integral: $W = \int_{V=V_0}^{V} P dV = P\int_{V=V_0}^{V} dV = P(V - V_0) = P\Delta V$. Heat can be expressed in terms of the heat capacity at constant pressure ($đQ = MC_P dT$ or $đQ = nc_P dT$). All three terms in the first law are generally nonzero ($\Delta U = Q - W$).
- Free expansion – there is no pressure ($P = 0$). No work is done ($W = 0$). The change in internal energy equals the heat exchanged ($\Delta U = Q$).
- Isotherm – temperature is constant. If the relationship between pressure and volume is known for fixed temperature, the work integral can be performed by expressing the pressure in terms of the volume (or, equivalently, if a graph is given, from the area under the P-V curve). Heat is generally exchanged even if temperature is constant. However, if the potential energy of the molecules is constant, then the change in internal energy will be zero ($\Delta U = 0$) and the heat exchange will equal the work done ($Q = W$).
- Adiabat – no heat is exchanged ($Q = 0$). If the relationship between pressure and volume is known for the adiabat, the integral can be performed by expressing the pressure in terms of the volume. The change in internal energy equals the negative of the work done ($\Delta U = -W$).
- Cycle – the net internal energy change for a complete cycle is zero ($\Delta U = 0$), but internal energy may change for parts of the cycle. The heat exchange equals the work done ($Q = W$).

Example. A system completes the thermodynamic cycle illustrated below, where $V_A = 20 \text{ m}^3$, $V_B = 4.0 \text{ m}^3$, $P_B = 8.0 \text{ kPa}$, and $P_C = 32 \text{ kPa}$. Determine the work done by the system and the net heat exchanged between the system and its surroundings for the complete cycle.

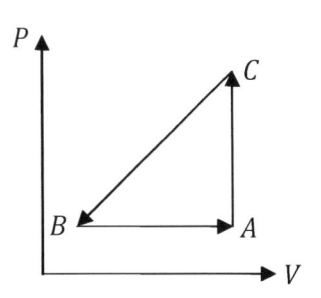

The net work equals the area of the triangle: $W_{net} = -\text{base} \times \frac{\text{height}}{2} = -\frac{(V_A - V_B)(P_C - P_A)}{2} = -(20 \text{ m}^3 - 4.0 \text{ m}^3)(32 \text{ kPa} - 8.0 \text{ kPa})/2 = -190 \text{ kJ}$. The overall minus sign represents that more work was done during the contraction than during the expansion. For a complete cycle, $\Delta U_{net} = 0$. Therefore, the net heat exchange is: $Q_{net} = \Delta U_{net} + W_{net} = -190 \text{ kJ}$. Since it is negative, the system lost heat to its surroundings.

Example. A (non-ideal) gas with n moles expands from initial volume V_0 to final volume $3V_0$ at a constant pressure of P_0, changing its temperature from T_0 to $3T_0/2$. The pressure then increases at constant volume, changing the temperature of the gas to $2T_0$. The specific heat capacities of the gas at constant pressure and volume, respectively, are $2R$ and $3R/2$. Derive an equation for the net internal energy change of the gas in terms of these symbols.

Since pressure is constant for the isobar, the work done along the isobar is: $W = \int_{V=V_0}^{V} P_0 dV = P_0 \int_{V=V_0}^{V} dV = 2P_0V_0$. No work is done along the isochor, so the net work is: $W_{net} = 2P_0V_0$. The heat added to the system along the isobar and isochor is: $Q = n(2R)(T_0/2) + n(3R/2)(T_0/2) = 7nRT_0/4$. The net internal energy change is therefore: $\Delta U_{net} = Q_{net} - W_{net} = 7nRT_0/4 - 2P_0V_0$.

Example. Determine the change in internal energy for a system along an adiabat for which $PV^2 = 200 \text{ J} \cdot \text{m}^3$, from initial volume $V_0 = 10 \text{ m}^3$ to final volume $V = 50 \text{ m}^3$.

The work integral can be performed by expressing the pressure in terms of the volume:

$$W = \int_{V=V_0}^{V} P dV = \int_{V=V_0}^{V} \frac{200 \text{ J} \cdot \text{m}^3}{V^2} dV = 200 \text{ J} \cdot \text{m}^3 \int_{V=V_0}^{V} \frac{dV}{V^2} = -200 \text{ J} \cdot \text{m}^3 \left(\frac{1}{V} - \frac{1}{V_0}\right)$$

$$W = -200 \text{ J} \cdot \text{m}^3 \left(\frac{1}{50 \text{ m}^3} - \frac{1}{10 \text{ m}^3}\right) = 16 \text{ J}$$

No heat is exchanged because the process is adiabatic. Thus, the net internal energy change is: $\Delta U_{net} = Q_{net} - W_{net} = -16 \text{ J}$.

Mechanical equivalent of heat: A block sliding down a frictionless incline loses gravitational potential energy and gains the same amount of kinetic energy – i.e. gravitational potential energy is transformed into kinetic energy, conserving the total energy of the block. If instead a block slides with an initial velocity down an incline for which there is friction, it loses more gravitational potential energy than it gains in kinetic energy – the difference equals the nonconservative work done by the force of friction. Mechanical energy is lost, yet the total energy of the system and surroundings must be conserved. Where does this lost mechanical energy go? From the first law, the nonconservative work may be transformed into an increase in internal energy of the block, or there may be an exchange of thermal energy (heat) between the block and incline, which increases the internal energy of the incline. This increased internal energy of the block and incline can be measured: In particular, each will be warmer.

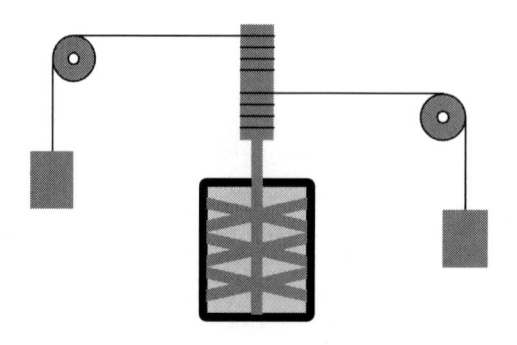

Joule demonstrated that mechanical energy can be transformed into heat, and vice-versa, conserving the total energy of the system and surroundings. His experiment demonstrated the mechanical equivalent of heat – i.e. how much mechanical energy in Joules corresponds to how much heat in calories. The significance of his experiment is that heat reveals itself as another form of energy. Joule placed a horizontal set of paddles connected to a vertical axle inside an insulated cylinder of water. He wound two cords around the axle and passed them over pulleys to suspend two weights. When the system is released, the weights fall, causing the paddles to rotate. The friction between the paddles and the water increases the temperature of the water. Measuring the initial and final gravitational potential energy of the blocks, translational kinetic energy of the blocks, and rotational kinetic energy of the axle and paddlewheel (and pulleys), a significant amount of mechanical energy is lost (much more than can be accounted for by sources of error). Joule equated this lost mechanical energy to heat transferred to the water. The heat capacity of water and measured temperature rise of the water show that this lost mechanical energy is indeed transferred to the water as heat.

Conceptual Questions

The selection of conceptual questions is intended to enhance the conceptual understanding of students who spend time reasoning through them. You will receive the most benefit if you first try it yourself, then consult the hints, and finally check your answer after reasoning through it again. The hints and answers can be found, separately, toward the back of the book.

1. The metal ring illustrated below on the left has a gap. What happens to the size of the gap as the metal ring is heated? Explain. Answer both in terms of angle (from the center) and arc length.

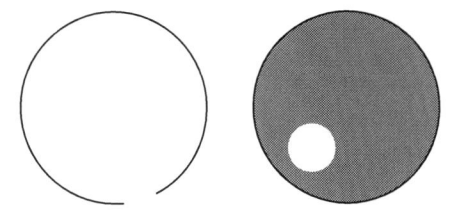

2. The metal disc illustrated above on the right has a circular hole. What happens to the size of the hole as the metal disc is heated? Explain.

3. A bimetallic strip consists of two strips of different metals that have been welded together. Explain why the bimetallic strip bends into the shape of an arc when it is cooled or heated. Also, what determines which way it will bend when it is heated? Repeat for when it is cooled.

4. A metal ring is a little too small to slip over a rod made of a different metal when both the ring and rod are at room temperature. However, the metal ring can be slipped over the metal rod if the metal ring is heated and the metal rod is cooled. Explain. What are the prospects for removing the metal ring from the metal rod afterward – without damaging either?

5. The coefficient of thermal expansion is negative for rubber. How does this impact the design of tires?

6. The density of water is maximum at a temperature of about 4° C. Consider a pond where the water has an initial temperature above 4° C, where the air temperature steadily drops until the pond freezes. Describe the temperature and density gradient of the pond that develops as it freezes, and explain why ice forms at the top.

7. Two coffee mugs are made from different materials. (A) Which coffee mug keeps hot coffee warm longer? Explain. (B) Which coffee mug has a handle that would be more comfortable to touch? Explain.

8. Two floors in the same room are made of different materials. Which floor would be more comfortable to walk on with bare feet? Explain.

9. Two outdoor walkways in the same garden are made of different materials. Which outdoor walkway would be more comfortable to walk on with bare feet during a hot summer day? Explain. Also describe the differences between the two outdoor walkways later that same evening.

10. A solid sample initially at temperature T_s is placed in a thermally insulated cup of liquid that is initially at temperature $T_\ell < T_s$. Neglect any heat exchanges between the system and surroundings and assume that the specific heat of the liquid exceeds that of the solid sample, $C_\ell > C_s$. (A) For a given specific heat of the liquid, C_ℓ, would a solid sample with a lower or higher specific heat, C_s, result in a higher equilibrium temperature? Explain. (B) For a given specific heat of the solid sample, C_s, would a liquid with a lower or higher specific heat, C_ℓ, result in a higher equilibrium temperature? Explain.

11. Identify the method of heat transfer that is primarily involved in each of the following processes: leftovers are warmed in a microwave oven, fingers holding a silver spoon feel cold while the spoon is dipped ice-cream, hands are warmed near a fireplace, a pot of tomato soup is cooked over a stove burner, a window painted with a pattern of metal ink is defrosted, and a sunbather receives a suntan.

12. If you hold a thin metal rod with your fingers and place the other end in contact with ice, your fingers will feel cold. Describe the transformation of heat that occurs. For example, does thermal energy travel from the ice to your fingers? Also, which method of heat transfer is involved?

13. If you touch a block of metal and a block of wood that have the same temperature, which will feel warmer to touch if their temperature is (A) less than, (B) equal to, and (C) greater than the temperature of your body? Explain your answers.

14. Given that water is a poor conductor of heat (which you can verify by comparing the thermal conductivity of water to good conductors of heat, such as metals), explain how it is possible to cook an egg quickly by boiling it in a pot of water on a stovetop.

15. (A) Explain why a good emitter of thermal radiation is also a good absorber of thermal radiation. (B) Explain why a good reflector of thermal radiation is also a poor emitter of thermal radiation.

16. Many people unfold a cardboard windshield cover and place it against the windshield on a hot summer day. (A) What is the purpose of this windshield cover? How does it work? (B) Some of these covers are highly reflective. What is the benefit of using a silvered cover rather than plain cardboard? (C) Would these windshield covers also be effective on a cold winter day? Explain.

17. A person stands in the center of a square room that has plane mirrors on opposite walls, holding a small object. Numerous images of the small object can be seen. How does the intensity of the brightest image seen compare to that of the second brightest image seen?

18. (A) In what sense do thermal conductors connected in series and parallel behave more like capacitors and less like resistors? (B) Does this mean that thermal conduction is a 'thermostatic' situation – since capacitive circuits are electrostatic? Explain.

19. Sketch an electric field map for an electric dipole and construct an analogous thermal map. How is heat flow similar to the flux of electric field lines possible, given that the electric dipole consists of a positive and negative charge, whereas all (absolute) temperatures are inherently positive. Precisely, what is the equivalent of electric charge on the thermal map?

Practice Problems

The selection of practice problems primarily consists of problems that offer practice carrying out the main problem-solving strategies or involve instructive applications of the concepts (or both). You will receive the most benefit if you first try it yourself, then consult the hints, and finally check your answer after working it out again. The hints to all of the problems and the answers to selected problems can be found, separately, toward the back of the book.

Thermal expansion

1. A solid cube of aluminum with 12-cm long edges at $20°$ C is heated to $50°$ C. The coefficient of linear expansion for aluminum is $2.4 \times 10^{-5}/°C$. Determine the increase in the cube's (A) volume, (B) surface area, and (C) edge length. Twelve aluminum wires, which are 12 cm in length at $20°$ C, are joined together to form the edges of a cube. The cube is heated to $50°$ C. Determine the increase in the cube's (D) volume, (E) surface area, and (F) edge length. Comment on the two sets of results.

2. A steel tape measure is designed for use at $20°$ C. The coefficient of linear expansion for steel is $1.1 \times 10^{-5}/°C$. On a very hot day, for which the temperature is $40°$ C, the tape measure is used to measure the length of a wooden pole. One end of the pole lies at the 0-cm mark and the other end lies at the 200-cm mark. Approximately, how long is the pole?

3. At 20° C, a brass ring has an inner radius of 4.00 cm and an aluminum rod has a radius that is 0.1 mm thicker than the inner radius of the brass ring. The coefficients of linear expansion for brass and aluminum are 1.9×10^{-5}/°C and 2.4×10^{-5}/°C, respectively. (A) To what temperature must the brass ring be heated – while the aluminum rod is kept at room temperature – in order to slip it over the aluminum rod? (B) Is it possible to remove the brass ring from the aluminum rod by cooling the system? If so, what cooling temperature must be reached in order to achieve this? If not, explain the problem.

4. The metal ring of Conceptual Question 1 is made of copper, for which the coefficient of linear expansion is 1.7×10^{-5}/°C. The ring has a radius of 15 cm and a gap arc length of 0.50 cm at 20° C. Determine the size of the gap in terms of (A) arc length and (B) angle measured from the center when the ring is heated to 80° C.

5. The metal disc of Conceptual Question 2 is made of copper, for which the coefficient of linear expansion is 1.7×10^{-5}/°C. At 20° C, the disc has a radius of 15 cm, the hole has a radius of 3 cm, and the center of the hole lies 10 cm from the center of the disc. Determine (A) the location and (B) the size of the hole when the disc is heated to 80° C.

Heat capacity and calorimetry

6. What is the temperature of the water in a pot if three times as much heat must be supplied to raise its temperature to its boiling point compared to how much heat must be released to lower its temperature to its freezing point at constant (standard) air pressure?

7. A pot contains 250 g of water at STP. The specific heat capacity of water is 4.186 kJ/kg/°C, and the latent heats of fusion and vaporization, respectively, for water are 3.3×10^5 J/kg and 2.3×10^6 J/kg. How much heat must be supplied/removed in order to completely (A) vaporize and (B) freeze the water at constant pressure?

8. A 200-g aluminum cube at 80° C is placed in a pot of 600 g of water at 20° C in a sealed, perfectly insulating container. The specific heat capacity of water is 4.186 kJ/kg/°C and of aluminum is 0.900 kJ/kg/°C. What is the equilibrium temperature of the system?

9. An ice cube at 0° C is placed in a pot of 300 g of water at 20° C in a sealed, perfectly insulating container. The specific heat capacity of water is 4.186 kJ/kg/°C and the latent heat of fusion is 3.3×10^5 J/kg. What fraction of the ice melts and what is the equilibrium temperature of the system if the ice cube has a mass of (A) 300 g and (B) 30 g?

Heat conduction

10. A copper conducting slab and aluminum conducting slab of the same length are placed in series. The slabs make thermal contact at the shared walls. The far end of the copper and aluminum slabs are at 20° C and 100° C, respectively. The remaining walls are thermally insulated. The thermal conductivities are 400 W/m/°C for copper and 240 W/m/°C for aluminum. What is the temperature of the shared wall?

11. Three conducting slabs of the same length are placed in series. The slabs make thermal contact at the shared walls. The far ends of the first and last slabs are at 20° C and 100° C, respectively. The remaining walls are thermally insulated. The thermal conductivity of the second slab is twice that of the first slab and three times that of the last slab. What are the temperatures of the shared walls?

12. A configuration of conducting slabs is constructed as illustrated below. The thermal conductivities in the legend are stated in multiples of 100 W/m/°C. The left and right walls are held at two different temperatures. (A) Determine the effective R-value of the configuration. (B) If the temperature of the left and right walls are 20° C and 100° C, respectively, determine the rate at which heat is transferred between the left and right walls.

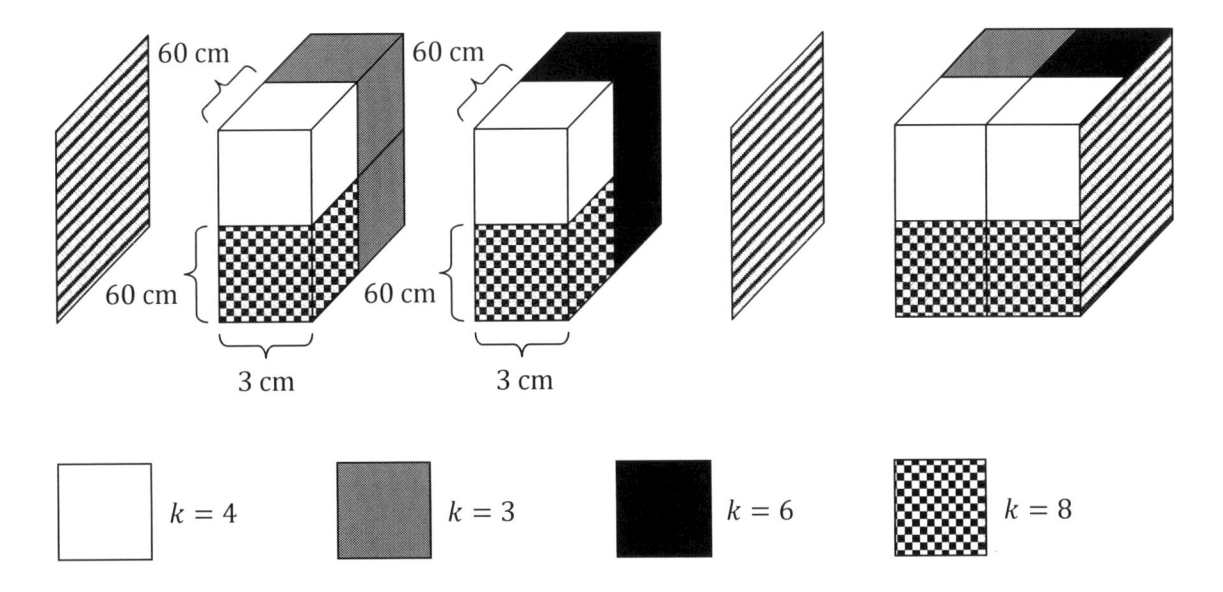

13. A cylindrical conducting shell with inner radius a, outer radius b, and length L has uniform thermal conductivity k. Derive an equation for the steady-state rate at which heat is transferred by thermal conduction across the conducting cylindrical shell (A) if the two ends have different temperatures such that heat flows along the body of the cylinder and (B) if the inner and outer walls have different temperatures such that heat flows radially across the cylindrical shell.

14. A cylindrical tub of water is left outside on a cold night. The temperature of the top layer of water is at its freezing point, while the air temperature is below the freezing point of water. As a result, a layer of ice forms at the top of the cylindrical tub. Assume that the temperature of the air and water (both the top layer of ice that forms and the layer of water just below it) remain constant. The thickness of the layer of ice will expand downward. (A) Justify that heat conduction is the most appropriate form of heat transfer for this problem. Also, draw and label a diagram, including the thickness of the frozen layer, x. Indicate the direction of heat flow on your diagram. (B) Conceptually and/or mathematically, justify that $đQ = L\rho A dx$. In addition, indicate what each of these five symbols represents, being clear which is which. (C) Derive an equation for the thickness of the layer of ice as a function of time.

The first law of thermodynamics

15. A 400-g copper cube is heated isobarically at 150 kPa, raising its temperature from 25 °C to 50 °C. The coefficient of linear expansion of copper is 17×10^{-6} /°C and the specific heat capacity of copper is 390 J/kg/°C. How much did the internal energy of the copper cube change? Also, did it increase or decrease?

16. A pot contains 5.0 kg of a particular liquid at a pressure of 100,000 N/m² and a temperature of 25 °C. The liquid has a specific heat capacity of 4,000 J/kg/°C, a heat of vaporization of 2,000,000 J/kg, a density of 1,500 kg/m³, and a boiling point of 125 °C. The liquid is isobarically heated until 40% of its mass has turned into gas at its boiling point. The density is 0.50 kg/m³ in the gaseous state. Define 'system' initially as liquid and finally as liquid plus the gas formed from boiling. (A) How much heat was supplied to the system? (B) How much work was done? Also, was it done on or by the system? (C) How much did the internal energy of the system change? Also, did it increase or decrease?

17. A system completes the thermodynamic cycle illustrated below, where $V_A = 50$ m³, $V_B = 10$ m³, $P_B = 4.0$ kPa, and $P_C = 20$ kPa. Determine the net heat exchanged between the system and its surroundings for the complete cycle. Also, was this heat absorbed or released by the system?

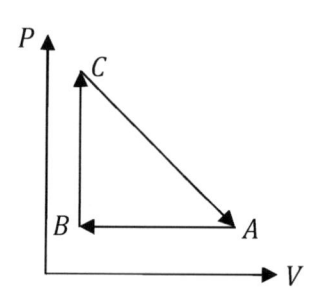

18. A system expands according to the equation $P^2V^3 = 8.0 \times 10^8$ J²m³, starting at $V_0 = 10$ m³ and ending at $V = 40$ m³. During the expansion, the internal energy of the system is given by $U = 3PV$. How much heat is exchanged between the system and its surroundings during this expansion? Also, was this heat absorbed or released by the system?

19. If a system expands along an adiabat from initial volume $V_0 = 10$ m³ to final volume $V = 50$ m³, it is found to obey the equation $PV^3 = 3,000$ J·m⁶. How much heat is exchanged between the system and its surroundings if instead the system expands isobarically to $V = 50$ m³ and then isochorically to the same final pressure? Also, was this heat absorbed or released by the system?

20. A fluid is governed by the equation $PV^{3/2} = aT^{1/2}$, where a is a constant. Express the internal energy change of the fluid along each of the following paths in terms of the initial pressure P_0 and initial volume V_0. (A) The fluid is heated from T_0 to $4T_0$ along an isobar. (B) The fluid is cooled from T_0 to $T_0/4$ along an isochor.

3 Ideal Gases in Thermal Physics

3.1 Definition of an Ideal Gas

An ideal gas: Experimentally, when we make measurements of pressure (P), volume (V), temperature (T), and mole number (n) for a wide variety of gases, we find that these variables tend to satisfy the same relationship if the density of the gas does not exceed a certain threshold. Since a low-density limit exists in which all gases tend to exhibit very similar behavior, with little to no regard for their chemical composition, we interpret this low-density behavior as that of an 'ideal gas.' Macroscopically, we define an ideal gas as a gas with low enough density that its pressure, volume, temperature, and mole number satisfy the ideal gas law $(PV = nRT)$.

Molecular model: We can also develop a molecular model that leads to the observed macroscopic behavior of low-density gases. Such a model is built from the following assumptions:
- An ideal gas consists of a very large number of identical molecules.
- The motion of the molecules is governed by Newton's laws.
- The distribution of molecular velocities is random and static (in thermal equilibrium).
- The molecules are very small compared to their average separation.
- The molecules collide elastically with one another and the walls of the container.
- Intermolecular forces are short-range, and therefore only affect collisions between molecules.
- The only significant internal forces are those exerted during collisions between molecules.
- The only significant external forces are exerted when molecules collide with the container.

Some immediate consequences of these assumptions are:
- The molecules travel with constant velocity between collisions: Since the net external force is zero, this follows from Newton's second law.
- Momentum and energy are both conserved during collisions between molecules: Momentum is conserved because the net external force is zero and energy is conserved because the collisions are elastic.
- The total volume of the molecules can be neglected in comparison with the volume of the container: This follows since the molecules are tiny compared to their average separation.
- The directions of the molecular velocities are randomly oriented. At any given instant, the same percentage of molecules are headed in any given direction (window of solid angle). While molecules may change direction as a result of collisions, there are so many molecules that these microscopic changes have no discernible effect on macroscopic averages.
- In thermal equilibrium, the distribution of molecular speeds is time-independent and uniform throughout the gas. There are so many molecules that speed changes during collisions do not affect macroscopic averages. The distribution of speeds is uniformly random throughout.

Relating the microscopic and macroscopic pictures: The molecular model for an ideal gas turns out to be consistent with the experimental observation that the macroscopic variables satisfy the ideal gas law ($PV = nRT$) for a sufficiently low-density gas. Let us consider, conceptually, how these microscopic assumptions relate to the macroscopic observations.

- The macroscopic observation of pressure results from the force imparted on the container when molecules collide with its walls. From Newton's third law, the force that a molecule exerts on the container is equal and opposite to the force that the container exerts on the molecule. The normal component of the force per unit area that the container exerts on the ideal gas is the pressure.
- The macroscopic observation of temperature corresponds to the average kinetic energy of the molecules of the ideal gas. In thermal equilibrium, the distribution of molecular speeds is space- and time-independent, meaning that the temperature is constant and uniform throughout.
- If the volume and mole number are held constant, an increase in temperature results in an increase in pressure, and vice-versa. As the temperature increases, the average speed of the molecules increases, and the molecules exert greater forces on the container when they collide with its walls.
- Since the molecules have negligible size, the volume of the molecules is not a factor when the volume of the container is being decreased during a compression. Indeed, the volume of a gas can be changed with ease in the laboratory through a large range of values, and if you consider how drastically its volume changed when it condenses to liquid, you can see just how tiny the volume of the molecules of a gas really is.
- If the temperature and mole number are held constant, pressure and volume need to have an inverse relationship. An increase in volume requires a decrease in pressure: The average force exerted on the walls of the container is the same, since temperature is held constant, but the normal component of force is effectively spread over a larger surface area if the volume is increased,[115] resulting in less pressure.
- If the pressure and mole number are held constant, volume and temperature must be proportional to one another. An increase in volume results in an increase in temperature: In order to hold the pressure constant, the average normal component of force exerted on the walls must increase as the volume increases, as it is effectively spread over a larger area, which means that the average speed of the molecules increases – i.e. the temperature increases.
- If volume and temperature are constant, more molecules results in more pressure. Since temperature is constant, the average speed of the molecules is constant, and since volume is constant, the average normal component of force exerted on the walls is effectively spread over the same surface area. Thus, more molecules results in more collisions with the walls, resulting in greater pressure.

---

[115] Volume and surface area are not proportional, in general, unless shape is preserved (and then there is generally an exponent involved in the proportionality). In fact, it is possible to have infinite surface area and finite volume – as in Gabriel's horn. Pressure involves the normal component of the force, however, so when you calculate how the normal component of the force is spread over the surface area, an increase in volume for a fixed average force does result in less pressure, regardless of the shape of the container.

- If pressure and volume are held constant, more molecules means less temperature. This time, more molecules implies more collisions with the walls, so in order for the pressure to remain constant, the average speed of the molecules must diminish – meaning temperature decreases.
- If pressure and temperature are held constant, more molecules results in more volume. Now there are more molecules to collide with the walls and the average speed of the molecules is held constant, so for pressure to also remain constant, the average normal component of force must effectively be spread over a greater surface area – i.e. volume must increase.

Dependence on density: There is very good agreement between the macroscopic model of an ideal gas and experimental observations of real gases, with a variety of chemical compositions, at low densities. However, the greater the density of the gas, the more it deviates from the ideal gas law and the more individualized it becomes. There are two obvious reasons for this dependence of the macroscopic behavior of the gas on its density:
- The molecules of the gas do occupy some fraction of the volume of the container. The volume of the molecules occupies a greater fraction of the volume of the container at high density than it does at low density. Thus, at high density, a gas does not compress as easily as the molecular model of an ideal gas predicts.
- The molecules of the gas do interact with one another, even in the absence of collisions. These intermolecular forces decrease with distance. The greater the average separation between molecules, the less effect these intermolecular forces have; the smaller the average separation between molecules, the greater their effect. These intermolecular forces are thus more relevant at high density. In particular, the attraction of molecules to other molecules decreases the average speed of molecules at the edges of the gas – i.e. those molecules colliding with the walls of the container – which decreases the pressure from expectations of the molecular model for an ideal gas.

Correction factors: The ideal gas law can be improved to accommodate higher-density gases by accounting for the volume of the molecules and the effect that intermolecular forces have on pressure. The virial expansion offers a means of adding a series of correction terms to the ideal gas law. The ideal van der Waals fluid is one such example, which accommodates not only higher-density gases, but also describes the behavior of liquids. We will consider the ideal van der Waals fluid in Sec. 6.3, where we also consider the ideal gas in greater detail (including its entropy and chemical potential).

3.2 The Ideal Gas Law

Boyle's law: For constant temperature and mole number, the pressure and volume of an ideal gas are observed to share a reciprocal relationship: $PV = P_0 V_0$.

Charles's law: For constant pressure and mole number, the temperature and volume of an ideal gas are observed be directly proportional: $V/T = V_0/T_0$. Of course, T is the absolute temperature in Kelvin.

Gay-Lussac's law: For constant volume and mole number, the pressure and temperature of an ideal gas are observed be directly proportional: $P/T = P_0/T_0$.

Ideal gas law: For constant mole number, combination of Boyle's law and Charles's law leads to:[116] $PV/T = P_0V_0/T_0$. Put another way, for constant mole number, PV/T equals a constant. From our conceptual considerations of how the molecular model for an ideal gas relates to macroscopic expectations for an ideal gas, we expect that this constant is proportional to the mole number. In general, an ideal gas obeys the relation $PV = nRT$, where R is the universal gas constant and T is the absolute temperature in Kelvin. The universal gas constant, R, is related to the Boltzmann constant, k_B, by $R = N_A k_B$, and the number of moles, n, is related to the number of molecules, N, by $N = N_A n$. Thus, $nR = nN_A k_B = Nk_B$ and the ideal gas law can alternatively be expressed as $PV = Nk_B T$. The difference is whether it is more convenient to work with the number of moles or number of molecules.

Ideal gas thermometer: The constant-volume gas thermometer is based on Charles's law (a special case of the ideal gas law): For a gas confined to a fixed volume, an increase in pressure results in an increase in temperature. In this case, the ratio of the temperature to the triple point temperature of water equals the ratio of the pressure to the triple point pressure of water. The constant-volume gas thermometer measures pressure for a fixed volume and has a scale that calibrates the measured pressure into a measurement of temperature. The device includes a bulb containing a gas and a manometer for measuring its pressure. It also includes a U-shaped tube filled with mercury, one side serving as a mercury reservoir. The height of the mercury reservoir is adjustable until the mercury level on the other side of the tube matches a reference height that ensures constant volume of the gas; the difference in heights indicates the pressure, which in turn indicates the temperature.

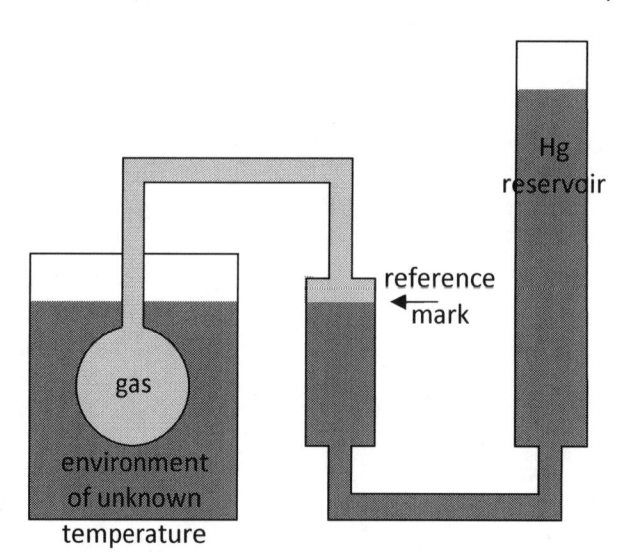

---

[116] All three laws are not independent. Any two of the three lead to the ideal gas law; the ideal gas law leads to all three. The ideal gas law also goes on to say what happens if mole number is not constant.

> **Example**. An ideal gas is initially at a temperature of 300 K. Its pressure is decreased by a factor of two as it expands to three times its original volume, while the mole number is held constant. What is the final temperature of the gas?
>
> According to the ideal gas law, initially $P_0 V_0 = nRT_0$ and finally $PV = nRT$. Dividing these two equations, $PV/P_0 V_0 = T/T_0$, or
>
> $$T = \left(\frac{P}{P_0}\right)\left(\frac{V}{V_0}\right) T_0 = \left(\frac{1}{2}\right)(3)(300 \text{ K}) = 450 \text{ K}$$

3.3 Pressure on an Ideal Gas

Colliding with a wall: Consider an elastic collision between a single molecule and a wall, where the wall is initially at rest ($\vec{v}_{w0} = 0$). Both linear momentum and mechanical energy are conserved for the system (molecule plus wall) for an elastic collision: $\vec{p}_{m0} + \vec{p}_{w0} = \vec{p}_m + \vec{p}_w$ and $K_{m0} + K_{w0} = K_m + K_w$. Resolving velocity, $\vec{v}$, into components that are normal (perpendicular) to the wall, (v_N), and tangential to the wall (v_T), these conservation laws give three equations:

$$m_m v_{mT0} + m_w v_{wT0} = m_m v_{mT} + m_w v_{wT}$$
$$m_m v_{mN0} + m_w v_{wN0} = m_m v_{mN} + m_w v_{wN}$$
$$\tfrac{1}{2} m_m v_{m0}^2 + \tfrac{1}{2} m_w v_{w0}^2 = \tfrac{1}{2} m_m v_m^2 + \tfrac{1}{2} m_w v_w^2$$

The mass of the molecule, m_m, is miniscule compared to the mass of a wall, m_w. In this extreme, the molecule will reflect off the wall – namely, $v_{mN} \approx -v_{mN0}$ and $v_{mT} \approx v_{mT0}$. The tangential component of momentum conservation then requires that $v_{wT} \approx v_{wT0} = 0$, whereas $m_w v_{wN} \approx 2 m_m v_{mN0}$ in order to conserve the normal component of momentum. Observe that v_{wN}/v_{mN0} and m_m/m_w are both very small. Thus, conservation of energy is conserved to first-order in m_m/m_w. That is, while $m_w v_{wN}$ makes a significant contribution to momentum conservation (since the enormous mass of the wall compensates for the wall's miniscule velocity), $\tfrac{1}{2} m_w v_w^2$ makes a negligible contribution (on the order of m_m/m_w) because now the smallness of the wall's velocity is squared compared to the wall's mass, which is not squared.[117]

The net result is that the molecule imparts a total momentum equal to twice the molecule's normal component of momentum $(2 m_m v_{mN0})$ to the wall, while its tangential component of momentum is conserved. That is, the wall receives an impulse equal to $\vec{J}_w = 2 m_m v_{mN0} \hat{n}$, where $\hat{n}$ is a unit vector normal to the wall (directed from the molecule toward the wall).

---

[117] We are not comparing apples to oranges; rather, we are comparing ratios. The ratio v_{wN}/v_{mN0} is about as small as the ratio m_m/m_w, since $m_w v_{wN} \approx 2 m_m v_{mN0}$. In the conservation of normal-component of momentum equation, $m_w v_{wN}/(m_m v_{mN0})$ is a significant ratio (≈ 2), when you divide every term in the equation by $m_m v_{mN0}$, whereas in the conservation of kinetic energy equation, $\tfrac{1}{2} m_w v_w^2/(\tfrac{1}{2} m_m v_{m0}^2)$ is a very insignificant ratio ($\approx 2 v_{wN}/v_{mN0}$), compared to other terms when you divide every term in the equation by $\tfrac{1}{2} m_m v_{m0}^2$.

Pressure on a plane wall: Consider a beam of N molecules, all with the same initial velocity ($\vec{v}_{m0}$), incident upon a plane wall at some angle, such that it has both normal and tangential components. If each molecule collides elastically with the wall, the net momentum imparted to the wall equals $\vec{p}_{net} = 2Nm_m v_{mN0}\hat{n}$.[118] The net force can then be expressed in terms of the particle current, dN/dt, through Newton's second law: $\vec{F}_{net} = d\vec{p}_{net}/dt = 2(dN/dt)m_m v_{mN0}\hat{n}$. The pressure exerted on the wall by the molecules, which equals the normal component of the force per unit area, is therefore equal to $P = F_N/A = 2(dN/dt)m_m v_{mN0}/A$.

Pressure on a rectangular box: Consider an ideal gas in a rectangular container, where the molecular distribution of velocities is random. Thus, the net force that the molecules exert on the walls of the container is zero – since for every molecule striking a wall in one direction with a given speed, there is on average a counterpart molecule striking the container in the opposite direction with the same speed. However, the net pressure exerted is not zero, as the sum of the normal components of the forces results in a net outward pressure. Note that intermolecular collisions have no effect on the pressure, as these net internal collision forces cancel out by Newton's third law and as the momentum and mechanical energies of the molecules is conserved for these collisions. So only the collisions between the molecules and the wall contribute to the pressure exerted on the container.

For convenience, we may setup our coordinate system with the origin at the center of the rectangular container and with each of the three coordinate axes perpendicular to one pair of walls. The pressure exerted on the $+x$ wall equals the pressure exerted on a wall by the beam of molecules considered previously, except that now the molecules do not have the same velocity so we use the average normal component (in this case, the x-component) of velocity: $P_{+x} = 2(dN_x/dt)m_m \overline{v_x}/A_x$, where we now interpret dN_x/dt as the collision frequency – i.e. the number of collisions that the molecules make with the $+x$ wall per unit time – and where A_x means the area of the $+x$ wall.[119]

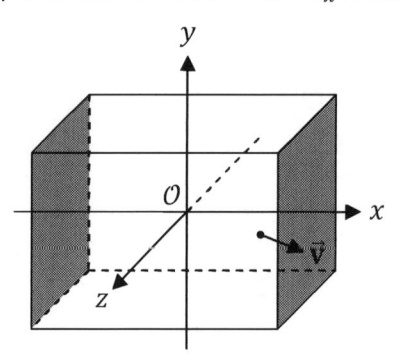

---

[118] We must presently assume that N is not so large that Nm_m is significant compared to m_w; otherwise, the wall's velocity grows more and more significant with each collision, affecting the net momentum more significantly. However, we shall be able to remove the restriction on the size of N when we apply the result to a closed container, since in that case for every particle striking one wall in one direction, there is another particle striking the opposite wall in the opposite direction, such that the net velocity of the container is always zero.

[119] When we considered the pressure exerted on an initially stationary plane wall, the wall developed a normal component of velocity in order to conserve momentum. For the pressure exerted on a closed container, however, the walls remain stationary, since any molecule that tends to impart momentum to a wall in one direction is compensated, on average, by another molecule that imparts the opposite momentum to an opposite wall.

When a molecule strikes the $+x$ wall, the x-component of its momentum reflects back toward the $-x$ wall, until it eventually collides with the $-x$ wall, at which time its x-component of momentum will reflect back toward the $+x$ wall.[120] Thus, a molecule imparts momentum equal to $2m_m v_x$ to the $+x$ wall with a frequency equal to one over the time it would take to complete a roundtrip in the absence of other collisions between the $\pm x$ walls.[121] The collision frequency, dN_x/dt, thus equals the number of molecules in the ideal gas, N, times the roundtrip frequency, $1/\Delta t_{rt}$. The roundtrip time is related to the average absolute value[122] of x-component of velocity by: $\Delta x_{rt} = 2x = \overline{v_x} \Delta t_{rt}$.[123] Thus, $dN_x/dt = N\overline{v_x}/2x$ and $P_{+x} = Nm_m \overline{v_x^2}/(A_x x)$. The denominator, $A_x x$, equals the volume of the container; Nm_m equals the mass of the ideal gas, M, and the ratio M/V is the density of the ideal gas, ρ. With these substitutions, $P_{+x} = \rho \overline{v_x^2}$.[124]

The pressure on the $+x$ wall equals the pressure on the $-x$ wall, such that $P_x = \rho \overline{v_x^2}$ for both $\pm x$ walls combined (twice the force and twice the area results in the same pressure). Similarly, the net pressure exerted on all six walls equals $P = \rho \overline{v_x^2}$ (again there is proportionally more force and area). Observe that $\overline{v_x^2} = \overline{v_y^2} = \overline{v_z^2}$ since the direction of the molecular velocities is completely random. From the Pythagorean theorem, $\overline{v_x^2} + \overline{v_y^2} + \overline{v_z^2} = \overline{v^2} = 3\overline{v_x^2}$. Thus, the pressure exerted on the walls (or, equivalently, on the ideal gas) is related to the root-mean-square (rms) speed by $P = \rho \overline{v^2}/3$.

Pressure on a spherical container: Now consider an ideal gas in a spherical, rather than rectangular, container. In this case, the radial component of momentum is normal to the container's wall, and thus makes the only contribution to the pressure. The pressure exerted on the container (or, equivalently, on the ideal gas) will now be:[125] $P = M\overline{v^2}/(AR_0)$, where R_0 is the radius of the container, and $A = 4\pi R_0^2$ is the surface area of the sphere. Since the volume of the sphere is $V = 4\pi R_0^3/3$, the denominator can be expressed as $AR_0 = 4\pi R_0^3 = 3V$. Thus, the pressure is $P = \rho \overline{v^2}/3$, which is identical to the result for a rectangular box.

---

[120] It does not matter that the molecule may collide with other molecules along the way to the opposite wall – since momentum is conserved for intermolecular collisions, the x-component of its momentum will eventually impart momentum equal to $2m_m v_x$ to that wall, regardless of such collisions. Similarly, collisions with the $\pm y$ and $\pm z$ walls do not affect the x-component of momentum, and thus do not prevent the x-component of momentum from eventually reaching the opposite x wall and imparting momentum equal to $2m_m v_x$ to that wall.

[121] Again, any collisions that occur in between the $\pm x$ walls do not affect the pressure exerted on the $\pm x$ walls.

[122] The average x-component of velocity is zero, since for every molecule with a positive x-component of velocity there is a counterpart with the same speed heading in the opposite direction. Technically, we don't want to average yet – we'll see another factor involved, and we'll actually want to average the product of these factors.

[123] Because average speed is defined as total distance over total time, unlike average velocity, which is net displacement over total time, the roundtrip distance (Δx_{rt}) needed here is the total distance traveled ($2x$).

[124] Two factors of $\overline{v_x}$ appeared in this derivation, leading to $\overline{v_x^2}$. Conceptually, we need to square and then average, not average and then square; we shall learn in the next chapter that this makes a difference, mathematically. For one, if you first average the velocities, you get zero by symmetry, so squaring first makes everything nonnegative before you average. But looking ahead, the final result depends on $\overline{v^2}$, which is the average square of the speed. Even in that case, squaring first and then averaging differs from averaging first and then squaring.

[125] This equation is obtained by jumping ahead to $P_{net,+x} = Nm_m \overline{v_x^2}/(A_x x)$ from the previous derivation, by comparison of the sphere to the $+x$ wall. Note that in spherical coordinates, the radial component of velocity is the velocity; just like $r^2 = x^2 + y^2 + z^2$.

Pressure on a general container: In general, the shape of the container does not matter. That is, the pressure exerted on an ideal gas is given by $P = \rho \overline{v^2}/3$, regardless of the geometry of the container. This result follows from the fact that pressure depends upon the normal component of the force exerted on the walls.

3.4 Energy of an Ideal Gas

Temperature of an ideal gas: Combining the equation for the pressure of an ideal gas obtained from the kinetic theory, $P = \rho \overline{v^2}/3$, with the ideal gas law, $PV = Nk_BT$, leads to $\rho \overline{v^2}/3 = Nk_BT/V$. Since density equals $\rho = M/V = Nm/V$, this becomes $m\overline{v^2}/3 = k_BT$. Recall that m is the molar mass (i.e. the mass of one mole), whereas M is the total mass of the gas and m_m is the mass of one molecule. The average translational energy[126] of one molecule is $\overline{K_{t,m}} = \frac{1}{2}m_m\overline{v^2} = 3k_BT/2$. Thus, we see that indeed absolute temperature is a measure of the average kinetic energy of the molecules of a gas – and we now have quantitative relationship between the two for the case of an ideal gas.

Translational kinetic energy of an ideal gas: The total translational kinetic energy of an ideal gas equals the average translational kinetic energy of one molecule times the number of molecules: $K_t = N\overline{K_{t,m}} = 3Nk_BT/2 = 3nRT/2$.

Equipartition of energy: At thermal equilibrium, the average kinetic energy of each molecule of an ideal gas must be divided equally, on average, among its independent degrees of freedom according to the theorem of equipartition of energy. Here, a degree of freedom refers to the number of independent ways that each molecule of the ideal gas is able to possess kinetic energy. Since the three dimensions of space are independent, translational kinetic energy includes three degrees of freedom – one degree of freedom for translation along each of the three dimensions. Thus, the average translational kinetic energy of a molecule of an ideal gas, $\overline{K_{t,m}} = 3k_BT/2$, is divided equally among translation along x, y, and z: $\overline{K_{t,mx}} = \frac{1}{2}m\overline{v_x^2} = k_BT/2$, $\overline{K_{t,my}} = \frac{1}{2}m\overline{v_y^2} = k_BT/2$, and $\overline{K_{t,mz}} = \frac{1}{2}m\overline{v_z^2} = k_BT/2$. That is, the average kinetic energy of each degree of freedom of each molecule is $\overline{K_{m,dof}} = k_BT/2$.

Total kinetic energy of an ideal gas: The average kinetic energy of a molecule of an ideal gas includes translational kinetic energy plus any rotational and/or vibrational kinetic energy that the molecule may have on average. In a monatomic ideal gas, the molecules are individual atoms, for which there is chiefly translational kinetic energy.[127] Hence, the average kinetic energy of a molecule of a monatomic ideal gas is $\overline{K_{mon,m}} = 3k_BT/2$ and the total kinetic energy of a monatomic ideal gas is $K_{mon} = 3Nk_BT/2 = 3nRT/2$.

---

[126] When we derived the equation for the pressure of an ideal gas, which we are now relating to temperature, we considered only the translation (straight-line motion) of the molecules, and not any possible rotations or vibrations of the molecules – as only translation affects the pressure. Hence, this is translational kinetic energy, and may not equal the total kinetic energy of the molecules (they won't be the same if the molecules rotate and/or vibrate).

[127] In the molecular model, we assumed that individual molecules have no internal structure.

Diatomic molecules, such as H_2, consist of pairs of atoms bound together covalently. The line joining the pair of atoms defines the axis of a diatomic molecule. The diatomic molecule can rotate in two independent directions – corresponding to the two dimensions perpendicular to its axis. For example, if the axis of the diatomic molecule is the z-axis, the two independent rotations include rotation about the x- and y-axes.[128] Diatomic molecules that have rotational kinetic energy, but no significant vibrational kinetic energy, have five degrees of freedom – three translational and two rotational. Thus, the average kinetic energy of a molecule of a diatomic ideal gas is $\overline{K_{dia,m}} = 5k_BT/2$ and the total kinetic energy of a monatomic ideal gas is $K_{dia} = 5Nk_BT/2 = 5nRT/2$.

If the diatomic molecules also have significant vibrational kinetic energy, there is a sixth degree of freedom, corresponding to vibrations along the axis of the molecule. Vibrational kinetic energy tends to be more relevant for very high temperatures. In this case, $\overline{K_{dia,m}^{high\,T}} = 3k_BT$ and $K_{dia}^{high\,T} = 3Nk_BT = 3nRT$.

At very low temperatures, diatomic molecules behave more like monatomic molecules – i.e. their rotational kinetic energy is not significant at very low temperatures. In this limit, $\overline{K_{dia,m}^{low\,T}} = 3k_BT/2$ and $K_{dia}^{low\,T} = 3Nk_BT/2 = 3nRT/2$.[129]

Polyatomic molecules consisting of three or more atoms, such as CO_2, are somewhat more complex that diatomic molecules. The main difference is that three or more atoms joined together form a two- or three-dimensional configuration, whereas a diatomic molecule is linear (a one-dimensional configuration). The rotational kinetic energy of a polyatomic molecule therefore tends to have three, rather than two, degrees of freedom. Therefore, at moderate temperatures, we predict that such polyatomic molecules will have an average kinetic energy of $\overline{K_{poly,m}} = 3k_BT$ and the polyatomic ideal gas to have a total kinetic energy of $K_{poly} = 3Nk_BT = 3nRT$. At high temperatures, the polyatomic molecules can be expected to have one or two additional vibrational modes, compared to diatomic molecules, depending upon whether the molecules are two- or three-dimensional structures.

Molecular structure	K.E. Degrees of freedom	Total Kinetic Energy
monatomic	3	$3nRT/2$
diatomic (moderate temperature)	5	$5nRT/2$
diatomic (high temperature)	6	$3nRT$
diatomic (very low temperature)	3	$3nRT/2$
polyatomic (moderate temperature)	6	$3nRT$
polyatomic (high temperature, 2D)	8	$4nRT$
polyatomic (high temperature, 3D)	9	$9nRT/2$
polyatomic (very low temperature)	3	$3nRT/2$

---

[128] Again, since the molecules are assumed to have negligible size in the kinetic theory, there is negligible kinetic energy for rotation about the axis of the molecule (in this case, about the z-axis).

[129] Any rotational or vibrational kinetic energy contributions must be 'activated' by collisions. The quantum mechanical energy levels of the molecules are temperature-dependent, and so the temperature of the ideal gas determines whether or not the energy transferred during collisions will be enough to 'activate' rotational and/or vibrational kinetic energies. At higher temperatures where rotational and/or vibrational kinetic energies are 'activated' during collisions, the collisions are inelastic rather than elastic; the difference in kinetic energy is used to raise the internal energy of the ideal gas.

Internal energy of an ideal gas: The internal energy of the ideal gas equals the total kinetic energy of the ideal gas plus any potential energies that the molecules may have. A monatomic ideal gas has purely kinetic energy because the molecules are assumed to be non-interacting except briefly during collisions – i.e. the molecular model did not provide for any potential energy terms for the molecules. The internal energy of a monatomic ideal gas is therefore $U_{mon} = 3Nk_BT/2 = 3nRT/2$.

Diatomic and polyatomic ideal gases also have the same internal energy as total kinetic energy unless they have vibrational energies – if they have vibrational energies, there are both kinetic and potential energies corresponding to the vibrations. Thus, at moderate temperatures a diatomic ideal gas has an internal energy of $U_{dia} = 5Nk_BT/2 = 5nRT/2$, while at high and very low temperatures, respectively, it has an internal energy of $U_{dia}^{high\,T} = 7Nk_BT/2 = 7nRT/2$ and $U_{dia}^{low\,T} = 3Nk_BT/2 = 3nRT/2$. At moderate temperatures, without looking at any additional complicating factors, we predict that a polyatomic ideal gas will have an internal energy of $U_{poly} = 3Nk_BT = 3nRT$.

The internal energy of an ideal gas only changes if temperature (or mole number) changes.

Molecular structure	Total Degrees of freedom	Internal Energy
monatomic	3	$3nRT/2$
diatomic (moderate temperature)	5	$5nRT/2$
diatomic (high temperature)	7	$7nRT/2$
diatomic (very low temperature)	3	$3nRT/2$
polyatomic (moderate temperature)	≥ 6	$\geq 3nRT$
polyatomic (high temperature, 2D)	≥ 10	$\geq 5nRT$
polyatomic (high temperature, 3D)	≥ 12	$\geq 6nRT$
polyatomic (very low temperature)	3	$3nRT/2$

3.5 Heat Capacities for an Ideal Gas

Specific heat capacity and molar specific heat capacity: The heat exchanged by the ideal gas and its surroundings can be expressed in terms of the heat capacities if either the pressure or the volume is held constant.[130] We have a choice between working with the specific heat capacities and mass, $đQ = MC_P dT$ or $đQ = MC_V dT$, or the molar specific heat capacities and mole number, $đQ = nc_P dT$ or $đQ = nc_V dT$. For an ideal gas, it is generally more convenient to work with the molar specific heat capacities, $đQ = nc_P dT$ or $đQ = nc_V dT$, because the internal energy involves the mole number rather than the total mass of the gas.[131]

---

[130] Observe that the heat capacity at constant pressure and the heat capacity at constant volume tend to be different for an ideal gas. If both the pressure and volume are constant, and mole number is constant, then temperature will also be constant – i.e. $dT = 0$, and therefore $đQ = 0$, in this case. Thus, in what follows we have in mind isobars where volume changes and isochors where pressure changes.

[131] Some texts use the same symbol for the specific heat and molar specific heat: $c_{P,V}$. In this case, the distinction must be understood from the context. Furthermore, it is not uncommon to call them both specific heats, or even heat capacities. You can easily sort these out, based on whether you see a factor of mass or mole number.

Molar specific heat capacity at constant volume: No work is done along an isochor, since the volume does not change: $đW = 0$. Therefore, assuming that the mole number of the gas is constant – i.e. molecules are not entering or exiting the container – the internal energy change equals the heat exchanged: $dU = đQ$. For an isochor, we must use the specific heat at constant volume: $dU = đQ = nc_V dT$. Thus, the molar specific heat capacity at constant volume for an ideal gas equals

$$c_V = \frac{1}{n}\frac{dU}{dT}$$

For a monatomic ideal gas, $c_V^{mon} = 3R/2$. For a diatomic ideal gas at moderate temperatures, $c_V^{dia} = 5R/2$. For a polyatomic ideal gas at moderate temperatures, our most straightforward prediction is $c_V^{poly} = 3R$.

Internal energy of an ideal gas: Previously, we found that the internal energy of an ideal gas depends only upon temperature. We also know that internal energy is path-independent. Thus, the internal energy of an ideal gas is given, in general, by the expression $dU = nc_V dT$, regardless of whether or not the volume is constant. (However, the expression for the heat exchange, $đQ$, is path-dependent, and so $nc_V dT$ is only appropriate for $đQ$ if volume is constant.) If temperature changes, the internal energy of an ideal gas definitely changes; however, it is possible for $đQ$ to be zero if temperature changes (as is the case of an adiabat – as we shall learn, an adiabat is not isothermal for an ideal gas).

Molar specific heat capacity at constant pressure: The work done by an ideal gas along an isobar depends upon the volume change: $đW = PdV$. Assuming constant mole number, the first law gives: $dU = đQ - đW$. For an isobar, we must use the specific heat at constant pressure: $dU = đQ - đW = nc_P dT - PdV$. The last term can be expressed in terms of temperature by implicitly differentiating both sides of the ideal gas law, holding pressure constant for the isobar: $PdV = nRdT$. Substituting this into the first law, $dU = nc_P dT - nRdT$. The internal energy change is given by $dU = nc_V dT$, even when volume changes, because it is path-independent and depends only on the temperature change. The first law then becomes $nc_V dT = nc_P dT - nRdT$. Therefore, $c_V = c_P - R$ or $c_P = c_V + R$.

Thus, the molar specific heat capacity at constant pressure for an ideal gas equals

$$c_P = R + \frac{1}{n}\frac{dU}{dT}$$

For a monatomic ideal gas, $c_P^{mon} = 5R/2$. For a diatomic ideal gas at moderate temperatures, $c_P^{dia} = 7R/2$. For a polyatomic ideal gas at moderate temperatures, our most straightforward prediction is $c_P^{poly} = 4R$.

Heat for other processes: It is important to realize that $đQ$ equals $nc_V dT$ only when volume is constant and $nc_P dT$ only when pressure is constant. For an adiabat, $đQ$ is zero. For other processes, such as for an isotherm, heat can be found by applying the first law: $đQ = dU + đW$. For an ideal gas, $dU = nc_V dT$ for any process and $đW = PdV$. For a finite volume change, work can be found by expressing pressure in terms of volume and integrating.

Adiabatic index: It is convenient to define the ratio of the specific heat at constant pressure to that at constant volume, $\gamma = c_P/c_V$, as it appears frequently in thermodynamic relations. It is symbolized by the lowercase Greek symbol gamma (γ), which goes by several names, including the adiabatic index, heat capacity ratio, ratio of specific heats, and isentropic expansion factor. In Sec. 3.7, we will see its role in an adiabatic expansion, and in later chapters we will explore the relationship between heat and entropy and see the connection between an adiabatic and isentropic process.

For a monatomic ideal gas, $\gamma_{mon} = 5/3$. For a diatomic ideal gas at moderate temperatures, $\gamma_{mon} = 7/5$. For a polyatomic ideal gas at moderate temperatures, our most straightforward prediction is $\gamma_{mon} = 4/3$.

Molecular structure	c_V	c_P	γ
monatomic	$3R/2$	$5R/2$	$5/3$
diatomic (moderate temperature)	$5R/2$	$7R/2$	$7/5$
diatomic (high temperature)	$7R/2$	$9R/2$	$9/7$
diatomic (very low temperature)	$3R/2$	$5R/2$	$5/3$
polyatomic (moderate temperature)	$\geq 3R$	$\geq 4R$	$\leq 4/3$
polyatomic (high temperature, 2D)	$\geq 5R$	$\geq 6R$	$\leq 6/5$
polyatomic (high temperature, 3D)	$\geq 6R$	$\geq 7R$	$\leq 7/6$
polyatomic (very low temperature)	$3R/2$	$5R/2$	$5/3$

3.6 Isothermal Expansion of an Ideal Gas

Equations of isotherms: The equation of an isotherm is $T = const.$ For a P-V diagram, however, it is more useful to express the equation of an isotherm in terms of pressure and volume. From the ideal gas law, for constant mole number, $T = const.$ implies that $PV = const.$ Put another way, $PV = P_0 V_0$.

Graphs of isotherms: On a P-V diagram, the equation $PV = const.$ is one branch of a hyperbola (just as $y = 1/x$ is a hyperbola in the xy plane, which is symmetric about the line $y = x$). The hyperbola has two asymptotes, which are the P- and V-axes. Thus, the family of isotherms for an ideal gas is a set of hyperbolas. No two isotherms overlap, of course, as the point of intersection would need to be multi-valued.

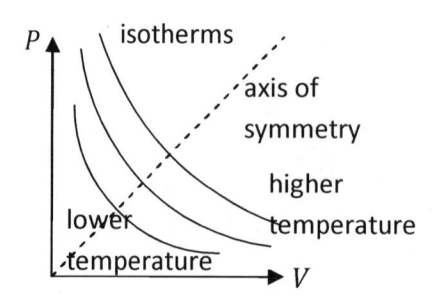

For an isothermal process, pressure and volume share an inverse relationship: As the gas expands, the pressure decreases, and as it contract, the pressure increases. Similarly, an increase in pressure results in a contraction, and a decrease in pressure results in an expansion. (Beware that these relationships may not hold for other processes; this is just if temperature is held constant.)

Work done along an isotherm: Work can be computed along an isotherm by expressing pressure in terms of volume and temperature. Temperature, unlike pressure, is constant for an isotherm and can therefore be pulled out of the work integral for an isotherm. For an ideal gas, $P = nRT/V$. For constant mole number, $W = \int_{V=V_0}^{V} PdV = \int_{V=V_0}^{V} (nRT/V)dV = nRT \int_{V=V_0}^{V} V^{-1} dV = nRT \ln(V/V_0)$. Using the ideal gas law, this can alternatively be expressed as $W = PV \ln(V/V_0)$. Since temperature is constant, $PV = P_0 V_0$, which means that the work done also equals $W = PV \ln(P_0/P)$.

Internal energy along an isotherm: Since $dU = nc_V dT$ for an ideal gas, internal energy is constant along an isotherm ($dU = 0$).

Heat exchanged along an isotherm: The heat added to or lost by the system along an isotherm can be found from the first law: $đQ = dU + đW$. Since $dU = 0$ for an isotherm, the heat added to the ideal gas equals the work done by the ideal gas: $Q = W = nRT \ln(V/V_0)$. (The specific heat capacity equations are not useful here, as neither pressure nor volume is constant along an isotherm.)

Example. A monatomic ideal gas (with constant n) completes the thermodynamic cycle illustrated below, where process $A \to B$ is isothermal and where $V_A = 2.0 \text{ m}^3$, $V_B = 8.0 \text{ m}^3$, and $P_A = 20.0 \text{ kPa}$. Determine the work done, heat exchange, and internal energy change for each process.

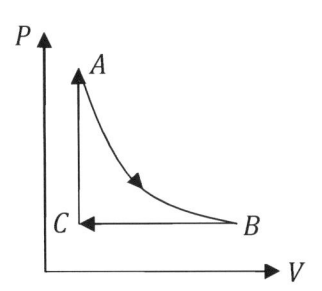

First, use the ideal gas law to determine the pressure at point B: $P_A V_A = P_B V_B$ along the isotherm, such that $P_B = P_A V_A / V_B = (20 \text{ kPa})(2 \text{ m}^3)/(8 \text{ m}^3) = 5.0 \text{ kPa}$.

Since $P_A V_A = PV$ along the isotherm, the work done for process $A \to B$ is:

$$W_{AB} = \int_{V=V_A}^{V_B} PdV = \int_{V=V_A}^{V_B} \frac{P_A V_A}{V} dV = P_A V_A \int_{V=V_A}^{V_B} \frac{dV}{V} = P_A V_A \ln\left(\frac{V_B}{V_A}\right) = 40 \text{ kJ} \ln(4) = 55.5 \text{ kJ}$$

The work done for the isobar $B \to C$ equals the area of the rectangle: $W_{BC} = \text{base} \times \text{height} = (V_C - V_B)P_B = (2.0 \text{ m}^3 - 8.0 \text{ m}^3) 5.0 \text{ kPa} = -30.0 \text{ kJ}$. No work is done along the isochor $C \to A$: $W_{CA} = 0$. The net work done is thus: $W_{net} = W_{AB} + W_{BC} + W_{CA} = 25.5 \text{ kJ}$.

The internal energy is constant along the isotherm: $\Delta U_{AB} = 0$. The internal energy decreases along the isobar and increases along the isochor, and these internal energy changes must be equal and opposite in order for the net internal energy change for the complete cycle to equal zero: $\Delta U_{net} = 0 = \Delta U_{AB} + \Delta U_{BC} + \Delta U_{CA} = 0 + \Delta U_{BC} + \Delta U_{CA} \Rightarrow \Delta U_{BC} = -\Delta U_{CA}$. The internal energy change is related to the temperature change by: $\Delta U_{BC} = nc_V(T_C - T_A) < 0$. For a monatomic ideal gas, and $c_V = 3R/2$. Therefore, $\Delta U_{BC} = 3nR(T_C - T_A)/2$. Using the ideal gas law, this can be expressed in terms of the pressure and volume as: $\Delta U_{BC} = 3(P_B V_A - P_A V_A)/2 = 3(2.0 \text{ m}^3)(5.0 \text{ kPa} - 20.0 \text{ kPa})/2 = -45.0 \text{ kJ}$. Hence, $\Delta U_{CA} = 45.0 \text{ kJ}$.

From the first law, the heat exchanged along the isotherm equals the work done along the isotherm (since $\Delta U_{AB} = 0$): $Q_{AB} = W_{AB} = 55.5 \text{ kJ}$. The heat exchanged along the isochor equals the internal energy change along the isochor (since $W_{CA} = 0$): $Q_{CA} = \Delta U_{CA} = 45.0 \text{ kJ}$. The net heat equals the net work (since $\Delta U_{net} = 0$): $Q_{net} = W_{net} = 25.5 \text{ kJ}$. Therefore, the heat exchanged along the isobar equals $Q_{BC} = Q_{net} - Q_{AB} - Q_{CA} = -75.0 \text{ kJ}$.

We can check this last answer using the first law: $Q_{BC} = \Delta U_{BC} + W_{BC} = -75.0 \text{ kJ}$. We can also check that it equals the heat for an isobar: $Q_{BC} = nc_P(T_C - T_A)$. For a monatomic ideal gas, $c_P = 5R/2$. Also, the temperatures can be expressed in terms of the pressures and volumes using the ideal gas law: $Q_{BC} = 5(P_B V_A - P_A V_A)/2 = 5(2.0 \text{ m}^3)(5.0 \text{ kPa} - 20.0 \text{ kPa})/2 = -75.0 \text{ kJ}$.

3.7 Adiabatic Expansion of an Ideal Gas

Equations of adiabats: The internal energy change equals the negative of the work done along an adiabat (since $Q = 0$): $dU = -đW$. For an ideal gas, $dU = nc_V dT$. Since $đW = PdV$, $nc_V dT = -PdV$. The differential temperature can be expressed in terms of pressure and volume by implicitly differentiating the ideal gas law: $d(PV) = d(nRT)$ or $PdV + VdP = nRdT$, where as usual we assume constant mole number. Putting all of this together: $nc_V dT = c_V(PdV + VdP)/R = -PdV$. This differential equation is separable – i.e. the two variables, pressure and volume, can be separated on different sides of the equation through some algebra, after which both sides can be integrated:

$$c_V PdV + c_V VdP = -RPdV$$
$$(c_V + R)PdV = -c_V VdP$$
$$\frac{c_P dV}{c_V V} = -\frac{dP}{P}$$
$$\gamma \int_{V=V_0}^{V} \frac{dV}{V} = -\int_{P=P_0}^{P} \frac{dP}{P}$$
$$\gamma \ln\left(\frac{V}{V_0}\right) = -\ln\left(\frac{P}{P_0}\right)$$
$$\ln\left(\frac{V}{V_0}\right)^\gamma = \ln\left(\frac{P_0}{P}\right)$$
$$PV^\gamma = P_0 V_0^\gamma$$

In this derivation, we have employed the fact that $c_P = c_V + R$ and $\gamma = c_P/c_V$. We also applied the logarithmic identities $\ln(x/y) = \ln x - \ln y$ and $a \ln x = \ln x^a$.

The equation for the adiabat of an ideal gas is thus $PV^\gamma = P_0 V_0^\gamma$, or $PV^\gamma = const.$ (Compare to the equation for an isotherm, $PV = const.$) Using the ideal gas law, the equation for an adiabat can alternatively be expressed as $TV^{\gamma-1} = T_0 V_0^{\gamma-1}$ or $TV^{\gamma-1} = const.$ by eliminating pressure, or as $P^{1-\gamma}T^\gamma = P_0^{1-\gamma}T_0^\gamma$ or $P^{1-\gamma}T^\gamma = const.$ by eliminating volume.

Graphs of adiabats: In a P-V diagram, an adiabatic curve is similar to, but a little different from, an isothermal curve (just as in the xy plane, the equation, $y = 1/x^a$, looks similar to a hyperbola, but it is not symmetric about the line $y = x$ and the curvature is a little different). Compared to an isotherm, an adiabat will have somewhat higher pressure at lower volume and somewhat lower pressure at higher volume because $\gamma = c_P/c_V = (c_V + R)/c_V > 1$.

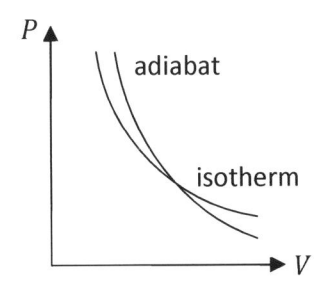

Work done along an adiabat: Work can be computed along an adiabat by expressing pressure in terms of volume and constants: Since $PV^\gamma = P_0 V_0^\gamma$ along an adiabat, $P = P_0 V_0^\gamma / V^\gamma$. For constant n,[132]

$$W = \int_{V=V_0}^{V} P\,dV = \int_{V=V_0}^{V} \frac{P_0 V_0^\gamma}{V^\gamma} dV = P_0 V_0^\gamma \int_{V=V_0}^{V} V^{-\gamma} dV = \frac{P_0 V_0^\gamma}{1-\gamma}\left(\frac{1}{V^{\gamma-1}} - \frac{1}{V_0^{\gamma-1}}\right)$$

$$W = \frac{P_0 V_0^\gamma}{\gamma - 1}\left(\frac{1}{V_0^{\gamma-1}} - \frac{1}{V^{\gamma-1}}\right)$$

Using the relation $PV^\gamma = P_0 V_0^\gamma$ for an adiabat, this can alternatively be expressed as:

$$W = \frac{P_0 V_0 - PV}{\gamma - 1} = \frac{nRT_0 - nRT}{\gamma - 1}$$

which is positive for an expansion (i.e. if $V > V_0$).

Internal energy change along an adiabat: No heat is exchanged between the system and its surroundings along an adiabat ($đQ = 0$). Therefore, the internal energy change equals the negative of the work done:

---

[132] Recall that $\gamma > 1$ for an ideal gas, so we don't need to treat the possibility of γ every equaling one.

3 Ideal Gases in Thermal Physics

$$\Delta U = -W = \frac{PV - P_0V_0}{\gamma - 1} = \frac{nRT - nRT_0}{\gamma - 1}$$

The internal energy change can alternatively be expressed in terms of the specific heat at constant volume: $\Delta U = nc_V(T - T_0)$.

Example. A monatomic ideal gas (with constant n) completes the thermodynamic cycle illustrated below, where process $A \to B$ is adiabatic and where $V_A = 2.0 \text{ m}^3$, $V_B = 8.0 \text{ m}^3$, and $P_A = 20.0$ kPa. Determine the work done, heat exchanged, and internal energy change for each process.

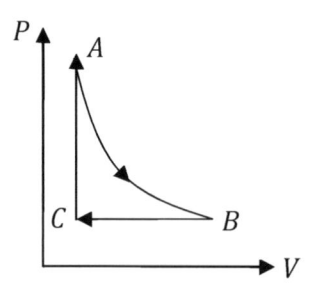

First, use the equation for the adiabat to determine the pressure at point B: $P_A V_A^\gamma = P_B V_B^\gamma$ along the adiabat, such that $P_B = P_A V_A^\gamma / V_B^\gamma = (20.0 \text{ kPa})(2.0 \text{ m}^3)^{5/3}/(8.0 \text{ m}^3)^{5/3} = 2.0$ kPa, since $\gamma = 5/3$ for a monatomic ideal gas.

Since $P_A V_A^\gamma = PV^\gamma$ along the adiabat, the work done for process $A \to B$ is:

$$W_{AB} = \int_{V=V_A}^{V_B} P\,dV = \int_{V=V_A}^{V_B} \frac{P_A V_A^\gamma}{V^\gamma}\,dV = P_A V_A^\gamma \int_{V=V_A}^{V_B} \frac{dV}{V^\gamma} = \frac{P_A V_A^\gamma}{\gamma - 1}\left(\frac{1}{V_A^{\gamma-1}} - \frac{1}{V_B^{\gamma-1}}\right)$$

$$W_{AB} = \frac{P_A V_A - P_B V_B}{\gamma - 1} = \frac{(20 \text{ kPa})(2 \text{ m}^3) - (2.0 \text{ kPa})(8 \text{ m}^3)}{5/3 - 1} = 36.2 \text{ kJ}$$

The work done along the isobar $B \to C$ equals the area of the rectangle: $W_{BC} = $ base $\times$ height $= (V_C - V_B)P_B = (2.0 \text{ m}^3 - 8.0 \text{ m}^3)\,2.0$ kPa $= -11.9$ kJ. No work is done along the isochor $C \to A$: $W_{CA} = 0$. The net work done is thus: $W_{net} = W_{AB} + W_{BC} + W_{CA} = 24.3$ kJ.

The internal energy change along the adiabat equals the negative of the work done (since $Q_{AB} = 0$): $\Delta U_{AB} = -W_{AB} = -36.2$ kJ. The internal energy along the isobar and isochor can be expressed in terms of the specific heat: $\Delta U_{BC} = nc_V(T_C - T_B)$ and $\Delta U_{CA} = nc_V(T_A - T_C)$. For a monatomic ideal gas, $c_V = 3R/2$. These internal energy changes can be expressed in terms of pressure and volume by using the ideal gas law: $\Delta U_{BC} = 3(P_B V_A - P_B V_B)/2 = 3(2.0 \text{ kPa})(2.0 \text{ m}^3 - 8.0 \text{ m}^3)/2 = -17.9$ kJ and $\Delta U_{CA} = 3(P_A V_A - P_B V_A)/2 = 3(2.0 \text{ m}^3)(20.0 \text{ kPa} - 2.0 \text{ kPa})/2 = 54.0$ kJ. As a check, $\Delta U_{net} = \Delta U_{AB} + \Delta U_{BC} + \Delta U_{CA} \approx 0$, since the net internal energy change for the cycle must be zero.

There is no heat exchange along the adiabat: $Q_{AB} = 0$. The heat exchange along the isochor equals the internal energy change (since $W_{CA} = 0$): $Q_{CA} = \Delta U_{CA} = 54.0$ kJ. The net heat exchange equals the net work (since $\Delta U_{net} = 0$): $Q_{net} = W_{net} = 24.3$ kJ $= Q_{AB} + Q_{BC} + Q_{CA} = 0 + Q_{BC} + 54.0$ kJ. Therefore, the heat exchange along the isobar is $Q_{BC} = -29.7$ kJ.

We can check this last answer using the first law: $Q_{BC} = \Delta U_{BC} + W_{BC} = -29.8$ kJ. We can also check that it equals the heat for an isobar: $Q_{BC} = nc_P(T_C - T_B)$. For a monatomic ideal gas, $c_P = 5R/2$. Also, the temperatures can be expressed in terms of the pressures and volumes using the ideal gas law: $Q_{BC} = 5(P_B V_A - P_B V_B)/2 = 5(2.0 \text{ kPa})(2.0 \text{ m}^3 - 8.0 \text{ m}^3)/2 = -29.8$ kJ.

Conceptual Example. It is instructive to compare the previous two examples. In both cases, the initial volume and pressure and final volume are the same, and the cycles look very similar; the only difference is that in one case the curved path is an isotherm and in the other case it is an adiabat. An immediate difference is that the final pressures vary somewhat: The final pressure is 5.0 kPa for the isotherm and 2.0 kPa for the adiabat. The net work is very similar for both cases, as the area under the isotherm and adiabat are nearly equal: It is 25.5 kJ for the isothermal case and 24.3 kJ for the adiabatic case; it is slightly larger in the isothermal case. These results are expected, since the isotherm and adiabat started in the same place, and the adiabat must curve downward more sharply. The net heat equals the net work, of course, since the net internal energy change is zero for a complete cycle.

The internal energy change was -45.0 kJ for the isobar and 45.0 kJ for the isochor in the isothermal case, compared to -17.9 kJ for the isobar and 54.0 kJ for the isochor in the cycle with the adiabat. These numbers represent a drastic difference between the two cycles. Similarly, the heat exchange was -75.0 kJ for the isobar and 45.0 kJ for the isochor in the cycle with the isotherm, compared to -29.8 kJ for the isobar and -54.0 kJ for the isochor in the adiabatic case.

The reason for these large differences between internal energy changes and heat exchanges in what otherwise appear to be very similar cycles is outlined as follows. The isotherm and adiabat appear as very similar curves in the two P-V diagrams, and therefore the work done in both cases is very similar. However, the internal energy change and heat exchange for an isotherm differ greatly from that of an adiabat, regardless of the similarity of the P-V curves. Whereas $\Delta U = 0$ for the isotherm, $\Delta U = -36.2$ kJ for the adiabat, which is a significant fraction of the energy involved; similarly, while $Q = 0$ for the adiabat, $Q = 55.5$ kJ for the isotherm. As a result, the isobars and isochors must differ substantially for the two cycles, in terms of internal energy and heat, in order to compensate for the differences in internal energy and heat for the isotherm and adiabat.

Bulk modulus for an ideal gas: A modulus is a property of the elasticity of a material. Young's modulus describes the elasticity of a material that is stretched along a single length, such as a stretched rod; shear modulus describes the elasticity of a material for which the shape is distorted by a shear; and bulk modulus describes the elasticity of a material that is compressed from all sides such that its volume decreases from all directions. For reasonable deformations, elastic modulus equals the ratio of the stress to the strain, where the stress relates to the pressure causing the deformation and the strain quantifies the fractional degree of deformation. For bulk modulus, B, the stress equals the change in applied pressure exerted on the material and the strain equals the fractional change in volume:

$$B = -\frac{dP}{dV/V} = -V\frac{dP}{dV}$$

Conceptually, the minus sign reflects that an increase in pressure ($dP > 0$) decreases the volume ($dV < 0$).

The value of the bulk modulus depends upon the nature of the thermodynamic process being applied. Let us consider an adiabatic compression of an ideal gas, corresponding to the adiabatic bulk modulus.[133] For an adiabatic compression, $PV^\gamma = P_0 V_0^\gamma$, such that:

$$B = -V\frac{d}{dV}\left(\frac{P_0 V_0^\gamma}{V^\gamma}\right) = -P_0 V_0^\gamma V \frac{d}{dV} V^{-\gamma} = \gamma P_0 V_0^\gamma V V^{-\gamma-1} = \frac{\gamma P_0 V_0^\gamma}{V^\gamma} = \gamma P$$

Speed of sound in an ideal gas: The speed of sound in a medium equals the squareroot of the ratio of an appropriate elastic property of the medium to a corresponding inertial property of the medium. As a sound wave propagates through a gas, pressure of the gas oscillates in space and time, creating regions of compression (pressure greater than the equilibrium gas pressure creating a region of greater density – i.e. the increased pressure contracted a local volume) and regions of rarefaction (pressure lower than the equilibrium gas pressure creating a region lower density – i.e. the decreased pressure expanded a local volume). Thus, the bulk modulus is the appropriate measure of elasticity, describing the pressure and volume changes associated with these compressions and rarefactions. The corresponding inertial property is density (mass being a measure of inertia, and the volume in the denominator giving rise to the correct units for the speed of sound). For a sound wave traveling through an ideal gas, the underlying thermodynamic process is approximately adiabatic:

$$v_s = \sqrt{\frac{B}{\rho}} = \sqrt{\frac{\gamma P}{\rho}} = \sqrt{\frac{\gamma RT}{m}}$$

where the ideal gas law and definition of density were used in the last step. Note that m is the molar mass of the gas, not the total mass of the gas.

> **Important Distinction.** Don't confuse the root-mean-square (average) speed of the molecules of an ideal gas, $\sqrt{\overline{v^2}}$, with the speed of sound in an ideal gas, v_s. The sound wave does not travel with the same speed as the molecules. However, the two speeds are close, as ultimately sound propagates via the motion and collisions of the gas molecules. Compare the two speeds:[134]
>
> $$\sqrt{\overline{v^2}} = \sqrt{\frac{3P}{\rho}} = \sqrt{\frac{3RT}{M}} \quad , \quad v_s = \sqrt{\frac{\gamma P}{\rho}} = \sqrt{\frac{\gamma RT}{m}}$$

---

[133] The adiabatic bulk modulus is the reciprocal of the adiabatic compressibility, κ_S, defined in Chapter 6.
[134] Recall that we found the pressure on an ideal gas to be $P = \rho\overline{v^2}/3$.

The average speed of the molecules is a little greater than the speed of sound in the gas, as $\sqrt{3} > \sqrt{\gamma}$, which is to be expected as the propagation of sound involves interactions between the molecules. For a monatomic ideal gas, $\gamma = 5/3$, and for various temperature ranges for diatomic and polyatomic ideal gases, γ is somewhat smaller, meaning that the gas molecules tend to travel about $\gtrsim \sqrt{2}$ faster than the sound wave.

Conceptual Questions

The selection of conceptual questions is intended to enhance the conceptual understanding of students who spend time reasoning through them. You will receive the most benefit if you first try it yourself, then consult the hints, and finally check your answer after reasoning through it again. The hints and answers can be found, separately, toward the back of the book.

1. The molecules of a real gas do interact with one another, which is one of the reasons that the ideal gas equation applies better to low-density gases than high-density gases. Assuming that the molecules of a real gas attract one another, how does this attraction affect the pressure of the real gas compared to the pressure calculated in the kinetic theory of ideal gases?

2. Explain how the ideal gas law incorporates Boyle's law, Charles's law, and Gay-Lussac's law.

3. How does the speed of smell compare to the average speed of air molecules? Explain.

4. Rank the following ideal gases in terms of the average speed of their molecules, assuming that each ideal gas has the same temperature: Ne, N_2, He, S, and Cl_2. Explain how you determined this.

5. How does the average speed of ozone (O_3) molecules compare to the average speed of diatomic oxygen (O_2) molecules at the same temperature? Explain.

6. Earth's atmosphere consists of five layers. The innermost layer is called the troposphere. The troposphere extends above Mount Everest and contains 90% of the air molecules. The next layer is the stratosphere, where ozone molecules absorb the sun's UV radiation. The middle layer is called the mesosphere, where molecules absorb very little radiation. Next is the thermosphere, which is the thickest layer. The outermost layer is called the exosphere, which is the sparsest.

   With increasing altitude, temperature decreases in the troposphere, increases in the stratosphere, decreases in the mesosphere (until reaching its minimum temperature), increases dramatically in the thermosphere, and maxes out in the exosphere. Explain this temperature variation. Also, explain why you would freeze in the exosphere, despite the extremely high temperatures.

7. Which is larger – the molar specific heat capacity at constant pressure or constant volume? Why?

8. As an ideal gas expands adiabatically, does its temperature increase or decrease? Why?

Practice Problems

The selection of practice problems primarily consists of problems that offer practice carrying out the main problem-solving strategies or involve instructive applications of the concepts (or both). You will receive the most benefit if you first try it yourself, then consult the hints, and finally check your answer after working it out again. The hints to all of the problems and the answers to selected problems can be found, separately, toward the back of the book.

Kinetic theory of ideal gases

1. During a rainstorm, an average of 80 raindrops strike a vertical window each second. The raindrops have an average mass of 4.0 g and an average speed of 10 m/s. The rain falls at an angle of 30° to the vertical. The window has a width of 100 cm and a height of 200 cm. What pressure does the rain exert on the window?

2. An ideal gas consisting of helium atoms is initially at a temperature of 40 °C. Its pressure is increased by a factor of three as it compressed to half its original volume, while the mole number is held constant. What are the initial and final root-mean-square speeds of the molecules?

3. Determine the root-mean-square speed of the molecules for each of the following types of ideal gases at 300 K: Ne, N_2, He, S, and Cl_2.

4. The speed of sound in a monatomic ideal gas is measured to be 453 m/s at room temperature. Of which element is the gas composed?

5. Show that $\frac{\gamma}{\gamma-1} = \frac{c_P}{R}$ for an ideal gas.

6. Show that $T_0 V_0^{\gamma-1} = TV^{\gamma-1}$ and $P^{1-\gamma}T^{\gamma} = P_0^{1-\gamma}T_0^{\gamma}$ along an adiabat for an ideal gas.

7. In analogy with the equation that we derived for the adiabatic bulk modulus, derive an equation for the isothermal bulk modulus.

The first law of thermodynamics applied to ideal gases

8. An ideal polyatomic gas (at moderate temperature) resides in a cylindrical container. The radius of the cylinder is 50 cm. A movable piston controls the length of the cylindrical container, which is initially $20/\pi$ m. The pressure is initially 60 kPa. The piston is pulled, adiabatically increasing the length of the cylinder to $30/\pi$ m. The piston is pushed, isothermally decreasing the length of the cylinder until the pressure returns to 60 kPa. Finally, the piston is pulled, this time isobarically increasing the length of the cylinder to $20/\pi$ m. (A) Determine the pressure of the gas after the adiabatic expansion. (B) Determine the volume of the gas after the isothermal compression. (C) Determine the work done, heat exchanged, and internal energy change for each process.

9. A monatomic ideal gas contracts adiabatically until $V = V_0/3$, isobarically until $V = V_0/6$, and then isothermally until $V = V_0/24$. Express the net change in the internal energy of the ideal gas in terms of P_0, V_0, and appropriate constants only.

10. A diatomic ideal gas consisting of 80 moles completes the thermodynamic cycle illustrated below, where $V_A = 25$ m^3, $V_B = 5.0$ m^3, $P_B = 10$ kPa, and $P_C = 30$ kPa. Determine the work done, heat exchanged, and internal energy change for each process.

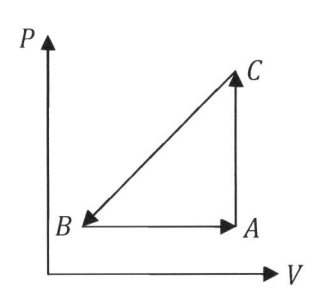

11. A monatomic ideal gas expands adiabatically until its volume is doubled, then contracts isothermally to its original volume. Determine by what factor each of the following changes: (A) its pressure, (B) its temperature, and (C) its internal energy.

12. A monatomic ideal gas expands according to the equation $P^a V^b = const.$, where a and b are constants. Derive equations for the work done, heat exchanged, and internal energy change in terms of the initial pressure and the initial and final volume only.

4 Probability Distribution Functions

4.1 Probability and Statistics

Probability: Some events are deterministic. For example, when two billiard balls collide, given their masses and initial velocities, their final velocities can be determined by conserving momentum and treating the collision as elastic. However, many events are not deterministic. For example, if a coin is flipped into the air multiple times, the outcome will not be the same each time. If the coin is flipped once, the outcome is uncertain – as it can land with either heads or tails up. For a random (or stochastic) process, the best prediction that can be made is to assign probabilities to the possible outcomes. A fair coin has a 50% chance of landing heads or tails up.

The set of possible outcomes can be used to determine the probability that each outcome will result. It is first necessary to introduce some notation. Let X_i refer to a specific outcome and N_x be the total number of possible outcomes. For example, flipping two coins at once, X_i may refer to the outcome where the first coin comes up heads (H) and the second coin comes up tails (T). Flipping two coins simultaneously, there are $N_x = 4$ possible outcomes: $X_1 = (H, H)$, $X_2 = (H, T)$, $X_3 = (T, H)$, and $X_4 = (T, T)$.

If the outcome of the event is random, every individual outcome, X_i, is equally probable. In this case, the probability, $P(X_i)$, that an event will result in a particular individual outcome, X_i, equals the inverse of the total number of possible outcomes, N_x:

$$P(X_i) = 1/N_x$$

For example, if both coins are fair in the example of flipping two coins at once, $P(H, H) = ¼$, $P(H, T) = ¼$, $P(T, H) = ¼$, and $P(T, T) = ¼$. Observe that $0 \leq P(X_i) \leq 1$ and $\sum_{i=1}^{N_x} P(X_i) = 1$. It is important to note that although the probabilities for individual outcomes, $P(X_j)$, are all equal in a random distribution, the probabilities for distinct outcomes are, in general, different.

Some of the possible outcomes may be equivalent. Indeed, if we are counting the total number of coins that land heads up when two coins are flipped simultaneously, two of the outcomes are equivalent: $X_2 = (H, T)$ and $X_3 = (T, H)$. Let Y_j denote a distinct outcome and N_y be the number of distinct outcomes. Obviously, $N_y \leq N_x$. In the example where two fair coins are flipped simultaneously, there are $N_y = 3$ distinct outcomes: $Y_1 = 0H$, $Y_2 = 1H$, and $Y_3 = 2H$.

Let Ω_j equal the total number of outcomes, X_i, that are equivalent to outcome Y_j.[135] Put another way, Ω_j is the number of ways that a distinct outcome Y_j can result. Continuing the two-coin example, $\Omega_1 = 1$, $\Omega_2 = 2$, and $\Omega_3 = 1$.

---

[135] The uppercase Greek letter omega looks like a horseshoe (Ω).

Assuming that each individual outcome is equally likely, the probability, $P(Y_j)$, that an event will result in a particular distinct outcome, Y_j, is computed as the ratio of the number of ways that the distinct outcome, Y_j, can occur to the total number of possible outcomes, N_x:

$$P(Y_j) = \frac{\Omega_j}{N_x}$$

Note that $0 \leq P(Y_j) \leq 1$ and $\sum_{i=1}^{N_x} P(Y_j) = 1$ – i.e. all of the probabilities must be between 0% and 100%, and the total probability that there is *some* outcome is 100%. In the two-coin example, $P(0H) = ¼$, $P(1H) = ½$, and $P(2H) = ¼$.

> **Important Distinction.** For an event with random outcomes, all of the outcomes are equally likely, which means that the probability, $P(X_i)$, that any individual outcome, X_i, will occur is equal for all of the possible outcomes. However, the probability, $P(Y_j)$, that a distinct outcome, Y_j, will occur may be different for each possible outcome. It is important to distinguish between an individual outcome, X_i, and a distinct outcome, Y_j. For example, when flipping two fair coins simultaneously, if the first coin comes up heads and the second comes up tails, this is a particular individual outcome, (H, T). Compare with a single coin landing heads up, which refers to a particular distinct outcome, $1H$. The probability, $P(H, T)$, that the first coin will be heads and the second tails is ¼, whereas the probability, $P(1H)$, that one of the two coins will be heads is ½ because it includes two possible outcomes – (H, T) and (T, H).

Probability distribution: The set of probabilities, $\{P(Y_j)\}$, relating to the distinct outcomes, $\{Y_j\}$, for an event is called the probability distribution function for the event, and $P(Y_j)$ is the probability distribution function.[136] For the example of flipping two fair coins simultaneously, there are three possible distinct outcomes – $0H$, $1H$, and $2H$. The odds of both coins being tails is 1 in 4, one being heads is 2 in 4, and both being heads is 1 in 4. This is the probability distribution, which only provides the odds for the possible distinct outcomes: The probability distribution function describes what is expected to happen *on average*, if the event is repeated a large number of times, but it does *not* state what *will* happen for a given number of events.

> **Conceptual Example.** It is important to realize that the probability distribution function does not state what will happen, but only what is expected to occur on average if the event is repeated a large number of times. Try flipping two pennies at once and tabulating the number of heads that come up for each flip.[137] For every four flips, you will not always see $0H$ occur once, $1H$ occur twice, and $2H$ occur once. In fact, if you repeat this enough times, you should eventually see $2H$ occur four times in a row! However, if you flip two pennies at once 100 times, you will probably see $0H$ occur about 25 times, $0H$ occur about 50 times, and $0H$ occur about 25 times. The more you repeat the event, the better the results tend to agree with the probability distribution.

---

[136] The brace, {}, denotes a set: $P(Y_j)$ is the probability for a specific event and $\{P(Y_j)\}$ is the set of probabilities.

[137] Throw two pennies in the air simultaneously and count how many heads you see. Throw the two pennies in the air and count the heads again. Repeat this several times and tabulate the outcomes.

Discrete probability distribution: An event that has a discrete set of possible outcomes has a discrete probability distribution.[138] In this case, the number of possible outcomes, N_x, can be counted, the set of possible outcomes, $\{X_i\}$, can be tabulated and grouped according to distinct outcomes, $\{Y_j\}$, and the number of ways, Ω_j, that each distinct outcome can result can be tabulated in order to determine the probability distribution, $\{P(Y_j)\}$. For example, tossing a six-sided die, there is a discrete set of possible outcomes.

Continuous probability distribution: An event that has a continuous set of possible outcomes has a continuous probability distribution. For example, if a child is playing a game of Pin the Tail on the Donkey, after being blindfolded and spun in a circle, it is possible for the child to be initially headed along *any* displacement vector lying in the horizontal plane. If the event is spinning the blindfolded child and the outcome is the initial direction, the outcome can be any angle between 0° and 360° (defined relative to some coordinate axis). In this case, there are an infinite number of possible outcomes, and the distribution is continuous – not discrete. The probability that any particular outcome will result is infinitesimal – $P(X_i) = 0$ – because $N_x = \infty$. However, the probability that an event will lead to a specific range of outcomes, such as $P(X_{min} \leq X \leq X_{max})$, is generally nonzero.

Law of large numbers: As an event is repeated a large number of times (assuming there are no factors to change the likelihood between trials), the distribution of actual outcomes approaches the probability distribution function. For example, if you flip a fair coin once, the probability for each outcome is 50%, but only one outcome will result for one flip, so the probability distribution function will definitely be wrong – by 50%! Suppose that the result is H, then you flip the coin again and the result is H again. The probability distribution is still way off! Suppose that you have now flipped the coin 10 times. By this time, you may very well have seen H occur 7 times and T occur 3 times – but the probability distribution function predicts 5 occurrences of H and 5 occurrences of T. Suppose after flipping the coin 100 times, H has occurred 54 times and T has occurred 46 times. If you try this, your results will very likely be somewhat different. However, you will probably observe the law of large numbers: The more times you repeat the event, in general, the closer the results will be to 50% occurrences of both H and T.

Statistics: The analysis of numerical data is referred to as statistics. Deterministic data leads to the deduction of precise relationships. For example, if the circumference of a variety of objects is plotted as a function of their diameters, the data will reveal a linear relationship with a slope approximately equal to π (since $C = \pi D$). Such data is deterministic. The relationship will not be perfectly linear, the slope will not be exactly equal to π, and the vertical intercept will not be exactly zero, however. Statistical analysis does not, in general, result in exact relationships due to inherent experimental errors. Rather, statistical analysis yields predictions in the form of expectation values, quantifies the uncertainties, and establishes quantitative confidence levels. Not all data is deterministic; it can also be probabilistic.

---

[138] Here discrete has the sense that the outcomes are distinct, separate, and discontinuous. It is a mistake to interpret the distinction between discrete and continuous to be that of a finite and infinite set: A discrete set can, in fact, be infinite. For example, the set of positive integers, $\{1,2,3,\cdots\}$, is both discrete and infinite.

Probabilistic data leads to statistical distributions, which are similar to probability distributions. The difference is that probability distributions are theoretical predictions that are based on theoretical assumptions and the application of probability theory, whereas statistical distributions are actual results obtained by performing experiments. For example, a statistical distribution can be made by repeatedly flipping two pennies at once and tabulating the number of times that outcomes of $0H$, $1H$, and $2H$ occur. According to the law of large numbers, the greater the number of data points obtained – i.e. the more the experiment is repeated – the greater the statistical distribution function will match the probability distribution function. Therefore, when the theory is not yet known, the statistical distribution function can provide great insight toward the development of the theory.

Deterministic data inevitably has a probabilistic component to it. Experimental errors can be classified as systematic and random. A systematic error favors a discrepancy in one direction. For example, if a ball is dropped from different heights and the time of descent is measured for each time, the value of gravitational acceleration obtained as the slope of a graph of $2h$ as a function of t^2 (since $h = \frac{1}{2}gt^2$, h is quadratic in t, but linear in t^2 with a slope equal to $\frac{1}{2}g$ – or, equivalently, $2h$ is linear in t^2 with a slope of g) can be expected to be small due to air resistance: That is, the neglect of air resistance in the formula $h = \frac{1}{2}gt^2$ systematically leads to values of time, t, which are larger than theory predicts. A random error has equal chance of missing in either direction and tends to fill out a Gaussian distribution (Sec. 4.4). For example, if the height from which the ball is dropped is measured with a meterstick, the error associated with the limitation of the measuring device – i.e. the actual height is apt to lie between the lines, which results in an error whether or not the distance between the lines is estimated – is random. The deviations associated with random errors tend to fill out a Gaussian statistical distribution, which is probabilistic in nature. That is, a deterministic set of data featuring random errors tends to statistically deviate from the actual value in a probabilistic way.

Statistical distribution functions for probabilistic data can be analyzed using the techniques of analyzing probability distribution functions – since the statistical distribution function and corresponding probability distribution function are expected to agree better and better for larger and larger numbers. For example, the same techniques used to compute expectation values for probability distribution functions can be applied to obtain statistical averages – and for large data sets the expectation values should agree with corresponding statistical averages.

Weighting factors: Let's face it, the universe is not as fair as we would like. A physical coin may not be perfectly fair – i.e. it is possible that either H or T is slightly more likely. If there are N_x possible outcomes, some of the outcomes may be more likely than others. In this case, a weighting factor must be incorporated into the probabilities, instead of assuming equal likelihood for all outcomes. Such weighting factors are not necessary for standard coins, but there are situations where weighting factors are quite significant. For example, rather than using a six-sided die in the shape of a cube, one could roll a ball with 6 colors painted on its surface, in which some of the colors occupy significantly more surface area than other colors. In this case, the color occupying the greatest surface area is most likely to lie at the top, and the weighting factors can be obtained by measuring the surface areas for each color. Some very natural problems involve weighting factors. For example, in order to find the center of mass of a system of objects, each object must be weighted by its mass – since, e.g., it would be absurd to expect the center of mass of the earth-moon system to lie halfway between the earth and the moon.

4.2 Discrete Probability Distributions

Permutation: The reordering of the elements of a set is called a permutation. For example, a set of three distinct elements – labeled as 1, 2, and 3 – can be permuted 6 ways: (1,2,3), (1,3,2), (2,1,3), (2,3,1), (3,1,2), and (3,2,1). It is also sometimes useful to permute the elements into a subset. For example, the three distinct elements of the previous set can be permuted into 6 different two-member subsets: (1,2), (1,3), (2,1), (2,3), (3,1), and (3,2). Conceptually, a permutation serves as a means for efficiently counting a large number of possibilities. Following are some formulas for counting some common classes of permutations:

- If the N elements of a set are all distinct, there are $N!$ permutations of the full set. [139]
- If the N elements of a set are all distinct, there are $\frac{N!}{(N-r)!}$ ways to permute the elements of the set into a subset with r elements, where $1 \leq r \leq N$.
- If the N_1 elements of one set and the N_2 elements of another set are all distinct, there are $\frac{N_1! N_2!}{(N_1-r_1)!(N_2-r_2)!}$ ways to permute the elements into a subset that consists of r_1 elements from the first set and r_2 elements from the second set, where $1 \leq r_{1,2} \leq N_{1,2}$.
- If there are N elements in a set, in which some elements may not be distinct, the number of permutations of the full set is $\frac{N!}{N_1! N_2! N_3! \cdots N_r!}$, where there are r distinct elements (where $1 \leq r \leq N$) and N_i is the number of elements of kind i (where $1 \leq i \leq r$). [140]

Example. Find the permutations for a set with four distinct elements.
There are $4! = 24$ permutations in all: (1,2,3,4), (1,2,4,3), (1,3,2,4), (1,3,4,2), (1,4,2,3), (1,4,3,2), (2,1,3,4), (2,1,4,3), (2,3,1,4), (2,3,4,1), (2,4,1,3), (2,4,3,1), (3,1,2,4), (3,1,4,2), (3,2,1,4), (3,2,4,1), (3,4,1,2), (3,4,2,1), (4,1,2,3), (4,1,3,2), (4,2,1,3), (4,2,3,1), (4,3,1,2), (4,3,2,1).

Example. Two six-sided dice are rolled simultaneously. How many permutations are there for the possible outcomes – where an outcome refers to the set of two dice?
Each die is a set of 6 possible outcomes, and one outcome is selected from each die. Thus, there are $\frac{6! 6!}{(6-1)!(6-1)!} = 36$ permutations (as in the popular casino game called craps). [141]

Example. How many permutations can be made by selecting 2 coins from a set of 6 distinct coins?
There are $6!/(6-2)! = 30$ ways to permute the 6 distinct coins into a subset of 2 coins.

---

[139] Obviously, the number of members in the set, N_s, must be a nonnegative integer. In this case, the factorial, $N_s!$, means $N_s! = N_s(N_s - 1)!$, where $0! \equiv 1$. For example, $4! = 4 \cdot 3! = 12 \cdot 2! = 24 \cdot 1! = 24 \cdot 0! = 24$, or $4! = 4 \cdot 3 \cdot 2 \cdot 1$. (The factorial can also be expressed in terms of an integral – called the gamma function – which is also valid for nonnegative fractions.)

[140] Observe that this agrees with $N!$ if all of the members of the set are distinct. In that case, $r = N$ and $N_i = 1$.

[141] It is a common mistake to think of this as 6 possible outcomes from which 2 must be selected. Indeed, $\frac{6!}{(6-2)!} = 30$ gives the incorrect answer. Study the distinction between this and the next example.

Example. Seven six-sided die are showing 3, 6, 3, 2, 5, 2, and 3. How many permutations of this set are there? How many begin with the 5?

These 7 dice are showing 4 distinct numbers, including three 3's, one 6, two 2's and one 5. The number of permutations of the full set of 7 dice is $\frac{7!}{3!1!2!1!} = 420$. If the 5 is placed in the first position, the remaining 6 dice can be permuted in the other positions a total of $\frac{6!}{3!1!2!} = 60$ ways.

Combination: Elements can be selected from a set and grouped into a subset either as a permutation or as a combination. The distinction is that the reordering of a subset yields a new permutation, whereas the order of the elements is irrelevant for a combination. For example, a set of three distinct elements – labeled as 1, 2, and 3 – can be grouped into 3 different combinations of two-member subsets: (1,2), (1,3), and (2,3). Compare this to the 6 different permutations that would be possible: For example, for a combination, (1,2) and (2,1) are no different, whereas for a permutation they are. Conceptually, a combination is a selection of elements, whereas a permutation is an arrangement of elements; both are useful for counting large numbers of possibilities (which one is useful depending upon whether or not order is important). The fundamental rule for combinations is:

- If the N elements of a set are all distinct, the number of ways to select a subset with r elements in which order is not important, where $1 \leq r \leq N$, is:

$$\binom{N}{r} \equiv \frac{N!}{r!(N-r)!}$$

The notation for the number of combinations, $\binom{N}{r}$, is read as, "N choose r." Conceptually, $\binom{N}{r}$ represents the number of combinations that can be made by choosing r elements from a set of N distinct elements. Observe that $\binom{N}{r}$ is equal to $\binom{N}{N-r}$. For example, $\binom{5}{3} = \binom{5}{2} = \frac{5!}{3!2!} = 10$.

Example. How many combinations can be made by selecting 2 coins from a set of 6 distinct coins?

There are $\frac{6!}{2!(6-2)!} = 15$ ways to combine the 6 distinct coins into a subset of 2 coins.[142,143]

Example. Two six-sided dice are rolled simultaneously. How many combinations are there for the possible outcomes – where an outcome refers to the subset of two dice?

Each die is a set of 6 possible outcomes, and one outcome is selected from each die. Observe that $\binom{6}{2}$ is not the correct answer, since with two separate dice it is possible to obtain repeated numbers – i.e. rolling doubles, as in (3,3). There are 6 possible doubles. The total number of combinations is thus $\binom{6}{2} + 6 = \frac{6!}{2!4!} + 6 = 21$.

---

[142] Compare this to the 30 permutations that are possible. When computing the number of ways to do something, it makes all the difference whether or not order is important.

[143] Study the distinction between this and the next example. For the coins, once the first coin is removed, there are only 5 coins remaining to choose from. For the dice, both dice provide 6 possible outcomes.

Binomial expansion: A binomial is an algebraic sum[144] of two terms of the form $(x + y)$. The expansion of $(x + y)^N$ – a binomial raised to a power – is called a binomial expansion. Let us limit ourselves to powers, N, that are integers, since the present section is on discrete probability distributions. Pascal's triangle can be observe by multiplying this out for a handful of cases:

$$(x + y)^0 = 1$$
$$(x + y)^1 = x + y$$
$$(x + y)^2 = x^2 + 2xy + y^2$$
$$(x + y)^3 = x^3 + 3x^2y + 3xy^2 + y^3$$
$$(x + y)^4 = x^4 + 4x^3y + 6x^2y^2 + 4xy^3 + y^4$$
$$(x + y)^5 = x^5 + 5x^4y + 10x^3y^2 + 10x^2y^3 + 5xy^4 + y^5$$
$$(x + y)^6 = x^6 + 6x^5y + 15x^4y^2 + 20x^3y^3 + 15x^2y^4 + 6xy^5 + y^6$$

Observe that the binomial coefficients are related by various triangles: For example, the 5 of $5x^4y$ plus the 10 of $10x^3y^2$ add up to the 15 of $15x^4y^2$, which is midway beneath the other two. All of the coefficients are related through such triangles.

The coefficients of the binomial expansion can also be found using the notation for combinations. For example, the 15 of $15x^4y^2$ can be found from $\binom{6}{2} = \frac{6!}{2!4!} = 15$. Simply choose the exponent of x (or y, since the expansion is symmetric) from the power of the binomial, n: That is, the coefficient of x^r in $(x + y)^N$ equals $\binom{N}{r}$. For example, the coefficients of $(x + y)^4$ are $\binom{4}{0} = \frac{4!}{0!4!} = 1$, $\binom{4}{1} = \frac{4!}{1!3!} = 4$, $\binom{4}{2} = \frac{4!}{2!2!} = 6$, $\binom{4}{3} = \frac{4!}{3!1!} = 4$, and $\binom{4}{4} = \frac{4!}{4!0!} = 1$.

Discrete probability distributions: In order to determine the probability distribution function for a discrete system, the first step is to determine the total number of possible outcomes, N_x, by applying the techniques of permutations and/or combinations, as appropriate, for counting large numbers.[145] Next, determine the distinct outcomes, $\{Y_j\}$, which are possible. (If N_x is large, denote the distinct outcomes symbolically and focus on conceptually understanding them.) For each distinct outcome, Y_j, use appropriate techniques of permutations and/or combinations – or, for a small system, simply count them – to determine the number of ways, Ω_j, that Y_j can result. (Use a formula to express this symbolically if N_x is large.) The probability distribution function is then computed as $P(Y_j) = \Omega_j/N_x$.

Example. A bag contains six billiard balls, which are numbered 1, 2, 3, 4, 5, and 6. Determine the probability distribution function for the sum of the numbers of two randomly selected billiard balls.

In this case, the order of the billiard balls does not affect the distribution of sums.[146] There are $N_x = \binom{6}{2} = \frac{6!}{2!4!} = 15$ possible outcomes – one for each combination:

---

[144] As either number may be positive or negative, this also accommodates a 'difference' of two terms.
[145] If $N_x = \infty$, work with summation formulas that are analogous to those used for continuous systems.
[146] Yet, the next example, which is very similar, but critically different, requires a permutation rather than a combination. The distinction between these examples is very fundamental toward understanding probabilities.

$$(1,2) \ , \ (1,3) \ , \ (1,4) \ , \ (1,5) \ , \ (1,6)$$
$$(2,3) \ , \ (2,4) \ , \ (2,5) \ , \ (2,6)$$
$$(3,4) \ , \ (3,5) \ , \ (3,6)$$
$$(4,5) \ , \ (4,6)$$
$$(5,6)$$

These are the possible outcomes, X_i, not to be confused with the distinct outcomes, Y_j. The distinct outcomes are the sums of the numbers of the two billiard balls, which can be as small as $1 + 2 = 3$ and as large as $5 + 6 = 11$. There are thus $N_y = 9$ distinct outcomes: 3, 4, 5, 6, 7, 8, 9, 10, and 11. For a small system like this, it is easy to directly count the degeneracy of each outcome, Ω_j:

$$\Omega(3) = 1 \ , \ \Omega(4) = 1 \ , \ \Omega(5) = 2 \ , \ \Omega(6) = 2 \ , \ \Omega(7) = 3$$
$$\Omega(8) = 2 \ , \ \Omega(9) = 2 \ , \ \Omega(10) = 1 \ , \ \Omega(11) = 1$$

It is worthwhile to check that $\sum_{j=1}^{N_y} \Omega_j = N_x$, which in this case is 15. Using the formula $P(Y_j) = \Omega_j/N_x$, the probability distribution is found to be:

$$P(3) = 1/15 \ , \ P(4) = 1/15 \ , \ P(5) = 2/15 \ , \ P(6) = 2/15 \ , \ P(7) = 1/5$$
$$P(8) = 2/15 \ , \ P(9) = 2/15 \ , \ P(10) = 1/15 \ , \ P(11) = 1/15$$

Observe that $\sum_{j=1}^{N_y} P(Y_j) = 1$.

Example. Two six-sided dice are rolled simultaneously. Determine the probability distribution function for the sum of the numbers shown on the two dice.

The two dice used provide independent outcomes, which, when paired together, allow for doubles — like (1,1). If the first die yields 1 and the second yields 2, this results in the same sum as in the case where the first die yields 2 and the second die yields 1. So it is tempting to count combinations, thinking that order is not important. However, compare the states (1,2) and (2,1) to the state (1,1) and you should be able to convince yourself that two dice adding to 3 is twice as likely as two dice adding to 2. Thus, we obtain this factor of two correctly if we count permutations.

Thinking this way, there are $N_x = \frac{6!6!}{(6-1)!(6-1)!} = 36$ possible outcomes for the two independent dice paired together:

$$(1,1) \ , \ (1,2) \ , \ (1,3) \ , \ (1,4) \ , \ (1,5) \ , \ (1,6)$$
$$(2,1) \ , \ (2,2) \ , \ (2,3) \ , \ (2,4) \ , \ (2,5) \ , \ (2,6)$$
$$(3,1) \ , \ (3,2) \ , \ (3,3) \ , \ (3,4) \ , \ (3,5) \ , \ (3,6)$$
$$(4,1) \ , \ (4,2) \ , \ (4,3) \ , \ (4,4) \ , \ (4,5) \ , \ (4,6)$$
$$(5,1) \ , \ (5,2) \ , \ (5,3) \ , \ (5,4) \ , \ (5,5) \ , \ (5,6)$$
$$(6,1) \ , \ (6,2) \ , \ (6,3) \ , \ (6,4) \ , \ (6,5) \ , \ (6,6)$$

The distinct outcomes, Y_j, are the sums of the numbers shown on the two dice, which can be as small as $1 + 1 = 2$ and as large as $6+6 = 12$. There are thus $N_y = 11$ distinct outcomes: 2, 3, 4, 5, 6, 7, 8, 9, 10, 11, and 12. The degeneracy of each outcome, Ω_j, is:

$$\Omega(2) = 1, \quad \Omega(3) = 2, \quad \Omega(4) = 3, \quad \Omega(5) = 4, \quad \Omega(6) = 5, \quad \Omega(7) = 6$$
$$\Omega(8) = 5, \quad \Omega(9) = 4, \quad \Omega(10) = 3, \quad \Omega(11) = 2, \quad \Omega(12) = 1$$

Using the formula $P(Y_j) = \Omega_j/N_x$, the probability distribution is found to be:

$$P(2) = \frac{1}{36}, \quad P(3) = \frac{1}{18}, \quad P(4) = \frac{1}{12}, \quad P(5) = \frac{1}{9}, \quad P(6) = \frac{5}{36}, \quad P(7) = \frac{1}{6}$$
$$P(8) = \frac{5}{36}, \quad P(9) = \frac{1}{9}, \quad P(10) = \frac{1}{12}, \quad P(11) = \frac{1}{18}, \quad P(12) = \frac{1}{36}$$

As a check, $\sum_{j=1}^{N_y} \Omega_j = 36$ and $\sum_{j=1}^{N_y} P(Y_j) = 1$.

Histogram: The probability distribution, $P(Y_j)$, or, equivalently, the degeneracy of states, Ω_j, can be concisely tabulated in a plot as a function of the distinct outcomes, Y_j, in a plot called a histogram. The histogram below shows the degeneracy of states for the two-dice example.

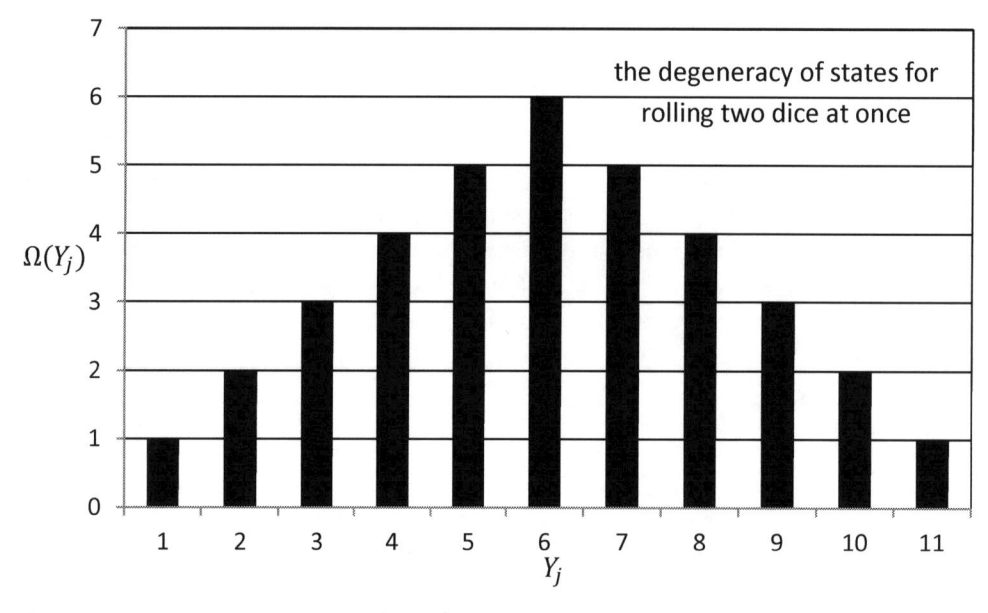

the degeneracy of states for rolling two dice at once

Grouped Outcomes: Sometimes an event is constructed as a group of events, such that the overall outcome is a group of individual outcomes. For example, when flipping two coins at once, the outcome for the event is the group of outcomes for each of the two coins. In this case, the probabilities for the overall outcome are related to the probabilities for the individual outcomes through arithmetic. This can be conceptually tricky, though, so it is important to think this through carefully and, where possible, check for consistency in the results.

The probability, $P\left(Y_j^1, Y_j^2, \cdots, Y_j^{N_o}\right)$, for combining N_o individual outcome with distinct events $\{Y_j^1, Y_j^2, \cdots, Y_j^{N_o}\}$, is related to the probabilities, $P(Y_j^k)$, for the individual outcomes, by multiplication: $P\left(Y_j^1, Y_j^2, \cdots, Y_j^{N_o}\right) = P(Y_j^1)P(Y_j^2)\cdots P(Y_j^{N_o})$. However, it is important to reason this out carefully, as numerous conceptual mistakes result in incorrect applications of this formula. In particular, the outcomes of individual events in the group may not be independent – i.e. they may be correlated, or one may affect the others.

For example, rolling a single die, the probability of getting any single outcomes, such as a 3, is $P_1(Y_j^1) = 1/6$. If two dice are rolled simultaneously, their outcomes are independent. Thus, for example, the probability of rolling snake-eyes (i.e. two 1's) is $P(1,1) = P_1(1)P_2(1) = (1/6)(1/6) = 1/36$, in agreement with the previous example. Compare this with the example of six billiard balls, numbered 1 thru 6, in a bag. In this case, the two probabilities are not independent because after the first ball is removed, only five billiard balls remain. The probability of selecting the 2 and the 5, for example, is $P(2,5) = P_1(2)P_2(5) = (1/6)(1/5) = 1/30$. Here, $P_1(2) = 1/6$ because there are 6 balls in the bag when the 2 is selected, but $P_1(5) = 1/5$ because there are 5 balls in the bag when the 5 is selected.

Probabilities do not always multiply, though; they can add, too. In particular, the probability for an outcome that involves different groupings of events is found by adding the relevant groups' probabilities together. For example, the probability of the sum of the two dice totaling 5, can be found by adding together the probabilities for paired events that sum to 5: $P(5) = P(1,4) + P(2,3) + P(3,2) + P(4,1) = P_1(1)P_2(4) + P_1(2)P_2(3) + P_1(3)P_2(2) + P_1(4)P_2(1) = 5/36$. Compare this to the probability that the numbers of the two billiard balls, numbered 1 thru 6, drawn randomly from a bag will add up to 5: $P(5) = P(1,4) + P(2,3) + P(3,2) + P(4,1) = P_1(1)P_2(4) + P_1(2)P_2(3) + P_1(3)P_2(2) + P_1(4)P_2(1) = 1/6$.

Example. Six fair coins are flipped at once. What is the probability that they will all land heads-up?
The probability that any individual coin will land heads-up is $P_1(H) = \frac{1}{2}$. The outcomes of the individual coins are independent. Therefore, $P(6H) = P_1^6(H) = (\frac{1}{2})^6 = 1/64$.[147]

Binomial distribution: Suppose that a grouped outcome consists of N_o individual outcomes, where the individual outcomes are independent and all of the individual events are identical – i.e. each event has the same probabilities, $P_1(Y_j)$. The probability, $P(rY_j)$, that r of the outcomes will equal Y_j is given by the binomial distribution (where $0 \leq r \leq N_o$):

$$P(rY_j) = \binom{N_o}{r}\left[P_1(Y_j)\right]^r \left[1 - P_1(Y_j)\right]^{N_o - r}$$

The factor $\left[P_1(Y_j)\right]^r$ gives the probabilities for the r individual events that yield the distinct outcome Y_j, $\left[1 - P_1(Y_j)\right]^{N_o-r}$ gives the probabilities that the remaining $N_o - r$ events yield a different outcome, and $\binom{N_o}{r}$ counts how many ways there are to obtain r of the Y_j's.

---

[147] The probability that 5 coins, e.g., will land heads-up is somewhat more complicated, as we will learn shortly.

> **Example.** Six fair coins are flipped at once. Determine the probability distribution for a given number of coins landing heads-up.
>
> The probability that any individual coin will land heads-up is $P_1(H) = \frac{1}{2}$, and the outcomes of the individual coins are independent. The probability distribution is given by the binomial distribution:
>
> $$P(rH) = \binom{6}{r}[P_1(H)]^r[1-P_1(H)]^{6-r} = \frac{6!}{r!(6-r)!}\frac{1}{2^r}\frac{1}{2^{6-r}} = \frac{6!}{r!(6-r)!}\frac{1}{2^6}$$
>
> $$P(0H) = \frac{6!}{0!\,6!}\frac{1}{2^6} = \frac{1}{64}$$
> $$P(1H) = \frac{6!}{1!\,5!}\frac{1}{2^6} = \frac{3}{32}$$
> $$P(2H) = \frac{6!}{2!\,4!}\frac{1}{2^6} = \frac{15}{64}$$
> $$P(3H) = \frac{6!}{3!\,3!}\frac{1}{2^6} = \frac{5}{16}$$
> $$P(4H) = \frac{6!}{4!\,2!}\frac{1}{2^6} = \frac{15}{64}$$
> $$P(5H) = \frac{6!}{5!\,1!}\frac{1}{2^6} = \frac{3}{32}$$
> $$P(6H) = \frac{6!}{6!\,0!}\frac{1}{2^6} = \frac{1}{64}$$
>
> It is a good idea to check that $\sum_{j=1}^{r} P(rH) = 1$.

4.3 Continuous Probability Distributions

Continuous probability distributions: An event that has a continuous set of possible outcomes has an infinite number of states ($N_x = \infty$). Hence, the probability that any particular outcome will result is infinitesimal – $P(X_i) = 0$. So instead of asking what the probability is of obtaining a particular outcome, a more relevant concern is the probability that the outcome of an event will lie within a specified finite range, $P(X_a \leq X \leq X_b)$.

The continuous limit: Consider a system with a discrete probability distribution, where the distinct outcomes, $\{Y_j\}$, are real numbers that are evenly spaced between two limits, Y_{min} and Y_{max}. For example, the distinct outcomes could be 0, 0.4, 0.8, 1.2, 1.6, and 2. In the limit that the number of distinct outcomes, N_y, approaches ∞, this discrete set becomes continuous. In this continuous limit, the sum $\sum_{j=1}^{N_y} P(Y_j) = 1$ becomes a Riemann sum, which can be expressed as an integral: $\int_{x=x_{min}}^{x_{max}} p(x)dx = 1$. Here, the continuous variable x is analogous to the discrete variable Y_j. In general, we can write $\int_{x=-\infty}^{\infty} p(x)dx = 1$, where $p(x)$ equals zero for $x < x_{min}$ or $x > x_{max}$. The integrand, $p(x)$, is termed the probability density.

Probability density: Continuous probability distributions are characterized by their probability densities, $p(x)$, for which $\int_{x=-\infty}^{\infty} p(x)dx = 1$. The probability for an outcome in some specified interval, $P(x_a \leq x \leq x_b)$, can be found by integrating the probability density. Thus, the goal for a continuous probability distribution is to determine the probability density, from which everything else can be found.

Probabilities: The probability of obtaining an outcome in the range $x_a \leq x \leq x_b$ equals the definite integral over the probability density for this range: $P(x_a \leq x \leq x_b) = \int_{x=x_a}^{x_b} p(x)dx$.

Example. A system has a probability density of $p(x) = A \sin x$, where A is a constant, for $0 \leq x \leq \pi$, and $p(x) = 0$ otherwise. What is the probability for an outcome in the range $\pi/6 \leq x \leq \pi/3$?

First, the numerical value of the constant A can be determined through normalization – i.e. by imposing the condition, $\int_{x=0}^{\pi} p(x)dx = 1$:

$$\int_{x=0}^{\pi} A \sin x \, dx = -A (\cos \pi - \cos 0) = 1$$

$$A = \tfrac{1}{2}$$

Now the probability for an outcome in the range $\pi/6 \leq x \leq \pi/3$ can be found via integration:

$$P\left(\frac{\pi}{6} \leq x \leq \frac{\pi}{3}\right) = \int_{x=\pi/6}^{\pi/3} A \sin x \, dx = -\tfrac{1}{2}\left(\cos\frac{\pi}{3} - \cos\frac{\pi}{6}\right) = -\frac{1}{2}\left(\frac{1}{2} - \frac{\sqrt{3}}{2}\right) = \frac{\sqrt{3}-1}{4}$$

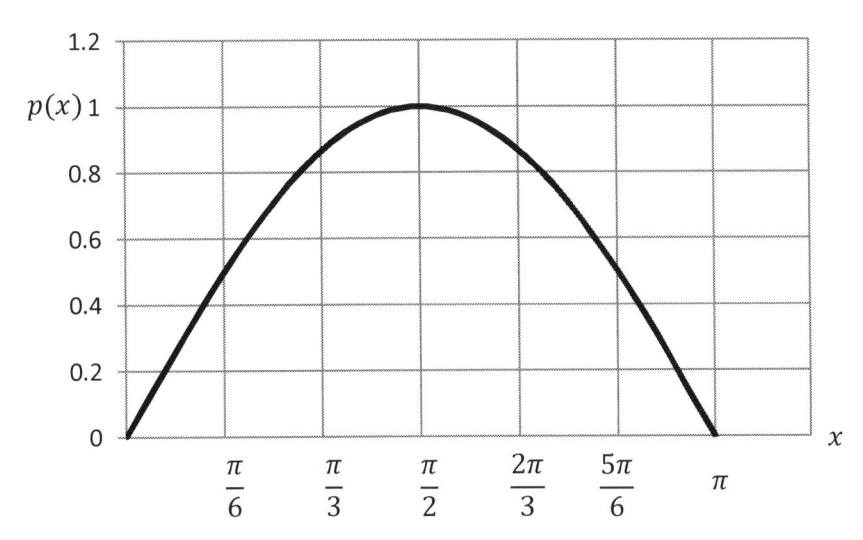

Probability density curve: Instead of the histogram which is made to represent the probabilities for a discrete probability distribution, for a continuous probability distribution, a plot of $p(x)$ is made as a function of x. The total area under the probability density curve must equal 1, and the area under the curve in the range $x_a \leq x \leq x_b$ equals $P(x_a \leq x \leq x_b)$. The probability density curve for the previous example is illustrated above.

Important Distinction. Whereas a histogram provides the probability, $P(Y_j)$,[148] of obtaining distinct outcome Y_j for a discrete probability distribution, a probability density curve provides the probability density, $p(x)$, as a function of x for a continuous probability distribution. Note the distinction: The vertical coordinate of a probability density curve does not tell you the probability of obtaining the outcome x – that would be zero (infinitesimal, since the number of possible outcomes is infinite)! Rather, the area under a section of the curve gives the probability of obtaining an outcome in a particular interval ($x_a \leq x \leq x_b$).

In the case of a discrete probability distribution, all of the vertical values are less than one because the total probability must be 100%: $\sum_{j=1}^{N_y} P(Y_j) = 1$. However, in the case of a continuous probability distribution, it is quite possible – and often occurs – for some of the vertical values to exceed one. In the continuous case, the total area under the curve must be 100%: $\int_{x=-\infty}^{\infty} p(x)dx = 1$. This allows for some of the vertical values to exceed one. For the continuous probability distribution, the vertical value is not a probability, but the probability density, $p(x)$.

Gaussian distribution: Many random variables in a very wide variety of applications of probability and statistics follow a Gaussian distribution (aka normal distribution). The probability density for the Gaussian distribution is:

$$p(x) = Ae^{-(x-\mu)^2/(2\sigma^2)}$$

where A, μ, and σ are constants. The probability density curve for the Gaussian distribution is shaped like a bell, which is symmetric about the line $x = \mu$. A probability density curve for a Gaussian distribution with $\mu = 3$ is illustrated below.

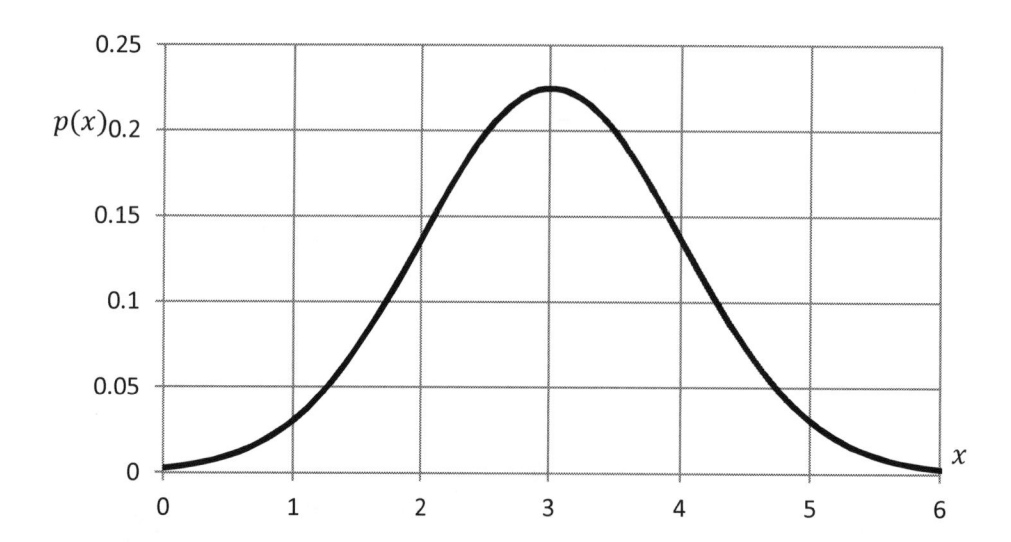

---

[148] Alternatively, a histogram may provide the degeneracy states, Ω_j, as a function of the distinct outcome, Y_j. In this case, the probabilities are found by simply dividing Ω_j by the total number of outcomes, N_x.

Normalization constant: The constant A is called the normalization constant. Its value is found to be

$$A = \frac{1}{\sigma\sqrt{2\pi}}$$

by imposing that $\int_{x=-\infty}^{\infty} p(x)dx = 1$, as we will see shortly. The normalization constant also equals the maximum height of the Gaussian, since this occurs at $x = \mu$.

Gaussian integral: The Gaussian probability distribution, $Ae^{-(x-\mu)^2/(2\sigma^2)}$, does not have an antiderivative that can be expressed algebraically. That is, it is not possible to find a function, $q(x)$, for which $q(x) = \int p(x)dx$, in general. The problem can be understood conceptually by considering the simple case, $\mu = 0$, for which the Gaussian has the form $p(x) = Ae^{-ax^2}$, where $a = 1/(2\sigma^2)$. A derivative of e^{-ax^2} with respect to x equals $-2axe^{-ax^2}$. So the trouble is that the antiderivative, $q(x)$, if it existed in algebraic form, would have to have the e^{-ax^2} part, while having some other factor to cancel the inevitable $1/x$ that is needed to cancel the x that comes from the chain rule when setting $p(x) = \frac{dq}{dx}$. Try it and you will better grasp the challenge. It can be proven that there is no such algebraic function $q(x)$. Nonetheless, the Gaussian can be integrated algebraically if the limits are $0 < x < \infty$ or $-\infty < x < \infty$. In these instances, there is a trick to performing the definite integral without actually knowing the antiderivative.

Consider the definite integral of the simple Gaussian, e^{-ax^2}, from $-\infty$ to ∞, which we shall denote by $\mathbb{I}$: $\mathbb{I} = \int_{x=-\infty}^{\infty} e^{-ax^2} dx$. The 'trick' is to square the integral:

$$\mathbb{I}^2 = \left(\int_{x=-\infty}^{\infty} e^{-ax^2} dx\right)\left(\int_{y=-\infty}^{\infty} e^{-ay^2} dy\right) = \int_{x=-\infty}^{\infty} e^{-ax^2}\left(\int_{y=-\infty}^{\infty} e^{-ay^2} dy\right) dx$$

$$\mathbb{I}^2 = \int_{x=-\infty}^{\infty}\int_{y=-\infty}^{\infty} e^{-ax^2} e^{-ay^2} dy\, dx = \int_{x=-\infty}^{\infty}\int_{y=-\infty}^{\infty} e^{-a(x^2+y^2)} dy\, dx$$

The separate integrals over the independent variables x and y can be merged in this way because the integrand is separable.[149] Since the variables x and y are independent – since the unmerged definite integrals are clearly independent – we can interpret this as a two-dimensional integral over the entire xy plane. This same integral can be performed in 2D polar coordinates rather than Cartesian coordinates. Since $r^2 = x^2 + y^2$ and $dA = dxdy = rdrd\theta$,

$$\mathbb{I}^2 = \int_{r=0}^{\infty}\int_{\theta=0}^{2\pi} e^{-ar^2} r\, dr\, d\theta$$

---

[149] Recall that in a double integral, when you perform the integration over y, you hold x constant. Also, note that the integral $\int_{x=-\infty}^{\infty} e^{-ax^2} dx$ is self-contained, such that the integration variable is a 'dummy' – i.e. it doesn't matter what you call it. However, when merging the two integrals, the two dummy variables need to be different.

Recall that values of (r, θ) for which $0 \leq r < \infty$ and $0 \leq \theta \leq 2\pi$ span the entire xy plane. The motivation for transforming to polar coordinates was to obtain the factor of r in the integrand that arose from $dA = r\, dr\, d\theta$: Whereas the function e^{-ar^2} has no algebraic antiderivative, the function re^{-ar^2} has a very natural antiderivative. Remember that the r that comes from the chain rule was the underlying problem with integrating e^{-ar^2}. The integral is now easily found to be:

$$\mathbb{I}^2 = 2\pi \int_{r=0}^{\infty} re^{-ar^2}\, dr = 2\pi \left[\frac{-e^{-ar^2}}{2a}\right]_{r=0}^{\infty} = \frac{\pi}{a} - \lim_{r \to \infty} \frac{\pi e^{-ar^2}}{a} = \frac{\pi}{a}$$

The original integral can now be found via a squareroot:

$$\mathbb{I} = \int_{x=-\infty}^{\infty} e^{-ax^2}\, dx = \sqrt{\frac{\pi}{a}}$$

Normalization: The Gaussian must be normalized – that is,

$$\int_{x=-\infty}^{\infty} p(x)\, dx = A \int_{x=-\infty}^{\infty} e^{-(x-\mu)^2/(2\sigma^2)}\, dx = 1$$

Transforming variables to $u \equiv x - \mu$, for which $du = dx$, the normalization integral becomes

$$A \int_{u=-\infty}^{\infty} e^{-u^2/(2\sigma^2)}\, du = 1$$

This is the same integral $\mathbb{I}$ that we considered previously, where $a = 1/(2\sigma^2)$. Therefore,

$$A \sqrt{\frac{\pi}{a}} = A\sigma\sqrt{2\pi} = 1$$

from which it follows that

$$A = \frac{1}{\sigma\sqrt{2\pi}}$$

Mean: The constant μ equals the mean – or average – as we will see in the following section. This is also easy to see conceptually, as the Gaussian distribution is symmetric about the line $x = \mu$. The outcome for a Gaussian distribution is expected to equal μ on average.

Standard deviation: The constant σ equals the variance – or standard deviation – as we will also learn in the next section. The standard deviation determines the width, and curvature, of the Gaussian. Since a wider Gaussian must be taller and a narrower Gaussian must be shorter – in order for the total area under the curve to always equal one – the standard deviation also affects the height of the Gaussian. Indeed, the height of the Gaussian is maximum at $x = \mu$, for which $p(x = \mu) = A = 1/(\sigma\sqrt{2\pi})$: As expected, the height, A, has an inverse relationship with the standard deviation, σ.

Example. Determine the width of the Gaussian distribution at half its maximum height.

The maximum height of the Gaussian is A, so half its height equals $A/2$. The two values of x for which the Gaussian equals half its height, which we shall denote by $x_\pm$, can be found by setting $p(x) = A/2$:

$$\frac{A}{2} = Ae^{-(x_\pm-\mu)^2/(2\sigma^2)}$$
$$e^{(x_\pm-\mu)^2/(2\sigma^2)} = 2$$
$$\frac{(x_\pm - \mu)^2}{2\sigma^2} = \ln 2$$
$$x_\pm = \mu \pm \sigma\sqrt{2\ln 2}$$

The full width at half the maximum is therefore:

$$FWHM = x_+ - x_- = 2\sigma\sqrt{2\ln 2}$$

Here we see explicitly that the standard deviation, σ, is proportional to the width of the Gaussian.

Probabilities: The probability, $P(x_a \leq x \leq x_b)$, that an outcome will lie in the range, $x_a \leq x \leq x_b$, can be found through the following definite integral:

$$P(x_a \leq x \leq x_b) = A \int_{x=x_a}^{x_b} e^{-(x-\mu)^2/(2\sigma^2)} dx$$

In general, this integral must be performed numerically.[150] A particularly important integral is $P(|x - \mu| \leq \sigma) = \int_{x=\mu-\sigma}^{\mu+\sigma} p(x)dx$, which establishes the conceptual significance of the standard deviation: From this numerical integral, it is found that there is a 68% chance that the outcome will lie within one standard deviation of the mean. Similarly, the integral $P(|x - \mu| \leq 3\sigma) = \int_{x=\mu-3\sigma}^{\mu+3\sigma} p(x)dx$ shows that there is a 90% chance that the outcome will lie within three standard deviations of the mean. Thus, the 3σ level expresses a level of confidence equal to 90%.

---

[150] The 'trick' that we applied when $x_a = -\infty$ and $x_b = \infty$ does not work in general. The reason relates to the transformation of $\mathbb{I}^2$ to polar coordinates: The Cartesian limits of integration represent a square in the xy plane, which leads to a complicated function in the limits of the polar coordinates – except for integrating an infinite plane (or the case $0 \leq x \leq \infty$, for which the infinite plane is cut in one-fourth – meaning that $\mathbb{I}$ is cut in half).

4.4 Most Probable, Average, and Root-Mean-Square Values

Average value of a function: The average value of a discrete function, $f(x)$, is given by its mean value:

$$\overline{f(x)} = \frac{\sum_{i=1}^{N} f_i(x_i)}{\sum_{i=1}^{N} 1} = \frac{1}{N} \sum_{i=1}^{N} f_i(x_i)$$

The reason for writing $\sum_{i=1}^{N} 1$ in the denominator, which simply equals N, is that it then easily generalizes in the case the average is weighted by weight factors, $\{W_i(x_i)\}$:[151]

$$\overline{f(x)} = \frac{\sum_{i=1}^{N} f_i(x_i) W_i(x_i)}{\sum_{i=1}^{N} W_i(x_i)}$$

The average value of a continuous function, $f(x)$, is given by the mean-value theorem of calculus:

$$\overline{f(x)} = \frac{\int_{x=x_a}^{x_b} f(x)dx}{\int_{x=x_a}^{x_b} dx} = \frac{\int_{x=x_a}^{x_b} f(x)dx}{x_b - x_a}$$

If the average is weighted by a weight function, $w(x)$, the average becomes:[152]

$$\overline{f(x)} = \frac{\int_{x=x_a}^{x_b} f(x)w(x)dx}{\int_{x=x_a}^{x_b} w(x)dx}$$

Expectation values: The expectation value (or expected value) of a function equals the average value of the function weighted by a probability distribution function. The expectation value for a discrete probability distribution is given by:[153]

$$\overline{f(x)} = \frac{\sum_{i=1}^{N} f_i(x_i) P_i(x_i)}{\sum_{i=1}^{N} P_i(x_i)} = \sum_{i=1}^{N} f_i(x_i) P_i(x_i)$$

---

[151] For example, the x-component of the center of mass of a system of point-like objects is weighted by mass: $x_{CM} = [\sum_{i=1}^{N} x_i M_i(x_i)] / [\sum_{i=1}^{N} M_i(x_i)] = [\sum_{i=1}^{N} x_i M_i] / M$.
[152] For example, for a one-dimensional distribution of mass along the x-axis, the center of mass is weighted by mass: $x_{CM} = \frac{1}{M} \int_{x=x_a}^{x_b} x dM$. In 1D, $dM = \lambda dx$, such that the total mass is $M = \int_{x=x_a}^{x_b} dM = \int_{x=x_a}^{x_b} \lambda dx$. Thus, the 1D center of mass integral can be expressed as $x_{CM} = \left(\int_{x=x_a}^{x_b} x\lambda dx \right) / \left(\int_{x=x_a}^{x_b} \lambda dx \right)$. The weighting function, λ, is the linear mass density, which does not cancel if the distribution of mass is non-uniform – i.e. if some regions of the object are more dense than others.
[153] The nature of the probability distribution (discrete or continuous) – and not the function $f(x)$, in the case of the expectation value – determines whether or not the discrete or continuous formula is appropriate.

since $\sum_{i=1}^{N} P_i(x_i) = 1$ for a complete set. The expectation value for a continuous probability distribution equals:

$$\overline{f(x)} = \frac{\int_{x=-\infty}^{\infty} f(x)p(x)dx}{\int_{x=-\infty}^{\infty} p(x)dx} = \int_{x=-\infty}^{\infty} f(x)p(x)dx$$

since $\int_{x=-\infty}^{\infty} p(x)dx = 1$.

Average outcome: The average (or mean) value of the outcome is the expectation value of the outcome – i.e. the average value equals the expectation value of x. For a discrete probability distribution:

$$\bar{x} = \sum_{i=1}^{N} x_i P_i(x_i)$$

If instead the probability distribution is continuous, the average outcome is:

$$\bar{x} = \int_{x=-\infty}^{\infty} xp(x)dx$$

Root-mean-square outcome: The average outcome is not always statistically useful. For example, the average value of a random distribution of vectors equals zero – since, in the large number limit, there will be as many vectors with a component along any given direction as there will be with its polar opposite, and therefore the addition of the complete set of vectors will be zero. We saw this when we explored the average velocity of a molecule in an ideal gas. If, as in the case of an ideal gas, the average value is zero, it may be more useful to compute the root-mean-square outcome – as the average value of the square of the outcome will be nonzero. The root-mean-square value is obtained by finding the expectation value of x^2 and square-rooting the result. If the probability distribution is discrete:

$$x_{rms} = \sqrt{\overline{x^2}} = \sqrt{\sum_{i=1}^{N} x_i^2 P_i(x_i)}$$

For a continuous probability distribution, the root-mean-square outcome is:

$$x_{rms} = \sqrt{\overline{x^2}} = \sqrt{\int_{x=-\infty}^{\infty} x^2 p(x)dx}$$

Most probable outcome: The most probable outcome, x_{mpo}, is simply the outcome for which the probability distribution, $P_i(x_i)$ or $p(x)$, is maximum. This can easily be found from a table or from a histogram or probability distribution curve. For a continuous probability distribution, the most probable outcome can also be found by setting a derivative of $p(x)$ with respect to x equal to zero. To insure that it is indeed an absolute maximum, also check that that the second derivative is negative and compare with the endpoints (since their slopes may not be zero, yet could have the greatest value).

Variance: The standard deviation (or variance) provides a measure of the spread of the outcomes. The standard deviation, σ_x, is smaller when the outcomes are more concentrated around the average (mean) value, and larger when the outcomes are less concentrated around the average outcome. The standard deviation can be found by taking the expectation value of $(\bar{x} - x)^2$ and square-rooting the result. Conceptually, this compares how close the outcome is to the average outcome, on average.[154] For a discrete probability distribution:[155]

$$\sigma_x = \sqrt{\sum_{i=1}^{N}(\bar{x} - x_i)^2 P_i(x_i)}$$

The standard deviation for a continuous probability distribution equals:

$$\sigma_x = \sqrt{\int_{x=-\infty}^{\infty}(\bar{x} - x)^2 p(x) dx}$$

Example. Find the average value, root-mean-square value, most probable value, and standard deviation for the Gaussian probability distribution.

The average outcome for the Gaussian distribution is found from the following integral:

$$\bar{x} = \int_{x=-\infty}^{\infty} x p(x) dx = \frac{1}{\sigma\sqrt{2\pi}} \int_{x=-\infty}^{\infty} x e^{-(x-\mu)^2/(2\sigma^2)} dx$$

It is convenient to change variables: $u \equiv x - \mu$, such that $du = dx$. In terms of u, the integral becomes:

---

[154] By definition, the expectation value of $(\bar{x} - x)$ would be exactly zero. Therefore, as in the case of the root-mean-square outcome, the standard deviation instead equals the square-root of the expectation value of $(\bar{x} - x)^2$, which is nonnegative.

[155] Compare with the statistical formula for the standard deviation of a discrete set of data, which is not weighted with probability factors. In that case, the average value is found by summing x_i and dividing by N, while the standard deviation has a denominator of $N - 1$, rather than N.

$$\bar{x} = \frac{1}{\sigma\sqrt{2\pi}} \int_{u=-\infty}^{\infty} (u+\mu)e^{-u^2/(2\sigma^2)} du = \frac{1}{\sigma\sqrt{2\pi}} \int_{u=-\infty}^{\infty} ue^{-u^2/(2\sigma^2)} du + \frac{\mu}{\sigma\sqrt{2\pi}} \int_{u=-\infty}^{\infty} e^{-u^2/(2\sigma^2)} du$$

Normalization requires that

$$\frac{1}{\sigma\sqrt{2\pi}} \int_{u=-\infty}^{\infty} e^{-u^2/(2\sigma^2)} du = 1$$

such that

$$\bar{x} = \mu + \frac{1}{\sigma\sqrt{2\pi}} \int_{u=-\infty}^{\infty} ue^{-u^2/(2\sigma^2)} du$$

This integral is exactly zero since the integrand consists of an odd function, u, times an even function, $e^{-u^2/(2\sigma^2)}$, which means that the integrand is overall an odd function, and the limits are symmetric:

$$\int_{u=-\infty}^{\infty} ue^{-u^2/(2\sigma^2)} du = 0$$

Therefore, $\bar{x} = \mu$, as expected since previously we described the constant μ as the mean.

The root-mean-square outcome of the Gaussian distribution is:

$$x_{rms} = \sqrt{\int_{x=-\infty}^{\infty} x^2 p(x) dx} = \sqrt{\frac{1}{\sigma\sqrt{2\pi}} \int_{x=-\infty}^{\infty} x^2 e^{-(x-\mu)^2/(2\sigma^2)} dx}$$

The same substitution as before, $u \equiv x - \mu$, yields:

$$x_{rms} = \sqrt{\frac{1}{\sigma\sqrt{2\pi}} \int_{u=-\infty}^{\infty} (u+\mu)^2 e^{-u^2/(2\sigma^2)} du}$$

$$x_{rms} = \sqrt{\frac{1}{\sigma\sqrt{2\pi}} \int_{u=-\infty}^{\infty} u^2 e^{-u^2/(2\sigma^2)} du + \frac{2\mu}{\sigma\sqrt{2\pi}} \int_{u=-\infty}^{\infty} ue^{-u^2/(2\sigma^2)} du + \frac{\mu^2}{\sigma\sqrt{2\pi}} \int_{u=-\infty}^{\infty} e^{-u^2/(2\sigma^2)} du}$$

By symmetry, again the middle integral is zero, and by normalization, the last term is μ^2:

$$x_{rms} = \sqrt{\mu^2 + \frac{1}{\sigma\sqrt{2\pi}} \int_{u=-\infty}^{\infty} u^2 e^{-u^2/(2\sigma^2)} du}$$

An interesting mathematical trick can be applied in order to evaluate the remaining integral: Starting with the normalized Gaussian integral, take a partial derivative with respect to σ, on both sides, holding the independent variable u constant during the differentiation:[156]

$$\frac{\partial}{\partial \sigma} \int_{u=-\infty}^{\infty} e^{-u^2/(2\sigma^2)} du = \frac{\partial}{\partial \sigma}(\sigma\sqrt{2\pi})$$

$$\frac{1}{\sigma^3} \int_{u=-\infty}^{\infty} u^2 e^{-u^2/(2\sigma^2)} du = \sqrt{2\pi}$$

Thus, the root-mean-square value for the Gaussian distribution is:

$$x_{rms} = \sqrt{\mu^2 + \sigma^2}$$

Conceptually, it should make sense that $\sigma^2 = x_{rms}^2 - \mu^2$ based on the definition of the standard deviation as the square-root of the expectation value of $(\bar{x} - x)^2$.

The most probable outcome of the Gaussian distribution can be found from a derivative:

$$\left.\frac{dp}{dx}\right|_{x=x_{mpo}} = \left[\frac{d}{dx} \frac{1}{\sigma\sqrt{2\pi}} e^{-\frac{(x-\mu)^2}{2\sigma^2}}\right]_{x=x_{mpo}} = 0$$

$$\frac{2(x_{mpo} - \mu)}{\sigma\sqrt{2\pi}} e^{-\frac{(x_{mpo}-\mu)^2}{2\sigma^2}} = 0$$

$$x_{mpo} = \mu$$

which is indeed the absolute maximum (as the endpoints are easily seen to be zero).

The standard deviation of the Gaussian distribution is given by an integral:

$$\sigma_x = \sqrt{\int_{x=-\infty}^{\infty} (\bar{x} - x)^2 p(x) dx} = \sqrt{\frac{1}{\sigma\sqrt{2\pi}} \int_{x=-\infty}^{\infty} (\mu - x)^2 e^{-\frac{(x-\mu)^2}{2\sigma^2}} dx}$$

The substitution $u \equiv x - \mu$ transforms this integral to:

---

[156] If you want to be boring, the technique of integrating by parts will suffice to yield the same result.

$$\sigma_x = \sqrt{\frac{1}{\sigma\sqrt{2\pi}} \int_{u=-\infty}^{\infty} u^2 e^{-\frac{u^2}{2\sigma^2}} du}$$

Having previously performed this same integration,

$$\sigma_x = \sqrt{\sigma^2} = \sigma$$

This agrees with our previous description of the constant σ as the standard deviation.

4.5 Maxwell-Boltzmann Distribution of Speeds for an Ideal Gas

Ideal atmosphere: The atmosphere is not an ideal gas in the sense that it does not satisfy all of the assumptions of the ideal gas model. In particular, the ideal gas model assumes that there are no external forces except when molecules collide with the container. This does not apply to the atmosphere, where the gravitational field has a significant effect on the distribution of molecules:[157] The external gravitational field creates a gradient in the density of molecules, with greater density near the surface. To good approximation, we may treat the atmosphere as an ideal gas except for the effect of the external gravitational field. This means that any equation that we derive for the atmosphere will also apply to an ideal gas in the limit that the gravitational field vanishes. This turns out to be a useful technique for deriving the distribution of speeds for an ideal gas: The effect that an external gravitational field has on the distribution of various quantities is relatively intuitive, which serves as a guide for working out the mathematical details; and once this is accomplished, gravity can be 'turned off' by setting g equal to zero, yielding the distributions for an ideal gas.

Pressure in a fluid at rest in a uniform gravitational field with uniform density: As a prelude to calculating the pressure distribution for the atmosphere, we first treat a container of fluid – i.e. liquid or gas – in a uniform gravitational field assuming that density is uniform throughout. This assumption of uniform density best applies to liquids, as they are nearly incompressible – i.e. the distribution of pressure leads to a significant distribution of density for a low-density gas, but virtually no distribution of density for a liquid or very high-density gas. We also assume that the fluid and container are in static equilibrium.[158]

---

[157] A container of gas also has an external gravitational field acting on its molecules, yet we can still treat such a gas as an ideal gas as long as the height of the container is small enough that the distribution of molecules is virtually independent of height. If the height is instead significant, the gas can be treated as an ideal gas, corrected using the distribution functions for a gas in an external gravitational field.

[158] This means that the fluid is not flowing – i.e. there is no current. Microscopically, of course, the fluid cannot be expected to be at rest – there will be inherent motions of the molecules (namely, Brownian motion).

Consider a differential volume element shaped like a right-circular cylinder with a vertical axis, located somewhere within the fluid. Since the fluid is in equilibrium, the net force acting on this differential volume element must be zero – which means that the net horizontal component of force and net vertical component of force must separately both be zero. The external gravitational field pulls downward vertically, which causes a gradient in the vertical distribution of pressure, but which does not affect the horizontal component of pressure. Thus, the pressure is symmetric horizontally and this symmetry causes the net horizontal component of force to be zero. The vertical component of force must also be zero, but in this case not by symmetry of pressure. The upward force exerted on the bottom face of the differential volume element is PdA. There are two downward forces exerted on the differential volume element: One downward force is from the pressure exerted on the top face, $(P + dP)dA$, and the other is from the weight of the differential volume element, $dW = gdm = \rho g dV = \rho g dA dy$, where ρ is the density of the fluid and dy is the vertical thickness of the differential volume element. The dP is needed in $(P + dP)dA$ in order for the net vertical component of force to equal zero:

$$PdA = (P + dP)dA + \rho g dA dy$$
$$dP = -\rho g dy$$

Since the gravitational field and density were both assumed to be uniform, integration yields

$$P - P_0 = -\rho g(y - y_0) = \rho g h$$
$$P(h) = P_0 + \rho g h$$

where h is the depth of the fluid.[159] When h is measured from the top of the fluid, P_0 is the pressure at the top of the fluid. For example, for a column of liquid, P_0 is the air pressure at the top of the column.

Pascal's principle: Consider the equation for the pressure in a fluid at a depth h, where both the density of the fluid and gravitational acceleration are independent of height: $P(h) = P_0 + \rho g h$. If the pressure P_0 at zero depth ($h = 0$) is increased by some amount – e.g. 1 atm – the pressure everywhere in the fluid increases by this same amount. Conceptually, this means that any pressure that is applied to the fluid is transmitted throughout the fluid and to the walls of the container – and not only that, but the effect of changing the pressure at one point changes the pressure everywhere in the fluid undiminished.

Atmospheric pressure: The density of the atmosphere is significantly non-uniform, and is a function of altitude. The value of gravitational acceleration is also a function of altitude. However, we will assume that gravitational acceleration is relatively uniform in comparison to the variation of density, which will provide a reasonably approximate behavior. With these assumptions, we may borrow the result $dP = -\rho g dy$, which we derived for the case of uniform density, but now integration will lead to a different pressure density, since in this case density is not constant.

---

[159] The use of the letter h, rather than d or D, to represent depth in the equation $P = \rho g h$ is almost universal. It is important to realize that h represents depth, though, and not height. Many students make mistakes by thinking, "How high is it from the bottom?" instead of thinking, "How far down is it from the top of the fluid level?"

As we remarked earlier, we will apply the ideal gas equation, $PV = Nk_BT$, to the atmosphere. Here, we have opted to work with the total number of particles, N, rather than the number of moles, n. (Recall that $Nk_B = nR$.) We can express pressure in terms of density using the ideal gas law:

$$P = \frac{Nk_BT}{V} = \frac{Nk_B\rho T}{M} = \frac{k_B\rho T}{m}$$

where M is the average mass of one molecule (and not the mass of one mole, m – i.e. $m = M/N$). Now we can write density in terms of pressure and integrate the equation $dP = -\rho g dy$:

$$\int_{P=P_0}^{P} \frac{dP}{P} = -\int_{y=y_0}^{y} \frac{mgdy}{k_BT}$$

$$\ln\left(\frac{P}{P_0}\right) = -\frac{mg}{k_BT}(y - y_0) = -\frac{mgH}{k_BT}$$

$$P(H) = P_0 e^{-\frac{mgH}{k_BT}}$$

In the context of the atmosphere, we define $y_0 = 0$ to correspond to sea level, such that P_0 is the atmospheric pressure at sea level. Thus, H is altitude relative to sea level (and not depth).[160] Thus, atmospheric pressure exponentially decays with increasing altitude.[161]

Number density: It is useful to work with number density, n_V, which is also called the concentration of molecules: $n_V = dN/dV$. When the molecules are distributed uniformly, this reduces to the number of molecules per unit volume ($n_V = N/V$). For the atmosphere, however, n_V is non-uniform, and is a function of altitude: $n_V = n_V(y)$.[162] The convenience of the number density lies in its relationship with the formulas for computing average values.

Distribution of number density: The number density, n_V, can be expressed as a function of some variable, which we denote by x: $n_V = n_V(x)$. Here, the variable x serves as a general variable, which may be altitude, y, or speed, v, for example. For a given range of outcomes, $x_a \leq x \leq x_b$, the distribution of number density, $n_V(x)$, gives the number of molecules, per unit volume, corresponding to $x_a \leq x \leq x_b$. In this way, we see that the distribution of number density, $n_V(x)$, is related to the probability distribution function, $p(x)$. Compare the total number of molecules per unit volume, N_V, in the range $x_a \leq x \leq x_b$,

---

[160] This is contrary to the case of uniform density, for which $P(h) = P_0 + \rho g h$, where the convention is to work with depth rather than height. For the atmosphere, however, it is much more convenient to use altitude. Notice the sign change associated with this distinction: Compare $+\rho g h$ with $-mgH/(k_BT)$, which follows from the definitions of depth as $h = -(y - y_0)$ and altitude relative to sea level as $H = y - y_0$.

[161] Observe that in the limit that g approaches zero, the pressure becomes the constant value, P_0. As expected, the distribution of pressure (and density) becomes uniform if the external field is removed.

[162] The variable y represents altitude, the difference $y - y_0$ is altitude with respect to some reference altitude, y_0, and H is altitude with respect to sea level (i.e. $H = y - y_0$ if y_0 corresponds to sea level).

$$N_V(x_a \leq x \leq x_b) = \int_{x=x_a}^{x_b} n_V(x)dx$$

to the probability of finding a molecule for which $x_a \leq x \leq x_b$:

$$P(x_a \leq x \leq x_b) = \int_{x=x_a}^{x_b} p(x)dx$$

If the range of outcomes, $x_a \leq x \leq x_b$, is extended to include all possible outcomes, these become:

$$\int_{x=-\infty}^{\infty} n_V(x)dx = \frac{N}{V} \quad , \quad \int_{x=-\infty}^{\infty} p(x)dx = 1$$

That is, these are equivalent normalization conditions. We thus see that averages may be computed in terms of $n_V(x)$ rather than $p(x)$. For example, the average value of x is

$$\bar{x} = \int_{x=-\infty}^{\infty} xp(x)dx = \frac{\int_{x=-\infty}^{\infty} xn_V(x)dx}{\int_{x=-\infty}^{\infty} n_V(x)dx}$$

This has the form of a weighted average, where $n_V(x)$ is the weighting factor.

Law of atmospheres: In terms of the number density, n_V, the ideal gas law can be expressed as $P = n_V k_B T$. Thus, the distribution of pressures that we derived for the atmosphere assuming that g is approximately constant,

$$P(H) = P_0 e^{-\frac{mgH}{k_B T}}$$

can be expressed in terms of the number density as

$$n_V(H) = n_{V0} e^{-\frac{mgH}{k_B T}}$$

where n_{V0} is the concentration of molecules at sea level. The exponential decay of the number density with increasing altitude is known as the law of atmospheres. Identifying mgH as the gravitational potential energy per molecule, U_m, for an approximately uniform gravitational field, the law of atmospheres can be expressed as

$$n_V(U_m) = n_{V0} e^{-\frac{U_m}{k_B T}}$$

Example. Derive an expression for the average gravitational potential energy of a molecule in the atmosphere, assuming that g and T are approximately constant.[163]

The average value of the gravitational potential energy of a molecule, $\overline{U_m}$, can be found by integrating the number density over the altitude:

$$\overline{U_m} = \frac{\int_{H=0}^{\infty} U_m n_V(H) dH}{\int_{H=0}^{\infty} n_V(H) dH} = \frac{\int_{H=0}^{\infty} mgH n_{V0} e^{-\frac{mgH}{k_BT}} dH}{\int_{H=0}^{\infty} n_{V0} e^{-\frac{mgH}{k_BT}} dH} = mg \frac{\int_{H=0}^{\infty} H e^{-\frac{mgH}{k_BT}} dH}{\int_{H=0}^{\infty} e^{-\frac{mgH}{k_BT}} dH}$$

Strictly speaking, the atmosphere is not infinite, but this integral still gives a good approximation. The bottom integral is easily computed:

$$\int_{H=0}^{\infty} e^{-\frac{mgH}{k_BT}} dH = \left[-\frac{k_BT}{mg} e^{-\frac{mgH}{k_BT}} \right]_{H=0}^{\infty} = \frac{k_BT}{mg}$$

The top integral can be found from our trick of applying a partial derivative to both sides:

$$\frac{\partial}{\partial a} \int_{H=0}^{\infty} e^{-aH} dH = \frac{\partial}{\partial a}\left(\frac{1}{a}\right)$$

$$\int_{H=0}^{\infty} H e^{-aH} dH = \frac{1}{a^2}$$

where the two minus signs cancel and where $a \equiv mg/k_BT$. Therefore, the average gravitational potential energy of a molecule in the atmosphere is found to be:

$$\overline{U_m} = mg \frac{\left(\frac{k_BT}{mg}\right)^2}{\frac{k_BT}{mg}} = k_BT$$

The average gravitational potential energy of a molecule in the atmosphere is independent of both g and m, and depends only upon temperature. It has the same form as the various average energies of an ideal gas discussed in Sec. 3.4, differing only in the overall factor.

---

[163] If you want to treat the general case, you must express g as a function of altitude as well as temperature, T. From Newton's law of universal gravitation, the dependence of gravitational acceleration on altitude is $g = GM_p/r^2$, where M_p is the mass of the planet and r is the distance from its center of mass. Furthermore, gravitational potential energy is not equal to mgh for a varying gravitational field, but to $-GM_p/r$, relative a point at infinity (the reference point is where potential energy is zero, and $-GM_p/r \to 0$ as $r \to \infty$). This expression is found by integrating the equation for gravitational force – Newton's law of universal gravitation – since work equals the negative change in potential energy for a conservative force and work is defined by a line integral.

Average potential and kinetic energy: In the previous example, we found that the law of atmospheres,

$$n_V(H) = n_{V0}e^{-\frac{mgH}{k_BT}}$$

leads to an average gravitational energy of a molecule equal to

$$\overline{U_m} = \frac{\int_{H=0}^{\infty} U_m n_V(H)dH}{\int_{H=0}^{\infty} n_V(H)dH} = k_BT$$

The result, $\overline{U_m} = k_BT$, differs only by an overall factor of 2 from the average kinetic energy per degree of freedom of one molecule of an ideal gas (Sec. 3.4): $\overline{K_{m,dof}} = k_BT/2$. The average translational kinetic energy of a molecule of an ideal gas associated with the x-component of velocity[164] is

$$\overline{K_{mx}} = \frac{\int_{v_x=0}^{\infty} K_{mx} n_V(v_x)dv_x}{\int_{v_x=0}^{\infty} n_V(v_x)dv_x} = \frac{k_BT}{2}$$

Notice that the integral expression for $\overline{K_{mx}}$ has a form that is similar to $\overline{U_m}$, and the result only differs by a factor of 2. Also, realize that while we have written integral expressions for both $\overline{K_{mx}}$ and $\overline{U_m}$, and while we have results for both ($\overline{K_{mx}} = k_BT/2$ and $\overline{U_m} = k_BT$), we have not yet derived an expression for $n_V(v_x)$. However, though we do not yet know the form of $n_V(v_x)$, this comparison suggests that $n_V(v_x)$ is very similar in form to $n_V(H)$. We shall now show that this is indeed the case.

This should not actually be surprising, with the theorem of equipartition of energy (see Sec. 3.4) in mind. From this perspective, we should expect $n_V(K_{mx})$ and $n_V(U_m)$ to have identical form:[165]

$$n_V(U_m) = n_{V0}e^{-\frac{U_m}{k_BT}} \quad , \quad n_V(K_{mx}) = n_{V0}e^{-\frac{K_{mx}}{k_BT}}$$

We shall now show that this deduction is correct. In terms of the integration variables H and v_x,

$$n_V(H) = n_{V0}e^{-\frac{mgH}{k_BT}} \quad , \quad n_V(v_x) = n_{V0}e^{-\frac{mv_x^2}{2k_BT}}$$

All we need to do to show if our expression for $n_V(K_{mx})$ works is substitute $n_V(v_x)$ into the integral expression for $\overline{K_{mx}}$ and see if the result is $k_BT/2$:

---

[164] It would be imprecise to label K_{mx} as "the x-component of kinetic energy of one molecule," since kinetic energy is a scalar – not a vector. That is, unlike a vector, kinetic energy does not have 'components.' It might be tempting to think this way, since the magnitude of a vector is related to the components through addition in quadrature: Namely, since $v = \sqrt{v_x^2 + v_y^2 + v_z^2}$, it follows that $K_m = K_{mx} + K_{my} + K_{mz}$. However, these are not at all vector-like components, since these terms cannot be found from K_m using expressions like $K_m \cos\varphi \sin\theta$.

[165] The argument is crucial: Note the distinctions between $n_V(K_{mx})$, $n_V(U_m)$, $n_V(H)$ and $n_V(v_x)$.

$$\overline{K_{mx}} = \frac{\int_{v_x=0}^{\infty} K_{mx} n_V(v_x) dv_x}{\int_{v_x=0}^{\infty} n_V(v_x) dv_x} = \frac{m}{2} \frac{\int_{v_x=0}^{\infty} v_x^2 n_{V0} e^{-\frac{mv_x^2}{2k_BT}} dv_x}{\int_{v_x=0}^{\infty} n_{V0} e^{-\frac{mv_x^2}{2k_BT}} dv_x} = \frac{m}{2} \frac{\int_{v_x=0}^{\infty} v_x^2 e^{-\frac{mv_x^2}{2k_BT}} dv_x}{\int_{v_x=0}^{\infty} e^{-\frac{mv_x^2}{2k_BT}} dv_x}$$

The bottom integral involves a Gaussian distribution, for which we have previously (Sec. 4.3) derived the result that $\int_{x=-\infty}^{\infty} e^{-ax^2} dx = \sqrt{\frac{\pi}{a}}$. Since the Gaussian distribution is symmetric, $\int_{x=0}^{\infty} e^{-ax^2} dx = \frac{1}{2}\sqrt{\frac{\pi}{a}}$. Therefore, the bottom integral results in

$$\int_{v_x=0}^{\infty} e^{-\frac{mv_x^2}{2k_BT}} dv_x = \frac{1}{2}\sqrt{\frac{2\pi k_BT}{m}}$$

The top integral is related to the bottom integral:

$$\int_{v_x=0}^{\infty} v_x^2 e^{-\frac{mv_x^2}{2k_BT}} dv_x = -\frac{\partial}{\partial a} \int_{v_x=0}^{\infty} e^{-av_x^2} dv_x = -\frac{\partial}{\partial a}\left(\frac{1}{2}\sqrt{\frac{\pi}{a}}\right) = \frac{1}{4a}\sqrt{\frac{\pi}{a}}$$

where $a \equiv m/(2k_BT)$. Substituting the results of these integrals into the original expression,

$$\overline{K_{mx}} = \frac{m}{2} \frac{\frac{1}{4a}\sqrt{\frac{\pi}{a}}}{\frac{1}{2}\sqrt{\frac{\pi}{a}}} = \frac{m}{4a} = \frac{m}{4} \frac{2k_BT}{m} = \frac{k_BT}{2}$$

which is indeed the correct result, justifying the form of $n_V(v_x)$ and hence $n_V(K_{mx})$.

Boltzmann distribution law: The fact that $n_V(U_m)$ and $n_V(K_{mx})$ have the same form, as expected from the theorem of equipartition of energy, can be stated more generally in what is known as the Boltzmann distribution law. For a general type of energy, E, the distribution of number density has the form,

$$n_V(E) = n_{V0} e^{-\frac{E}{k_BT}}$$

This holds for E equal to either U_m or K_{mx}, and other forms of energy, too. Note that although $n_V(U_m)$ and $n_V(K_{mx})$ have the same structure, the fact that K_{mx} is quadratic in v_x while U_m is linear in H leads to a different overall factor between $\overline{U_m}$ and $\overline{K_{mx}}$. Thus, while the Boltzmann distribution law says that all energy distributions have the same form, the Boltzmann distribution law does not say that all average energies are the same.

Maxwell-Boltzmann distribution: When the distribution of number density is expressed as a function of the speed, v, of the molecules – i.e. $n_V(v)$ – it is called the Maxwell-Boltzmann distribution. Recall that the distribution of number density in terms of the x-component of velocity is

$$n_V(v_x) = n_{V0} e^{-\frac{mv_x^2}{2k_B T}}$$

As the three dimensions of space constitute independent directions, $n_V(v_y)$ and $n_V(v_z)$ must have the same form. The number of molecules per unit volume with x-components of velocity in the range $v_a \leq v_x \leq v_b$ is

$$N_V(v_a \leq v_x \leq v_b) = \int_{v_x=v_a}^{v_b} n_V(v_x) dv_x$$

Compare this to the total number of particles with speeds in the range $|v_a| \leq v \leq |v_b|$:

$$N_V(|v_a| \leq v \leq |v_b|) = \int_{v=|v_a|}^{|v_b|} n_V(v) dv$$

Since speed, v, is the magnitude of velocity $\left(v = \|\vec{v}\| = \sqrt{v_x^2 + v_y^2 + v_z^2}\right)$, the integral over speeds can be combined with the integral over solid angle (since v is related to the components of velocity – v_x, v_y, and v_z – the same way that r of spherical coordinates is related to the components of position – x, y, and z):

$$\int_{v=|v_a|}^{|v_b|} \int_{\theta=0}^{\pi} \int_{\varphi=0}^{2\pi} n_V(v) dv d\theta d\varphi = \int_{v_z=v_a}^{v_b} \int_{v_y=v_a}^{v_b} \int_{v_x=v_a}^{v_b} n_V(v_x) n_V(v_y) n_V(v_z) dv_x dv_y dv_z$$

These are analogous to volume elements in spherical and Cartesian coordinates: $dV = r^2 \sin\theta\, d\theta d\varphi = dx dy dz$. Therefore, by comparison, we conclude that (where $\propto$ means "proportional to")

$$n_V(v) dv \propto v^2 n_V(v_x) n_V(v_y) n_V(v_z) dv_x dv_y dv_z$$

$$n_V(v) dv \propto v^2 e^{-\frac{mv_x^2}{2k_B T}} e^{-\frac{mv_y^2}{2k_B T}} e^{-\frac{mv_z^2}{2k_B T}} dv \quad \Rightarrow \quad n_V(v) dv \equiv A v^2 e^{-\frac{mv^2}{2k_B T}} dv$$

The overall factor, A, can be found from normalization:

$$\frac{N}{V} = N_V(0 \leq v < \infty) = \int_{v=0}^{\infty} n_V(v) dv = \int_{v=0}^{\infty} A v^2 e^{-\frac{mv^2}{2k_B T}} dv$$

We have previously computed an integral of this same form:

$$\frac{N}{V} = \int_{v=0}^{\infty} Av^2 e^{-\frac{mv^2}{2k_BT}} dv = \int_{v=0}^{\infty} Av^2 e^{-av^2} dv = \frac{A}{4}\sqrt{\frac{\pi}{a^3}} = \frac{A}{4}\sqrt{\frac{8\pi k_B^3 T^3}{m^3}} = \frac{A}{2}\sqrt{\frac{2\pi k_B^3 T^3}{m^3}}$$

The normalization constant turns out to be

$$A = \frac{2^{1/2} m^{3/2} N}{\pi^{1/2} k_B^{3/2} T^{3/2} V}$$

The Maxwell-Boltzmann distribution can then be expressed as

$$n_V(v) = \frac{2^{1/2} m^{3/2} N}{\pi^{1/2} k_B^{3/2} T^{3/2} V} v^2 e^{-\frac{mv^2}{2k_BT}}$$

This and other results that are independent of g apply to ideal gases in general, and are not limited to the atmosphere (since the atmosphere becomes an ideal gas – i.e. the external field vanishes – in the limit that g approaches zero).

Example. Determine the average, root-mean-square, and most probable speeds for the Maxwell-Boltzmann distribution.

The average speed of a molecule following the Maxwell-Boltzmann distribution is

$$\bar{v} = \frac{\int_{v=0}^{\infty} v n_V(v) dv}{\int_{v=0}^{\infty} n_V(v) dv} = \frac{V}{N} \int_{v=0}^{\infty} v n_V(v) dv = \frac{AV}{N} \int_{v=0}^{\infty} v^3 e^{-av^2} dv$$

where $a = m/(2k_B T)$. This integral can be reduced using the partial derivative trick:

$$\bar{v} = -\frac{AV}{N} \frac{\partial}{\partial a} \int_{v=0}^{\infty} v e^{-av^2} dv$$

This last integral can be performed by substituting $u \equiv v^2$, for which $du = 2v dv$:

$$\int_{v=0}^{\infty} v e^{-av^2} dv = \frac{1}{2} \int_{u=0}^{\infty} e^{-au} du = -\left[\frac{e^{-au}}{2a}\right]_{u=0}^{\infty} = \frac{1}{2a}$$

The average speed of a molecule is therefore:

$$\bar{v} = -\frac{AV}{N}\frac{\partial}{\partial a}\left(\frac{1}{2a}\right) = \frac{AV}{2a^2 N} = \frac{AV}{2N}\left(\frac{2k_B T}{m}\right)^2 = \frac{m^{\frac{3}{2}}}{2^{\frac{1}{2}}\pi^{\frac{1}{2}}k_B^{\frac{3}{2}}T^{\frac{3}{2}}}\left(\frac{2k_B T}{m}\right)^2$$

$$\bar{v} = \sqrt{\frac{8\, k_B T}{\pi\, m}}$$

The root-mean-square speed is

$$v_{rms} = \sqrt{\overline{v^2}} = \sqrt{\frac{\int_{v=0}^{\infty} v^2 n_V(v)\,dv}{\int_{v=0}^{\infty} n_V(v)\,dv}} = \sqrt{\frac{V}{N}\int_{v=0}^{\infty} v^2 n_V(v)\,dv} = \sqrt{\frac{AV}{N}\int_{v=0}^{\infty} v^4 e^{-av^2}\,dv}$$

This integral is found to be:

$$\int_{v=0}^{\infty} v^4 e^{-av^2}\,dv = -\frac{\partial}{\partial a}\int_{v=0}^{\infty} v^2 e^{-av^2}\,dv = -\frac{\partial}{\partial a}\left(\frac{1}{4}\sqrt{\frac{\pi}{a^3}}\right) = \frac{3}{8}\sqrt{\frac{\pi}{a^5}}$$

Therefore, the root-mean-square speed is

$$v_{rms} = \sqrt{\frac{AV}{N}\frac{3}{8}\frac{\pi^{1/2}}{a^{5/2}}} = \sqrt{\frac{2^{1/2}m^{3/2}}{\pi^{1/2}k_B^{3/2}T^{3/2}}\frac{3}{8}\frac{2^{5/2}\pi^{1/2}k_B^{\frac{5}{2}}T^{\frac{5}{2}}}{m^{5/2}}}$$

$$v_{rms} = \sqrt{\frac{3k_B T}{m}}$$

The most probable speed is obtained by setting a derivative of $n_V(v)$ with respect to v equal to zero (and checking the sign of the second derivative to see if it is indeed a minimum – for some distributions, it would not be – and checking the endpoints, since an endpoint could be an absolute minimum, lower than a relative endpoint found by setting the first derivative equal to zero):

$$\left.\frac{d}{dv}\right|_{v=v_{mps}} n_V(v) = A\left.\frac{d}{dv}\right|_{v=v_{mps}}\left(v^2 e^{-av^2}\right) = -2Av_{mps}e^{-av_{mps}^2} + 2aAv_{mps}^3 e^{-av_{mps}^2} = 0$$

$$v_{mps} = av_{mps}^2$$

$$v_{mps} = \sqrt{\frac{1}{a}} = \sqrt{\frac{2k_B T}{m}}$$

Observe that v_{rms} is a little smaller than $\bar{v}$, which itself is a little smaller than v_{mps}.

4.6 Mean-Free Path and Molecular Flux

Collisions in an ideal gas: The molecules of an ideal gas interact with one another and with the walls of the container through collisions. The nature of these collisions is very significant to the kinetic theory of gases because these collisions are ultimately responsible for the equilibrium state that the gas reaches. The intermolecular collisions of an ideal gas – for which the Maxwell-Boltzmann distribution applies – are most simply modeled for the case of a low-density gas, for which the following assumptions (in addition to those stated for an ideal gas in Sec. 3.1) tend to be most applicable:

- Each molecule spends a small fraction of its time interacting with other molecules in collisions.
- The intermolecular collisions occur almost exclusively in pairs – i.e. three or more particles seldom interact with one another simultaneously.[166]
- The average intermolecular separation is large compared to average deBroglie wavelength – $\lambda = h/p = h/(mv)$, where p is momentum and h is Planck's constant – of the molecules.[167]

Some immediate consequences of these assumptions are:

- The molecules spend most of their time traveling according to their own inertia – i.e. moving in straight lines with constant speeds, for which the distribution may be modeled by the Maxwell-Boltzmann distribution.[168]
- Each intermolecular collision may be treated as a collision between two molecules. That is, very seldom do three or more molecules collide together simultaneously.
- The collisions may be treated classically (i.e. applying classical mechanics, rather than quantum mechanics) since the deBroglie wavelength is negligible.[169]

Collision frequency: The rate at which intermolecular collisions occur – i.e. the average number of collisions occurring per unit time – is called the collision frequency, f_c (aka collision rate). The collision frequency, f_c, for a molecule is independent of its past – i.e. it does not depend upon how much time has elapsed since its last collision (just as the probability that a fair six-sided die will land with the side with 2 dots facing up is $1/6$, regardless of what the prior rolls were). Qualitatively, we do expect the collision frequency, f_c, to depend upon the average speed of the molecules: It is reasonable to predict that the faster the molecules travel, on average, the greater the collision frequency will be.

Mean-free time: The reciprocal of the collision frequency, f_c, is called the mean-free time, τ (aka relaxation time or collision time[170]): $\tau = 1/f_c$. Conceptually, the mean-free time, τ, represents the time that each molecule spends, on average, between collisions.[171]

---

[166] Of course, at any given moment, there will be numerous molecules engaged in collisions. However, virtually all of these collisions will each involve just two molecules.
[167] A small deBroglie wavelength – represented by the Greek symbol lambda (λ) – thus implies a large momentum.
[168] Of course, the molecular speeds are far from constant, since every molecule undergoes numerous collisions. The speeds are, however, assumed to be relatively constant during the interval between collisions.
[169] Nonetheless, if you are deriving scattering cross sections, you should consider the quantum effects.
[170] However, 'collision time' is confusing, since τ is the average time spent *between* (and not *during*) collisions.
[171] The mean-free time is represented by the Greek symbol tau (τ).

Collision probability: For a duration, Δt, the ratio $\Delta t/\tau = f_c \Delta t$ represents the collision probability – i.e. the chance of having a collision in the time interval, Δt.[172]

Mean-free path: The distance that a molecule travels, on average, between collisions is called the mean-free path, $\bar{\ell}$. The mean-free path, $\bar{\ell}$, is related to the average speed, $\bar{v}$, of the molecules between collisions and the mean-free time, τ, by $\bar{\ell} = \bar{v}\tau$.

Scattering cross section: The probability that two molecules will interact in a collision can be determined from the scattering cross section, σ.[173] The quantitative calculation of the scattering cross section, σ, which can become quite involved for common nuclear collisions, provides a numerical value of area that conceptually represents an effective target area in the following sense: If an incident molecule intercepts this effective target area, there will be an interaction between the two molecules. Calculation of this effective target area is generally rather involved when the force of interaction varies with distance – e.g. Rutherford's bombardment of a thin gold foil with alpha particles. It is very simple, on the other hand, for hard geometric objects like billiard balls, which either hit or miss. The low-density ideal gas can be treated, to good approximation, with this geometric hit-or-miss situation.

Scattering between hard spheres: The scattering cross section for two hard spheres with radii R_1 and R_2 equals $\sigma = \pi(R_1 + R_2)^2$ because the two hard spheres will hit if they are on course for their centers to be separated by a distance less than $R_1 + R_2$ and they will miss otherwise, so the effective target area is $\pi(R_1 + R_2)^2$. For identical hard spheres, $R_1 = R_2 \equiv R$, the scattering cross section reduces to $\sigma = 4\pi R^2$, which equals the surface area of each sphere.

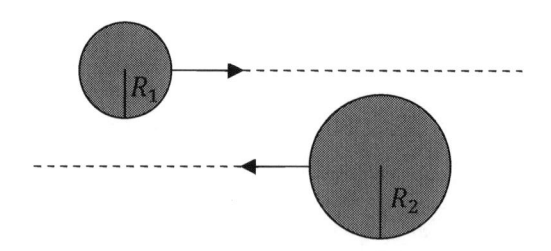

Mean-free values for a low-density ideal gas: A single classical, hard molecule of radius R travels in a straight line with constant speed between collisions, with an average speed of $\bar{v} = \sqrt{8k_B T/(\pi m)}$ (see the previous example). For a single-component gas, all of the molecules are identical, so as one molecule proceeds forward with constant velocity, it will collide with any other molecule that it intercepts in a cylinder of cross-sectional area equal to the effective target area $\sigma = 4\pi R^2$. In a duration Δt, this cylinder will have length $\bar{v}\Delta t$ and volume $V = \sigma \bar{v}\Delta t$. This volume contains $\sigma \bar{v}\Delta t n_V$ molecules, which represent potential collisions that will occur in time interval Δt.

---

[172] If the time interval, Δt, is larger than the mean-free time, τ, the 'collision probability' exceeds unity. This means that, on average, if $\Delta t > \tau$, the molecule will experience multiple collisions – hence the term 'collision frequency.'

[173] The Greek symbol sigma (σ), which looks like a sideways 6, is best drawn in a clockwise fashion (which is opposite to the manner in which most students would intuitively draw it): That is, draw the circular part first, with a clockwise motion, starting at the top, and when you close the circle, continue horizontally to the right. (If instead you draw the horizontal section first, the circular section usually does not close properly.)

Note that the other molecules are not at rest: The average speed $\bar{v}$ represents how fast one molecule moves relative to the walls of the container (assumed to be at rest), while two molecules move with an average relative speed $\bar{v}_r$ that is different from $\bar{v}$. Thus, a molecule does not actually have the potential to interact with $\sigma \bar{v} \Delta t n_V$ other molecules in a duration Δt, but instead has the potential to interact with $\sigma \bar{v}_r \Delta t n_V$ other molecules in a duration Δt. Therefore, the collision probability is

$$f_c \Delta t = \sigma \bar{v}_r \Delta t n_V$$

and the collision frequency is

$$f_c = \sigma \bar{v}_r n_V = 4\pi R^2 \bar{v}_r n_V$$

It is important to remember that $\bar{v}_r$ is not the average speed of the molecules from the Maxwell-Boltzmann distribution, but instead the average relative speed between molecules. However, average relative speed between molecules, $\bar{v}_r$ is related to the average speed of the molecules $\bar{v}$. If two molecules have velocities $\vec{v}_1$ and $\vec{v}_2$, their relative velocity equals their difference: $\vec{v}_r = \vec{v}_1 - \vec{v}_2$. The square of the magnitude of the relative velocity equals

$$v_r^2 = \vec{v}_r \cdot \vec{v}_r = (\vec{v}_1 - \vec{v}_2) \cdot (\vec{v}_1 - \vec{v}_2) = v_1^2 + v_2^2 - 2\vec{v}_1 \cdot \vec{v}_2$$

Taking the average of both sides results in

$$\overline{v_r^2} = \overline{v_1^2} + \overline{v_2^2} = 2\overline{v^2}$$

Note that $\overline{\vec{v}_1 \cdot \vec{v}_2}$ because $\vec{v}_1 \cdot \vec{v}_2 = 2 v_1 v_2 \cos\theta$, where θ is the angle between the two velocities, and the average value of $\cos\theta$ is zero. Since $\overline{v_r^2} = 2\overline{v^2}$, it follows that $\bar{v}_r \approx \sqrt{2}\bar{v}$, where the approximation comes from replacing the root-mean-square speed with the average speed.

In terms of the average molecular speed, the collision frequency is

$$f_c = 4\pi\sqrt{2} R^2 \bar{v} n_V = 16\pi R^2 n_V \sqrt{\frac{k_B T}{\pi m}}$$

from which the mean-free time is found to be:

$$\tau = \frac{1}{4\pi\sqrt{2} R^2 \bar{v} n_V} = \frac{1}{16\pi R^2 n_V}\sqrt{\frac{\pi m}{k_B T}}$$

The mean-free path equals

$$\bar{\ell} = \bar{v}\tau = \frac{1}{4\pi\sqrt{2} R^2 n_V}$$

Molecular flux: The number of molecules per unit area per unit time passing through a given surface is referred to as the molecular flux, Φ_N.[174] Although from the definition it might seem like molecular flux, Φ_N, should equal the collision frequency, f_c, per unit area, from a more careful analysis which follows we will see that they have roughly the same order of magnitude, but are not quite the same. The SI units of molecular flux are $m^{-2}s^{-1}$. The molecular flux would be a useful quantity to calculate, for example, if there is a small hole in a container that allows some of the molecules to escape.

Molecules striking a wall: Imagine an ideal gas in a container in the shape of a cube. We may setup a coordinate system with the origin at the center and each axis passing through the center of one of the square sides. Consider the flux of molecules, Φ_N, striking the $+z$ square wall – i.e. the number of molecules per unit area per unit time that strike the $+z$ wall.

Let us first consider the problem intuitively, which will provide only an order-of-magnitude – rather than exact – result. By symmetry, the number of molecules per unit volume headed in the direction of the $+z$ wall must be $n_V/6$. Naïvely, we may adapt the infinitesimal collision probability for intermolecular collisions to collisions with the $+z$ wall by setting the scattering cross section for hard spheres equal to the area of the wall, A (the actual target area), setting the average relative velocity $\bar{v}_r$ equal to $\bar{v}$ (since the wall is assumed to be stationary), and dividing by 6 (since 1 out of 6 molecules, on average, will be headed toward the $+z$ wall). This leads to a differential collision probability on the order of

$$f_c dt \sim \frac{A n_V \bar{v} dt}{6}$$

The number of particles striking the $+z$ wall per unit area per unit time is therefore on the order of

$$\Phi_N \sim \frac{n_V \bar{v}}{6}$$

As we shall learn shortly, the correct factor is actually 1/4, instead of 1/6. The correct procedure requires application of the distribution of velocities in the low-density ideal gas – including both the distribution of speeds and directions.

---

[174] The capital Greek symbol phi (Φ) is arguably the most awesome physics symbol since, if you turn it sideways, it looks like Darth Vader's spaceship. The lowercase and uppercase Greek letters φ and Φ may be pronounced with either a long 'i' sound – as in 'pie' – or a long 'e' sound – as in 'fee.' Well, you probably can't get away with that if you speak Greek: The two common pronunciations ultimately result from the difference between traditionalists who try to imagine how the ancients would pronounce words in their original Latin and Greek and modernists who Americanize the words that we borrow from other languages. For example, if you pronounce 'cacti' as 'kak-tie,' you are being inconsistent if you say 'fee' instead of 'fie.' Perhaps the ultimate example of Americanizing words from other languages belongs to the word 'alumnus.' At graduation, a single male is an alumnus, a single female is an alumna, a group of males are alumni, and a group of females are alumnae. In the original tongue, you would say 'alumni' with a long 'e' (alum-nee) and 'alumnae' with a long 'i' (alum-nigh); but in modern English, we have swapped the pronunciations of these plural forms. Nevertheless, 'phi' is pronounced both 'fee' and 'fie' so often presently that it would be absurd to claim that either way is correct or better. Whichever form you choose, though, at the very least you ought to have a reason to support your choice.

Let us consider the molecules just below the $+z$ wall – close enough that they have the potential to strike the wall in the infinitesimal time interval dt for their given speeds. These molecules will travel a distance vdt that is small compared to the mean-free path, such that we may focus on the collisions that the molecules make with the walls (because there is negligible probability that these particular molecules under consideration will collide with other molecules on the way to the wall). The z-components of their velocities equal $v\cos\theta$ (in spherical coordinates). It is precisely $v\cos\theta$ that determines whether or not a given molecule will strike the wall in the time interval dt given its present location. For one, we require that $0 < \theta \leq \pi/2$, otherwise the molecule is heading in the wrong direction; for another, molecules with larger values of θ require either closer proximity to the $+z$ wall or greater speed in order to reach the $+z$ wall.

Therefore, the collision probability, which we naïvely reasoned to be $An_V \bar{v} dt/6$ in our previous conceptual argument, should be modified as follows:

- Replace $\bar{v}$ by $v\cos\theta$ as the z-component of velocity determines how quickly the molecule is heading upward.
- Multiply by the Maxwell-Boltzmann distribution of speeds to account for the probability that the molecule will have enough speed to reach the $+z$ wall.
- Integrate over $v^2 \sin\theta \, dv d\theta d\varphi$ (the volume integral in spherical coordinate velocity space) instead of dividing by 6 to correctly determine the averages.
- Divide $n_V(v)$ by $4\pi v^2$, since the Maxwell-Boltzmann distribution of speeds already has this factor built-in, as we included the v^2 and angular integration (which equaled 4π) when we derived the Maxwell-Boltzmann distribution of speeds. Essentially, we presently need to undo the 4π from the Maxwell-Boltzmann distribution and redo the angular integration in order to determine the effect of the $\cos\theta$ factor from the z-component of velocity.
- Restrict θ to lie in the range $0 < \theta \leq \pi/2$ in order to ensure that the molecule is heading in the right direction.

The molecular flux, Φ_N, is then obtained – as before – by dividing by Adt:[175]

$$\Phi_N = \int_{v=0}^{\infty} \int_{\theta=0}^{\pi/2} \int_{\varphi=0}^{2\pi} \frac{v\cos\theta \, n_V(v)}{4\pi v^2} v^2 \sin\theta \, dv d\theta d\varphi = \frac{1}{4\pi} \int_{v=0}^{\infty} n_V(v) v dv \int_{\theta=0}^{\pi/2} \frac{\sin 2\theta}{2} d\theta \int_{\varphi=0}^{2\pi} d\varphi$$

$$\Phi_N = \frac{2\pi}{4\pi} \left[-\frac{\cos 2\theta}{4} \right]_{\theta=0}^{\pi/2} \int_{v=0}^{\infty} n_V(v) v dv$$

$$\Phi_N = \frac{1}{4} \int_{v=0}^{\infty} Av^3 e^{-av^2} dv = \frac{n_V \bar{v}}{4}$$

---

[175] Dividing by Adt to obtain the molecular flux from the integral for the collision probability, we see that the final result does not actually depend upon the size of the $+z$ wall. Furthermore, the result is independent of our original consideration of molecules close enough to the wall to reach it in the infinitesimal time interval dt, which was convenient for computing the collision probability – since the time interval cancels out in the expression for the molecular flux.

where we have used the double-angle trig identity $\sin 2\theta = 2 \sin\theta \cos\theta$. The integral of $Av^3 e^{-av^2}$ was encountered in the example from the previous section when we integrated the Maxwell-Boltzmann distribution to compute the average speed of the molecules of an ideal gas. The molecular flux can be expressed in terms of the temperature as

$$\Phi_N = n_V \sqrt{\frac{k_B T}{2\pi m}}$$

It is conceptually instructive to compare and understand the factors of ¼ in $\Phi_N = n_V \bar{v}/4$ and the relationship between the radiancy of electromagnetic energy escaping through a hole and the energy density of the enclosure, $R_T = c\rho_T/4$, for a blackbody cavity (see Sec. 6.4), which is essentially a photon gas.

Brownian motion: When very small particles, which are big enough to see (e.g. grains of pollen, with a size in the sub-millimeter regime), are dispersed in a liquid (such as water), the effect of the motion of molecules in the liquid can be seen: Namely, when the small particles (not to be confused with the much smaller molecules) are viewed through a microscope, they are seen to exhibit constant random zigzag motions. The constant random zigzag motions are the result of continual bombardment of the small particles by the molecules of the liquid. Careful observation of this Brownian motion — i.e. the motion of the small particles, which results from collisions from molecules in the liquid — has proven to be consistent with the kinetic theory for molecular motion. The molecules of the liquid transfer an average translational kinetic energy of $3k_B T/2$ to the small particles, according to the theorem of equipartition of energy (since there are three degrees of freedom), and the resulting distribution of speeds of the small particles reflects the distribution of speeds of the colliding molecules. The distribution of the small particles with depth in the liquid is very similar to the law of atmospheres:[176]

$$n_V(h) = n_{V0} e^{\frac{(\rho - \rho_\ell)Vgh}{k_B T}}$$

where ρ is the density of the small particles and ρ_ℓ is the density of the liquid. Note that what appears as exponential growth with increasing depth, h, is really exponential decay with increasing height, H.

---

[176] This equation is obtained in close analogy with the derivation of the law of atmospheres, simply accounting for the Archimedes' principle — i.e. the small particles are buoyed upward with a force that equals the weight of the liquid that is displaced by the volume of the small particles (since they are wholly submerged in the liquid). From Newton's second law, then, there is a vertical downward pull on each small particle equal to its weight, $mg = \rho V g$, in addition to a vertical upward buoyant force on each small particle equal to the weight of the liquid that it displaces, $m_\ell g = \rho_\ell V_\ell g = \rho_\ell V g$. Therefore, the net force on each small particle is $(\rho - \rho_\ell)V g$. Simply replace mg in the law of atmospheres with $(\rho - \rho_\ell)V g$ to obtain the distribution of n_V with height, and then change perspective from height to depth to obtain n_V.

Conceptual Questions

The selection of conceptual questions is intended to enhance the conceptual understanding of students who spend time reasoning through them. You will receive the most benefit if you first try it yourself, then consult the hints, and finally check your answer after reasoning through it again. The hints and answers can be found, separately, toward the back of the book.

1. Which tends to be larger – the number of permutations or the number of combinations? Why?

2. In roulette, a ball rolls into a spinning wheel and eventually comes to rest in one of 38 slots numbered 1 to 36 plus 0 and 00. Half of the first 36 are red and half are black, while the zero and double zero are green. The odds are based on 36 slots. As examples, if you bet $5 on 17 and the ball lands in the 17 slot, you receive $180 for your $5 bet, and if you bet $20 on black and the ball lands in a black slot, you double your wager. Justify that you might win money in the short run, but will definitely lose money in the long run (assuming that you play fairly).

3. You play roulette with the following strategy in mind. You start out betting $1 on red. If you lose, you bet $3 on red. If you lose again, you bet $7 on red. If you lose again, you bet $15 on red. This way, when you finally win, you will eventually gain $1, regardless of how many times you lose in a row. When you finally do win, you start over with a $1 bet on red. What's the problem with this strategy?

4. You go on a 1D random walk as follows. You flip a fair coin and walk one step north if it lands heads up and one step south if it lands tails up. Then you flip the coin again and decide whether to walk north or south in the same way. You do this repeatedly for a given number of steps. What will be the average net displacement if thousands of people go on such a random walk? Compare this to the average distance that they wind up from where they started. What analogy does this have with the average velocity of the molecules of an ideal gas?

5. Which of the following values are likely to occur most often – the average value, the most probable value, or the root-mean-square value? Explain. Which of these did we come across in the kinetic theory of ideal gases in the previous chapter? Why was this best-suited?

6. Is the water pressure greater at the bottom of a lake that is very large, but shallow or very small, but deep? Explain.

7. One glass is shaped like a right-circular cylinder, while another is shaped like a truncated right-circular cone. Water is poured into both glasses such that the water level is the same height in both glasses. In which glass is the pressure greatest at the bottom of the glass? Explain.

8. Would you expect for the region of air just above hot desert sand to follow the same trend as the law of atmospheres? Explain.

9. When two molecules collide, what will generally become of their speeds? In light of this, is it possible for the speeds of all of the molecules of an ideal gas to be the same?

Practice Problems

The selection of practice problems primarily consists of problems that offer practice carrying out the main problem-solving strategies or involve instructive applications of the concepts (or both). You will receive the most benefit if you first try it yourself, then consult the hints, and finally check your answer after working it out again. The hints to all of the problems and the answers to selected problems can be found, separately, toward the back of the book.

Permutations and combinations

1. Determine the number of permutations for the arrangement of letters in each of the following words: HEAT, ENERGY, PRESSURE, and STATISTIC.

2. How many 4-letter permutations are possible using letters from the word ENTROPY? How many of these begin with the letter Y? How many begin with the letters TR (in this order)?

3. How many 5-letter permutations can be made by choosing 3 letters from the word SECOND and choosing 2 letters from the word LAW?

4. How many permutations and combinations are possible by drawing 2 cards from a standard 52-card deck? How many permutations and combinations are possible by drawing one card from a standard 52-card deck and another card from a different standard 52-card deck? If your answers to these two questions are different, explain why.

5. How many permutations are possible for 10 coins that are flipped simultaneously?

Discrete probability distributions

6. What is the probability of rolling 5 fair six-sided dice and having all 5 dice land with the same side up?

7. What is the probability of being dealt a blackjack from a (full) standard deck of 52 cards?

8. What is the probability of being dealt a blackjack if you draw one card from a standard 52-card deck and another card from a different standard 52-card deck?

9. What is the probability of being dealt a full house from a standard deck of 52 cards?

10. Two fair six-sided dice are rolled simultaneously in the game of craps. Compute the most probable outcome, the average value, and the root-mean-square value for the sum of the dice.

Continuous probability distributions

11. An ideal gas is composed of sulfur atoms. Determine the most probable, average, and root-mean-square speeds of the molecules at 300 K.

12. The distribution of speeds of the molecules of a (non-ideal) gas is described by the following distribution function: $N(v) = av\left(b - \frac{v^2}{4}\right)$, where a and b are constants and all of the molecular speeds lie in the range $0 \leq v \leq 2\sqrt{b}$. (A) Sketch this distribution function and label key points. (B) Determine the most probable, average, and root-mean-square speeds. (C) Determine the probability that a molecule has a speed in the range $\sqrt{b}/2 \leq v \leq \sqrt{b}$.

13. Determine the mean inverse speed for the Maxwell-Boltzmann distribution. That is, find the expectation value (i.e. the average value) of $1/v$, denoted by $\overline{1/v}$. Compare this to the inverse of the average value of the Gaussian distribution, denoted by $1/\bar{v}$.

14. The molecules of an ideal gas have a diameter of 3 Å.[177] Determine the mean-free path at STP.

15. Show that the distribution of molecules in a gas centrifuge rotating with constant angular speed, ω_0, as a function of the molecule's radial distance from the axis, r, is $N(r) = N_0 e^{-mr^2\omega_0^2/(2k_BT)}$.

---

[177] One Angstrom (Å) equals 10^{-10} m.

5 Heat Engines and Refrigerators

5.1 Entropy and Reversibility

Thermodynamic processes: A system that begins and ends in states of equilibrium is a thermodynamic process. The initial and final states can be specified in terms of thermodynamic variables such as pressure and temperature, which are well-defined for the system during equilibrium. The interim states connecting the initial and final equilibrium states may or may not be in states of equilibrium, depending upon whether or not the process is quasistatic.

Quasistatic processes: A thermodynamic process that is carried out as a series of infinitesimal disturbances from equilibrium and that is carried out slowly enough that equilibrium is reached between consecutive disturbances is said to be quasistatic. Thermodynamic variables are well-defined for each equilibrium state of a quasistatic process. Each equilibrium state may be plotted on a P-V diagram, and the set of equilibrium states traces out a smooth curve. The area under this curve represents the work done by the system.

A thermodynamic process that is not quasistatic is irreversible, as it instead evolves a series of non-equilibrium states. However, a process that is quasistatic is not necessarily reversible. In particular, if a quasistatic process features frictional or dissipative forces, it will be irreversible.

Irreversible processes: There are two classes of irreversible processes: those which are not quasistatic and those which are quasistatic, but involve frictional or dissipative forces. A non-quasistatic process passes through a series of non-equilibrium states; spontaneous processes fall into this category. Thermodynamic variables, such as pressure and temperature, are ill-defined for the system during these non-equilibrium states since they tend to be non-uniform spatially and temporally. Therefore, a non-quasistatic process cannot be represented by a solid curve on a P-V diagram, as the non-equilibrium states do not have well-defined values for P. The initial and final equilibrium states can be plotted on a P-V diagram, but there is no well-defined path by which to connect them for a non-quasistatic process. Thus, it is not viable to think of the work done during a non-quasistatic process as the definite integral of pressure over volume – that applies to quasistatic processes only.

Reversible processes: A quasistatic process that can be carried out in reverse is said to be a reversible process. A quasistatic process for which there are no frictional or dissipative forces is reversible. Therefore, no physical process can be truly reversible. However, a thermodynamic process can be carried out quasistatically with very small changes over a long period of time, and to the extent that frictional and dissipative forces can be minimized, a physical process can be approximately reversible.

Important Distinction. A process that is carried out very slowly compared to the characteristic relaxation times of the system is quasistatic because there will be only temporary, very minute departures from equilibrium. If frictional and dissipative forces are approximately negligible during a quasistatic process, the process is also reversible. A process that is not quasistatic is irreversible, as is a quasistatic process where frictional or dissipative forces are present. Thermodynamic variables like temperature and pressure are well-defined for quasistatic processes, for which the system is in virtual equilibrium states throughout, but are not well-defined for non-quasistatic processes. Thus, the definite integral of pressure over volume – or, equivalently, the area under a P-V curve – only makes sense for a quasistatic process.

Conceptual Examples. Imagine a cylinder with one fixed wall and a moveable piston, which contains a gas that is initially in equilibrium. The gas has some initial temperature, T_0, an initial volume, V_0, given by the location of the piston, and is initially under pressure, P_0, from the enclosure. An adiabatic (zero heat exchange between the system and surroundings) process can be carried out if the walls of the cylinder, including the piston, are extremely well insulated. If the piston was locked in place and is suddenly unlocked, the gas – which is exerting pressure on all of the walls – will cause the piston to move in order to expand its volume until the piston reaches its limit. During this spontaneous adiabatic expansion, the system is constantly in non-equilibrium states between the initial and final states. Once the piston has come to a stop, the system will eventually come to equilibrium. The gas occupies a larger volume, $V_1 > V_0$, is under less pressure, $P_1 < P_0$, and drops its temperature, $T_1 < T_0$. This process is irreversible. If you shove the piston back to its initial position in an effort to 'reverse' the process, microscopically things won't be quite the same – in particular, the entropy will be greater.

The gas can instead be allowed to expand quasistatically as follows. Rather than locking and unlocking the piston, we can imagine that the cylinder is vertical with the moveable piston at the top. Initially, there is a pile of sand on the top of the piston that creates the initial pressure, P_0. We can remove a few grains of sand, patiently wait for equilibrium to set in again, remove a few more grains of sand, wait further, and repeat this many times until the piston has reached its final position. To the extent that friction between the piston and cylinder walls and dissipative forces may be negligible, this quasistatic adiabatic expansion can be reversible. In the reverse process, a few grains of sand are added little by little, each time waiting for equilibrium to be attained, slowly returning the piston to its initial position.

Other thermodynamic processes besides an adiabatic expansion may be carried out quasistatically. For example, this can be converted to an isothermal – rather than adiabatic – expansion by removing the bottom, fixed wall and setting the system instead on a base that is a very good conductor of heat, where the conducting base is at the top of a very large thermal reservoir held at constant temperature. The process will no longer be adiabatic since the system (i.e. the gas) will exchange heat with its surroundings (i.e. the thermal reservoir), but it will now be isothermal because the reservoir will regulate the temperature of the system through the heat exchanges.

This technique of adding or removing a few grains of sand at a time provides a concrete means of visualizing the more abstract notion of a quasistatic process for which the volume changes. An isochoric (i.e. constant volume) process may be imagined by making very slow adjustments of some other variable.

Entropy: A quantitative measure of the statistical disorder of the molecules of a substance is provided by a thermodynamic parameter called the entropy. Statistical disorder is characterized not just in terms of the geometric arrangement of the molecules, but also in terms of physical quantities such as spin angular momentum (for which some may be spin up or spin down). For a quasistatic process, the differential change in entropy is defined macroscopically in terms of the quasistatic differential heat exchange and temperature as đ$Q = TdS$. Like the expression đ$W = PdV$, the equation đ$Q = TdS$ does not make sense for a non-quasistatic process because, in that case, temperature is not a well-defined variable for the system.

It turns out that entropy is path-independent, like internal energy, but unlike work and heat. We thus represent the differential entropy, dS, like the differentials of other state variables – such as dU, dV, and dT – rather than path-dependent differentials – in particular, đQ and đW. This means that the change in entropy can be calculated for any process: Since the change in entropy is path-independent, for given initial and final equilibrium states, the change in entropy can be calculated for any reversible path connecting these endpoints – and will be the same as for any other path, including irreversible paths. In this way, the expression đ$Q = TdS$ can be employed to compute the entropy change even when the actual process under consideration is irreversible:[178] Just find an equivalent reversible path and integrate.

Microscopically, for a system in equilibrium the entropy of the system is defined in terms of the total number of microstates available to the system, Ω, as $S = k_B \ln \Omega$. The expression đ$Q = TdS$ provides a useful macroscopic measure of disorder for quasistatic processes in thermodynamics, while the relation $S = k_B \ln \Omega$ very literally quantifies the level of disorder microscopically for a system in equilibrium in statistical mechanics. The two definitions turn out to be equivalent – a notion that becomes most clear when comparing the postulates of thermodynamics (Chapter 6) with the fundamentals of the formulation of statistical mechanics.

Microstates and macrostates: The state of a system can be specified in terms of microscopic or macroscopic parameters. A microstate specifies the properties of the individual molecules of a system, whereas a macrostate provides information about coarse spatial and temporal averages. For example, a microstate might be specified as a tabulation of the positions, velocities, and spin angular momenta of gas particles, and a macrostate might be specified in terms of temperature, pressure, and volume.

Macroscopic and statistical definitions of entropy: The entropy of a microstate is calculated as $S = k_B \ln \Omega$ for a system in equilibrium, where Ω is the total number of microstates available for the given fixed conditions of the system (the volume or energy, for example, of the system may be fixed). The entropy change of a macrostate is calculated from the definition đ$Q = TdS$, applied to a reversible path between specified initial and final equilibrium states. Both the microscopic and macroscopic definitions of the entropy provide a measure of the statistical order of a system, and both formulations are equivalent (which, again, can be seen most clearly after studying the fundamentals of thermodynamics and statistical mechanics).

---

[178] If you want to calculate the entropy change, you should use a reversible path – not necessarily a quasistatic path. For a quasistatic irreversible path, you would need to account for the heat exchanges associated with the frictional and dissipative energy losses. So henceforth, if we are discussing how to compute entropy changes, we will speak of đ$Q = TdS$ for a reversible process (rather than merely a quasistatic process).

Statistically, it makes sense that the entropy, S, is proportional to the total number of available microstates, Ω, through a relation such as $S = k_B \ln \Omega$. Assuming that all of the available microstates are equally likely, the system is much more likely to be found in a more disordered state than in a more ordered state simply because there are many, many more disordered states than ordered ones. For example, suppose that you label each molecule in a room according to whether it is presently on the left or right half of the room. For this simple two-state system, the state of maximum order corresponds to every air molecule occupying the same half of the room, and the state of maximum disorder corresponds to half of the air molecules lying on each side. Even if there were a mere 100 air molecules in the room, out of the $2^{100} \approx 10^{30}$ possible microstates, just 2 are perfectly ordered, whereas there would be $\binom{100}{50} = \frac{100!}{50!50!} \approx 10^{29}$ microstates with half of the air molecules on each side. The odds of all 100 air molecules occupying on the same side of the room are approximately 2 in 1,000,000,000,000,000,000,000,000,000,000. This doesn't mean that exactly 50 air molecules will lie on each side, but it does mean that most of the time roughly the same number of air molecules will be found on each side. If there are Avogadro's number of air molecules, rather than a mere 100, the effect is far more extreme. The main idea is that the more microstates that are available to a system – given any constraints (e.g. if there are no outlets such as open doors or windows, then the total number of air molecules is fixed) – the greater the entropy of the system.

The factor of Boltzmann's constant, k_B, in $S = k_B \ln \Omega$, provides the correct units for entropy – Joules/Kelvin (J/K). The precise form of the statistical definition of the entropy – in particular, why there is a natural logarithm – is most clearly understood from studying the fundamentals of the formulation of statistical mechanics. In short, in statistical mechanics, the number of available microstates, Ω, depends upon parameters such as internal energy, U, volume, V, and mole numbers, n_i, of the various chemical components. In thermodynamics, the entropy, S, also depends upon U, V, and n_i. Once the relationship (called the fundamental relation) between S, U, V, and n_i is known, other tangible quantities – such as temperature and pressure – can be computed directly. The natural logarithm relationship between the entropy and the total number of microstates, $S = k_B \ln \Omega$, gives the correct behavior of physical quantities – like temperature and pressure – for real systems.

Following are a few examples of this 'correct behavior.' The expression $S = k_B \ln \Omega$ leads to a positive temperature with a limit that agrees with the third law of thermodynamics. It also leads to natural equilibrium conditions – such as equal temperatures for two systems in thermal equilibrium and equal pressures for two systems in mechanical equilibrium. Furthermore, in attaining thermal equilibrium, the direction of spontaneous heat flow is seen to be from higher to lower temperature. As a last example, it agrees with observed equations of state – such as $PV = nRT$ for a low-density gas.

The agreement between the statistical formula for the entropy, $S = k_B \ln \Omega$, and the macroscopic formula for the entropy change, $đQ = TdS$, for a reversible process also follows from the connection between the total number of available microstates, Ω, in statistical mechanics and the fundamental relation in thermodynamics. The details can be seen in subsequent chapters, but the essential point is that, in the formulation of equilibrium thermodynamics, the first law naturally arises from a handful of fundamental postulates as $dU = TdS - PdV + \sum_{i=1}^{r} \mu_i dN_i$. In order to identify this as the usual first law for a system with constant mole numbers, it is necessary to interpret the TdS term as the differential heat, $đQ$, added to the system. Since this comes from the formulation of equilibrium thermodynamics, we limit this definition of the entropy change, $đQ = TdS$, to quasistatic processes (and in application, it is much simpler if the process is also reversible).

5 Heat Engines and Refrigerators

The macroscopic definition of the entropy, S, associates an increase in entropy with heat added to the system and a decrease in entropy with heat lost to the surroundings for a reversible process, according to $đQ = TdS$ (since temperature, T, must be positive). For a reversible process, the relation $đQ = TdS$ reveals quite instructively the connection between temperature and heat: Adding heat to (or removing heat from) most precisely causes an increase (or decrease) in the entropy of the system, and does not necessarily cause a change in temperature (since the right-hand side involves a T, but not a dT). Heat corresponds more to entropy change than to temperature, and temperature change corresponds more to internal energy change than to heat.

The factor of temperature does have significance, though. It says that – for a reversible process – the same addition of heat will increase the entropy more for a low-temperature system than it will for a high-temperature system. For example, for an isothermal quasistatic addition of heat to a system, $Q = T\Delta S$, since temperature is constant, so for a given Q, a smaller T requires a larger ΔS.[179] Conceptually, it should make sense that it is much easier (i.e. less heat is required) to disturb the order of a low-temperature system than a high-temperature system.

Important Distinction. It is conceptually important not to confuse heat with temperature, or to associate addition of heat to a system with an increase in the temperature of the system. Though sometimes adding heat does increase the temperature, there are many common situations where this is not the case. For example, in general there is heat exchanged between the system and surroundings for an isothermal process, and temperature generally changes for an adiabatic process.

On the other hand, for a reversible process, the heat absorbed by a system from its surroundings is associated with the entropy change of the system. Similarly, temperature change corresponds more to internal energy change – e.g. see the equation that relates internal energy to temperature for an ideal gas (Sec. 3.4).

Entropy change: The change in entropy between initial and final equilibrium states can be found by integrating both sides of the equation, $đQ = TdS$, after isolating the entropy:

$$\Delta S = \int_i^f \frac{đQ}{T} \quad \text{(reversible path)}$$

It is important to realize that this integral applies only to a reversible path, but also that this integral can be applied to irreversible processes by finding a reversible path between the same initial and final equilibrium states. This integral only reduces to Q/T for a reversible isotherm. Note also that an isentropic (constant entropy) process is necessarily adiabatic, but that an adiabatic process is not necessarily isentropic. In particular, an irreversible adiabatic process will result in an increase in the overall entropy (system plus surroundings), as the equivalent reversible path will not be adiabatic.

---

[179] It is very important to realize that Q does not equal $T\Delta S$ in general. This is only true for an isothermal process. If the process is not isothermal, it is necessary to first express T as a function of Q and then integrate both sides of $dS = (1/T)đQ$. Even the equation $đQ = TdS$ does not apply in general, but is limited to quasistatic processes.

The total entropy change equals the sum of the entropy changes for the system and surroundings. The integral $\int_i^f \frac{dQ}{T}$ can be separately applied to the system and/or the surroundings. When speaking of the entropy change, ΔS, it is very important to clarify whether it relates to the system, surroundings, or the overall entropy change. In particular, the second law of thermodynamics involves the overall entropy (system plus surroundings).

Important Distinction. The equation $dQ = TdS$ implies that an isentropic process is adiabatic, but it does not imply that an adiabatic process is isentropic, since $\Delta S = \int_i^f \frac{dQ}{T}$ only applies to a reversible path. As it turns out, an irreversible adiabatic process results in an increase in the overall entropy.

Heat and entropy change: The relation $dQ = TdS$ for the differential heat exchange is analogous to the expression $dW = PdV$ for the differential work done. The work done over a finite volume change is obtained by expressing pressure as a function of volume and integrating both sides: $W = \int_{V=V_0}^{V} PdV$. In the case of heat, it is generally more practical to express temperature in terms of heat than to express temperature as a function of entropy: $\Delta S = \int_i^f \frac{dQ}{T}$. Both expressions, $dQ = TdS$ and $dW = PdV$ make sense only for quasistatic processes (and are most readily applied if the processes are also reversible), for which both temperature and pressure are well-defined. Also, the differentials for heat and work, dQ and dW, are both imperfect differentials, as the heat exchanged and work done for finite processes are both path-dependent, whereas the differentials for entropy and volume, dS and dV, are perfect differentials. In particular, the integral for the entropy change, $\Delta S = \int_i^f \frac{dQ}{T}$, is path-independent, and in this way the finite entropy change can be found even for irreversible processes just by finding an equivalent reversible process connecting the same initial and final equilibrium states.

First law of thermodynamics and entropy change: For a system with constant mole numbers, the differential form of the first law of thermodynamics is expressed as $dU = dQ - dW$. For a quasistatic process, this can also be written as $dU = TdS - PdV$ (and is applied most readily if the process is also reversible). This form of the first law, in terms of entropy and volume, is actually derived (as we shall see in Chapter 6) from a handful of postulates in thermodynamics, and proves to be of considerable use. It is important to realize that this equation cannot, in general, be integrated term by term (we will learn the correct technique for integrating the differential form of the first law in Chapter 6). This form of the first law expresses conservation of energy in terms of the entropy. This is interesting because, according to the second law of thermodynamics, whereas the overall energy is a strict conservation law, the overall entropy is not conserved in general.

Second law of thermodynamics and entropy change: According to the second law of thermodynamics, the overall entropy, $S_{overall}$, which equals the entropy of the system, S_{system}, plus the entropy of the surroundings, $S_{surroundings}$ – i.e. $S_{overall} = S_{system} + S_{surroundings}$ – cannot decrease: $\Delta S_{overall} \geq 0$, where $\Delta S_{overall} = \Delta S_{system} + \Delta S_{surroundings}$ is the change in the total entropy. Whereas the overall energy is governed by a strict equality, the overall entropy is governed by an inequality. Conceptually, the universe can only become more and more statistically disordered as time progresses.

5 Heat Engines and Refrigerators

The entropy changes of the system and surroundings can separately be calculated through integration: $\Delta S_{system} = \int_i^f \frac{dQ_{system}}{T_{system}}$ and $\Delta S_{surroundings} = \int_i^f \frac{dQ_{surroundings}}{T_{system}}$. It is possible for either ΔS_{system} or $\Delta S_{surroundings}$ to be negative – i.e. for the entropy of the system or surroundings to decrease – but not both. If $\Delta S_{system} < 0$, it follows that $\Delta S_{surroundings} \geq |\Delta S_{system}|$, and, likewise, if $\Delta S_{surroundings} < 0$, it follows that $\Delta S_{system} \geq |\Delta S_{surroundings}|$, such that $\Delta S_{overall} \geq 0$. That is, the entropy can be decreased in one part of the universe, at the expense of increasing the entropy more elsewhere – and the overall entropy must be a monotonically non-decreasing function of time. In the sense that $\Delta S_{overall}$ is always nonnegative for any time interval, $\Delta S_{overall}$ serves conceptually as a sort of arrow of time.[180]

The distinction between a reversible and irreversible process can be expressed very clearly in terms of the overall entropy: The overall entropy is constant for a reversible process and increases for an irreversible process:

$$\Delta S_{overall} = 0 \quad \text{(reversible)}$$
$$\Delta S_{overall} > 0 \quad \text{(irreversible)}$$

For a reversible process, it follows that the entropy changes of the system and surroundings are exact opposites:

$$\Delta S_{system} = -\Delta S_{surroundings} \quad \text{(reversible)}$$

That is, if the entropy of the system decreases for a reversible process, the entropy of the surroundings must increase by the same amount, and vice-versa. For an irreversible process, if the entropy of the system decreases, the entropy of the surroundings must increase even more than it decreases for the system (and similarly for the entropy of the system if the entropy of the surroundings decreases).

The fact that $\Delta S_{overall}$ equals zero for reversible processes does not imply that all reversible processes are adiabatic – which might seem to follow from $\Delta S = \int_i^f \frac{dQ}{T}$. Rather, for a reversible process,

---

[180] We have the freedom to travel through space at will, including motion both forward and backward along any coordinate, but evidently can only travel forward through time (yet the relative rate at which we travel forward through time is governed by Einstein's special theory of relativity). A consequence of moving only forward through time is that the overall entropy increases with the progression of time. Before you get carried away, though, it would be incorrect to associate the first law of thermodynamics – conservation of energy – with the ability to move forward and backward along spatial coordinates and the second law of thermodynamics – the non-decreasing overall entropy – with the restriction of moving forward through time. It may be tempting, as one expresses a strict equality and the other an inequality. However, from Noether's theorem, conservation of momentum is associated with translational symmetry in space, conservation of angular momentum is associated with rotational symmetry in space, and conservation of energy is associated not with space, but with translational symmetry in time. That is, from the perspective of the first law, there is a translational symmetry in time – all processes are 'reversible' as far as energy is conserved. From the second law, this translational symmetry in time is broken for irreversible processes.

$$\Delta S_{overall} = \int_i^f \frac{dQ_{system}}{T_{system}} + \int_i^f \frac{dQ_{surroundings}}{T_{system}} = 0 \quad \text{(reversible)}$$

As ΔS_{system} and $\Delta S_{surroundings}$ may each be nonzero even if $\Delta S_{overall}$ equals zero, heat may be exchanged between the system and surroundings for a reversible process. Therefore, a reversible process may or may not be adiabatic.

Important Distinction. The second law of thermodynamics speaks of the change in the overall entropy, $\Delta S_{overall}$, of the system plus the surroundings: $\Delta S_{overall} = \Delta S_{system} + \Delta S_{surroundings}$. It is the overall entropy change, $\Delta S_{overall}$, which is monotonically non-decreasing: $\Delta S_{overall} \geq 0$. For a reversible process, $\Delta S_{rev} = 0$, and for an irreversible process, $\Delta S_{irrev} > 0$, where ΔS_{rev} and ΔS_{irrev} both refer to the overall entropy change, $\Delta S_{overall}$. It is possible for the entropy of the system or surroundings to decrease, provided that the other increases at least as much: $\Delta S_{system} < 0 \Rightarrow \Delta S_{surroundings} \geq |\Delta S_{system}|$ and $\Delta S_{surroundings} < 0 \Rightarrow \Delta S_{system} \geq |\Delta S_{surroundings}|$, in order that $\Delta S_{overall} \geq 0$. For a reversible process, the equalities apply, while for an irreversible process, the inequalities apply.

Note that the second law of thermodynamics speaks of how much the overall entropy changes. Thus, it is not necessary to know the actual value of the entropy initially or finally, only how much it has changed (if at all) between the initial and final equilibrium states.

Also, since $\Delta S_{overall}$ is zero for a reversible process, from the relation $\Delta S = \int_i^f \frac{dQ}{T}$, it might be intuitive to want all reversible processes to be adiabatic, yet this is not the case – reversible processes can indeed involve exchanges of heat between the system and surroundings. Rather, the change in the overall entropy is related to the heat exchanges between the system and surroundings by: $\Delta S_{overall} = \int_i^f \frac{dQ_{system}}{T_{system}} + \int_i^f \frac{dQ_{surroundings}}{T_{system}}$. For a reversible process, $\Delta S_{system} = \int_i^f \frac{dQ_{system}}{T_{system}}$ must equal the negative of $\Delta S_{surroundings} = \int_i^f \frac{dQ_{surroundings}}{T_{system}}$. Thus, for a reversible process, either the entropy of the system or surroundings must decrease if the other increases, in order to preserve the overall entropy. So while the overall entropy equals zero for a reversible process, the separate entropies of the system and surroundings may change, such that a reversible process need not be adiabatic.

Isentropic and adiabatic processes: A thermodynamic process is isentropic if the entropy of the system does not change. For an adiabatic process, no heat is exchanged between the system and its surroundings. An isentropic process is inherently adiabatic, but an adiabatic process is only isentropic if it is reversible – since in that case the relation $\Delta S_{system} = \int_i^f \frac{dQ_{system}}{T_{system}}$ applies to the adiabat.

Reversible and irreversible adiabats: For an adiabat, $dQ_{system} = 0$. Therefore, for a reversible adiabat, $\Delta S_{system} = 0$ – i.e. a reversible adiabat is an isentrope. For an irreversible adiabat, we can conclude that $\Delta S_{system} > 0$. This follows because the overall entropy must increase – i.e. $\Delta S_{overall} > 0$ – for an irreversible process according to the second law of the thermodynamics, and for an adiabat the entropy can only increase in the system, as the system is thermally insulated from its surroundings.

Entropy changes for reversible processes: The change in the entropy of the system can be computed via integration: $\Delta S_{system} = \int_i^f \frac{dQ_{system}}{T_{system}}$. For a reversible process, the integration may be performed over the actual path. Following are the entropy changes for some common thermodynamic processes:

- Isentrope – the entropy of the system remains constant: $\Delta S_{system} = 0$.
- Reversible adiabat – this is the same as an isentrope: $\Delta S_{system} = 0$.
- Reversible isochor – since the volume of the system is constant, the heat absorbed by the system can be expressed in terms of the specific or molar heat capacity at constant volume: $đQ_{system} = MC_V dT$ or $đQ_{system} = nc_V dT$. If the factors MC_V or nc_V are constant for the isochor, integration yields: $\Delta S_{system} = \int_i^f \frac{MC_V dT}{T} = MC_V \ln(T/T_0)$ or $\Delta S_{system} = \int_i^f \frac{nc_V dT}{T} = nc_V \ln(T/T_0)$. If the temperature increases, heat is absorbed by the system and the entropy of the system increases (and the entropy of the surroundings decreases by this same amount), but if the temperature decreases, heat is released by the system (i.e. the heat absorbed by the system is negative) and the entropy of the system decreases (and, of course, the entropy of the surroundings increases by the same amount).
- Reversible isobar – this is similar to a reversible isochor, except for using the specific or molar heat capacity at constant pressure: $đQ_{system} = MC_P dT$ or $đQ_{system} = nc_P dT$. If MC_P or nc_P are constant for the isobar, integration yields: $\Delta S_{system} = MC_P \ln(T/T_0)$ or $\Delta S_{system} = nc_P \ln(T/T_0)$.
- Reversible isotherm – since the temperature is constant, it may be pulled out of the integral: $\Delta S_{system} = Q_{system}/T$. The heat capacity formulas are not generally useful for reversible isotherms, since neither pressure nor volume is generally constant along an isotherm. However, Q_{system} can be computed from the first law of thermodynamics, given other information about the specific system and process to which it is applied.

Example. Derive an expression for the entropy change of an ideal gas with constant mole number that expands to twice its initial volume during a reversible isothermal process.

For a reversible isotherm, temperature may be pulled out of the integral for entropy change since it is constant: $\Delta S_{system} = \int_i^f \frac{dQ_{system}}{T} = Q_{system}/T$. The heat absorbed (or released) by the system can be found from the first law: $Q_{system} = \Delta U_{system} + W_{system}$. The internal energy is constant for an isothermal process for an ideal gas, since $U_{system} = nc_V T$. Therefore, $Q_{system} = W_{system}$. The work done by (or on) the system can be found from integration:

$$W_{system} = \int_{V=V_0}^{V} PdV = \int_{V=V_0}^{V} \frac{nRT}{V} dV = nRT \int_{V=V_0}^{V} \frac{dV}{V} = nRT \ln\left(\frac{V}{V_0}\right) = nRT \ln 2$$

where temperature could come out of the integral since it is constant for the isotherm. The entropy change is then found to be $\Delta S_{system} = nR \ln 2$.

Reversible phase changes: Consider a substance for which the combination of pressure and temperature put it on the verge of undergoing a phase change on a *P-T* diagram: That is, the point (P,T) lies right next to the coexistence curve, such that a very minute change in pressure or temperature will result in a change of phase – e.g. from solid to liquid. If such a phase change is carried out reversibly, then the change in entropy will be approximately given by $\Delta S_{susbstance} = Q_{substance}/T$, as the temperature will be virtually constant. In this case, the heat absorbed or released by the system associated with the phase change is $Q_{substance} = ML$, where the latent heat of transformation, L, may be positive of negative, depending upon whether the substance absorbs or releases heat – e.g. heat is absorbed when melting from solid to liquid, but absorbed when freezing from liquid to solid. The entropy change is then $\Delta S_{susbstance} = ML/T$. The entropy of the substance increases ($\Delta S_{susbstance} > 0$) for a phase change from solid to liquid (melting), liquid to gas (boiling), or solid to gas (sublimation), for which the substance absorbs heat from its surroundings ($Q_{susbstance} > 0$). Likewise, the entropy of the substance decreases ($\Delta S_{susbstance} < 0$) for a phase change from gas to liquid (condensation), liquid to solid (freezing), or gas to solid (deposition), in which case the substance emits heat ($Q_{susbstance} < 0$). The entropy of the surroundings equals the exact opposite of the entropy of the system, since we are discussing a phase change that is reversible.

Entropy changes for irreversible processes: The change in the entropy of the system can be computed via integration even for an irreversible process: $\Delta S_{system} = \int_i^f \frac{dQ_{system}}{T_{system}}$. For an irreversible process, however, the integration must be performed over the equivalent reversible path – and not the actual path. The equivalent reversible path will have the same initial and final values of mole number, pressure, volume, and temperature. However, it is very important to realize that the work done and heat exchanged are generally different for the equivalent reversible path and the actual path. For example, $Q_{system} \neq 0$ for a reversible path that is equivalent to an irreversible adiabatic expansion, even though $Q_{system} = 0$ for the actual adiabatic expansion. It would therefore be incorrect to use đ$Q_{system} = 0$ in the integration, which would incorrectly result in $\Delta S_{system} = 0$, since đ$Q_{system} = 0$ corresponds to the actual path, rather than the equivalent reversible path. It is instead necessary to express đQ_{system} appropriately for the equivalent reversible path, which will correctly lead to $\Delta S_{system} > 0$. Following is a discussion of how to appropriately express đQ_{system} for some common sections of reversible paths:[181]

- Reversible adiabat – đ$Q_{system} = 0$ for an equivalent reversible adiabat, even though đQ_{system} is not zero for the actual irreversible process. Note that a reversible adiabat cannot, by itself, by equivalent to an irreversible process, but it may be possible to combine a reversible adiabat with another reversible process such that the combination is equivalent to the actual irreversible process.
- Reversible isochor – the heat capacity at constant volume applies: đ$Q_{system} = MC_V dT$ or đ$Q_{system} = nc_V dT$.

---

[181] A reversible path that is equivalent to the actual irreversible process may be a combination of reversible processes – such as a reversible isobar plus a reversible isotherm. This can be the case if, for example, neither a reversible isobar nor a reversible isotherm connects to both the initial and final equilibrium states, but joining a reversible isobar to a reversible isotherm is able to accomplish the task.

- Reversible isobar – the heat capacity at constant pressure applies: $đQ_{system} = MC_P dT$ or $đQ_{system} = nc_P dT$.
- Reversible isotherm – the heat exchange can be expressed in terms of the first law. For a system with constant mole numbers, $đQ_{system} = \Delta U_{system} + W_{system}$. Note that the heat capacity equations are not useful for a general isotherm, as generally neither pressure nor volume is constant for an isotherm.

> **Important Distinction.** When computing the entropy change for the system that undergoes an irreversible process, it is necessary to perform the integration, $\Delta S_{system} = \int_i^f \frac{đQ_{system}}{T_{system}}$, over the equivalent reversible path, and not over the actual path. Both the actual irreversible path and the equivalent reversible path will connect to the same initial and final equilibrium states, but path-dependent quantities – namely, work and heat – will generally be different for the equivalent reversible path and the actual irreversible path. So, for example, if the actual irreversible path is an adiabat, the heat exchanged between the system and surroundings will not be zero: Even though no heat is exchanged in the actual irreversible adiabatic process, heat must be exchanged for the equivalent reversible process – in accordance with the second law of thermodynamics (since the entropy of the surroundings must be zero in the case of an adiabat). Thus, the conditions of the equivalent reversible path will generally not match the actual conditions for the irreversible process. In the example of the irreversible adiabat, the system is, by definition, actually perfectly thermally insulated, yet in the equivalent reversible process, the system absorbs heat from its surroundings. The equivalent reversible process is equivalent in the sense of how much the entropy changes to produce the same final equilibrium state from the same initial equilibrium state – but not in terms of what is physically done to the system.

Equivalent reversible path: The strategy for calculating the entropy change for a system that undergoes an irreversible process involves the art of finding a reversible path – which may be a combination of two (or more) reversible processes – that is equivalent to the actual irreversible path. The technique for finding an equivalent reversible path involves the following steps:
- First, relate the values of P, T, n, and V for the final equilibrium state to the values of P_0, T_0, n_0, and V_0 for the initial equilibrium states using equations that apply to the system (e.g. the equation $PV = nRT$ would be useful for relating these variables in the case of an ideal gas).
- Next, examine the initial and final values to check if pressure, temperature, or volume remained constant during the actual process.[182] If so, this suggests using a reversible isobar, isotherm, or isochor – as appropriate – as part or all of the equivalent reversible path.
- If the previous step suggested using a particular reversible process – e.g. a reversible isotherm – relate the initial and final equilibrium states – using equations that apply to the system for the reversible process – to see if it is equivalent to the actual irreversible path.

---

[182] We are not checking that the values remain constant throughout the actual irreversible process, just whether or not the values are equal in the initial and final states (i.e. it is possible for a quantity such as temperature to increase and decrease during a process, so that its initial and final values are the same even though it was not constant). Only the endpoints matter, since the change in entropy is path-independent.

- If a reversible isobar, isotherm, or isochor will not work by itself – i.e. it could not reproduce the same final equilibrium state as the actual irreversible path from the same initial equilibrium state – it will be necessary to combine (at least) two reversible processes together to form an equivalent reversible path. To start with, try choosing two of the following: reversible isobar, isotherm, isochor, and adiabat.[183]
- When joining two reversible processes together, connect one of the reversible processes to the initial equilibrium state in a P-V diagram, and the other to the final equilibrium state. Solve for the point of intersection and algebraically check that the pressure, temperature, mole number, and volume of the system match for the two reversible processes at the point of intersection.

Free expansion: In a free expansion, the gas expands under zero pressure. This is a unique circumstance whereby volume changes, yet work is not done (since $P = 0$): $W = 0$. From the first law, $\Delta U = Q$. A free expansion is spontaneous, and is not quasistatic. Therefore, being irreversible, the overall entropy (system plus surroundings) increases during a free expansion. An equivalent reversible path must be found in order to compute the entropy change for the system.

Adiabatic free expansion: A free expansion can also be adiabatic – if the system is thermally insulated. For the pressure to be zero, the gas must be expanding into an evacuated chamber. For an adiabatic free expansion, $W = 0$ and $Q = 0$; and from the first law, $\Delta U = 0$ also. Volume increases, though there are no energy changes for the adiabatic free expansion. Yet, there is another effect: The total entropy of the system increases, as the free expansion is irreversible – and being adiabatic (no heat exchanged with the surroundings), the increase in entropy required by the second law of thermodynamics necessitates an increase in the entropy of the system. So even though the actual process is adiabatic, the equivalent reversible path involves a positive absorption of heat.

Example. Derive an expression for the entropy change of an ideal gas with constant mole number that undergoes an adiabatic free expansion to 5 times its initial volume.

Since an adiabatic free expansion is irreversible, in order to compute the entropy change via the integral $\Delta S_{system} = \int_i^f \frac{dQ_{system}}{T_{system}}$, we must find an equivalent reversible path. Not any reversible process will do: We must ensure that the reversible path connects the same initial and final equilibrium states. For initial pressure P_0, temperature T_0, mole number n_0, and volume V_0, we must ensure that both the actual irreversible adiabat and the equivalent reversible path that we use to perform the integration lead to the same final equilibrium states. We know that $n = n_0$ and $V = 5V_0$. It would not be satisfactory to find any reversible path that results in a volume increase of a factor of 5: The final pressure and temperature of the reversible path must also agree with those of the actual irreversible adiabat. So let us first determine the final pressure and temperature of the actual irreversible adiabat and then look for an equivalent reversible process (or, if necessary, combination of processes).[184]

---

[183] A reversible adiabat will not be equivalent to the actual irreversible process all by itself, but it may serve a useful role as part of an equivalent reversible path.

[184] The important thing is not to guess – you have to make sure that the final equilibrium values for P, T, n, and V are the same for both the actual irreversible path and the equivalent reversible path.

For the actual irreversible adiabat, we know that $Q_{irrev} = 0$ and that $W_{irrev} = 0$ (since it is a free expansion – i.e. it expands without pressure).[185] Therefore, from the first law, $\Delta U_{irrev} = 0$. The gas is ideal, for which $U_{irrev} = nc_V T$, which means that $T = T_0$. Although an adiabat is generally not also isothermal, this is a rare case – indeed, it is rare to have a process with zero pressure. The final pressure can be related to the initial pressure through the ideal gas law: $P_0 V_0/T_0 = PV/T$ implies that $P = P_0/5$, since $V = 5V_0$ and $T = T_0$. Therefore, the equivalent reversible path must result in a final pressure of $P_0/5$, temperature equal to T_0, mole number of n_0, and a volume increase to $5V_0$.

Since the actual irreversible adiabat turned was isothermal ($T = T_0$), we see if a reversible isotherm by itself will match the initial and final equilibrium states of the actual irreversible adiabat. A reversible isotherm with mole number held constant will guarantee that $T = T_0$ and $n = n_0$. According to the ideal gas law, a fivefold volume increase, $V = 5V_0$, will lead to the desired reduction in pressure: $P = P_0/5$. Therefore, a reversible isotherm can serve as the equivalent reversible path. Observe that $Q_{rev} \neq 0$, $W_{rev} \neq 0$, and $\Delta U_{rev} \neq 0$ for the reversible isotherm, completely unlike the actual irreversible adiabat.

Having found an equivalent reversible process, we can now integrate over the equivalent reversible path to find the entropy change. For the equivalent reversible isotherm, we use the first law of thermodynamics to determine the differential heat exchange: $đQ_{rev} = dU_{rev} + đW_{rev}$. The last term can be expressed as $đW_{rev} = PdV$ for a reversible process. For an ideal gas the first term can be written as $dU_{rev} = nc_V dT$, which equals zero since the equivalent reversible process is isothermal: $dT = 0 \Rightarrow dU_{rev} = 0$. Therefore, the entropy change for the system for both the equivalent reversible process and for the actual irreversible adiabat is:[186]

$$\Delta S_{system} = \int_i^f \frac{đQ_{rev}}{T} = \int_{V=V_0}^{V} \frac{PdV}{T} = \int_{V=V_0}^{V} \frac{nRdV}{V} = nR \ln\left(\frac{V}{V_0}\right) = nR \ln 5$$

Example. A calorimeter is perfectly thermally insulated from its surroundings, and consists of just two substances, which are initially thermally separated and in equilibrium, where $T_{20} > T_{10}$. The two substances are then placed in thermal contact. The mole numbers and volumes of the two substances remain constant. Derive an equation for the overall entropy change of the system.

---

[185] There is initial pressure and final pressure, but no pressure during the expansion. A wall might suddenly be removed to create a free expansion into a vacuum chamber, with perfect thermal insulation of the complete volume (original volume plus previously evacuated volume). In this way, the gas is under pressure prior the adiabatic free expansion, and is again under pressure after filling the entire volume and coming to equilibrium, but can expand freely for a brief period between these equilibrium states.

[186] Here, we expressed the ratio of pressure to temperature in terms of volume using the ideal gas law, so that we could integrate over the volume. Before you pull a variable out of the entropy change integral for an irreversible process, check to make sure it is constant for the equivalent reversible process – not for the actual irreversible path (e.g. temperature could be constant for an actual irreversible isotherm, but would not be for an equivalent path consisting of a reversible isobar and isochor). Also, if the equivalent reversible process consists of two (or more) sections – e.g. a reversible isotherm and a reversible isochor – separate the integral into the sum of integrals over the constituent reversible sub-processes.

The two substances will eventually come to thermal equilibrium, for which $T_1 = T_2 \equiv T_e$. This process occurs spontaneously – so it is definitely not quasistatic. It is an irreversible adiabatic process (adiabatic since the calorimeter is perfectly thermally insulated from its surroundings). The initial equilibrium states of the two systems are described by $P_{10}, T_{10}, n_{10}, V_{10}, P_{20}, T_{20} > T_{10}, n_{20}$, and V_{20}. The final equilibrium state is described by $P_1, T_e, n_1 = n_{10}, V_1 = V_{10}, P_2, n_2 = n_{20}$, and $V_2 = V_{20}$. We know that the mole numbers and volumes are constant for both substances, and that the final equilibrium temperatures of the two substances are equal. Since the problem did not specify the nature of the substances, we cannot calculate the final pressures, yet we can safely assume that some relationship exists for each substance, whereby the pressure of the substance could be determined from its temperature, mole number, and volume: $P_1 = P_1(T_e, n_1, V_1)$ and $P_2 = P_2(T_e, n_2, V_2)$.

Since the volumes of the two substances remains constant, we first try a reversible isochor (really, a pair of reversible isochors – one for each substance), where the mole numbers are held constant. This pair of reversible isochors satisfies $n_2 = n_{20}$ and $V_2 = V_{20}$. The isochoric process can be carried out until $T_1 = T_e$ and $T_2 = T_e$. Whatever equations of state, $P_1 = P_1(T_e, n_1, V_1)$ and $P_2 = P_2(T_e, n_2, V_2)$, govern the pressure of each substance, they will lead to the same final pressures, P_1 and P_2 as the actual irreversible adiabat. Therefore, these eversible isochors serve as equivalent reversible paths for the entropy changes of the two substances.

The entropy of the surroundings is constant, since the calorimeter is perfectly thermally insulated from its environment. Therefore, the overall entropy change equals the sum of the entropy changes for the two substances: $\Delta S_{overall} = \Delta S_1 + \Delta S_2$. The individual entropy changes can be found by integrating over the equivalent reversible isochors. For this calorimeter, the heat absorbed by the initially cooler substance (substance 1) equals the heat released by the initially warmer substance (substance 2), according to conservation of energy: $Q_1 = -Q_2$. For these isochors, the differential heat exchanges can be expressed in terms of the specific heat capacities at constant volume: $đQ_{1,2} = M_{1,2} C_{V1,2} dT_{1,2}$. The entropy changes of the two substances are:[187]

$$\Delta S_{1,2} = \int_i^f \frac{đQ_{1,2}}{T_{1,2}} = \int_i^f \frac{M_{1,2} C_{V1,2} dT_{1,2}}{T_{1,2}} = M_{1,2} C_{V1,2} \int_{T_{1,2}=T_{1,2;0}}^{T_e} \frac{dT_{1,2}}{T_{1,2}} = M_{1,2} C_{V1,2} \ln\left(\frac{T_e}{T_{1,2;0}}\right)$$

The overall entropy equals

$$\Delta S_{overall} = M_1 C_{V1} \ln\left(\frac{T_e}{T_{10}}\right) + M_2 C_{V2} \ln\left(\frac{T_e}{T_{20}}\right)$$

Conservation of energy for the calorimeter requires that

$$M_1 C_{V1}(T_e - T_{10}) = M_2 C_{V2}(T_{20} - T_e)$$

---

[187] Here, and elsewhere in this text, we have assumed the specific heat capacities to be relatively independent of temperature. This is not true in general, and for those substances that are integrated over temperature ranges for which the specific heat capacities are significantly variable, the functional dependence of the specific heat capacities needs to be known before the integrals can be performed.

> In Sec. 6.5, we will show that these two equations – one for the overall entropy change and the other for conservation of energy – leads to a positive increase in the overall entropy, regardless of the initial and final temperatures (so long as $T'_{20} \neq T'_{10}$), which is consistent with the second law of thermodynamics. Equivalently, the first and second laws of thermodynamics are only satisfied if heat spontaneously flows from the higher-temperature substance to the lower-temperature substance.

Reversible and irreversible cycles: Like the change in internal energy, ΔU, – but unlike the work done, W, or the heat exchanged, Q – the entropy change, ΔS, is path-independent. Therefore, like the change in internal energy, the entropy change for a system, ΔS_{system}, must be zero for a thermodynamic cycle (i.e. when the initial and final equilibrium states are identical states):

$$\Delta S_{system} = 0 \quad (thermodynamic\ cycle)$$

For a reversible cycle, this can be expressed as a closed integral (i.e. an integral where the initial and final states are identical):

$$\Delta S_{system} = \oint \frac{dQ_{rev}}{T} = 0 \quad (reversible\ cycle)$$

For an irreversible cycle, $\Delta S_{system} = 0$, but the closed integral expression does not apply to the irreversible path. However, *any* reversible cycle is equivalent to *any* irreversible cycle with the same initial (and therefore final) equilibrium state.

Although $\Delta S_{system} = 0$ for any thermodynamic cycle – reversible or not – the entropy change of the surroundings, and therefore also the overall entropy change, depends upon whether or not the cycle is reversible. From the second law of thermodynamics, the entropy of the surroundings must be constant for a reversible cycle in order for the overall entropy change to be zero:

$$\Delta S_{surroundings} = 0 \quad (reversible\ cycle)$$

For an irreversible cycle, the entropy of the surroundings must increase in order for the overall entropy to increase:

$$\Delta S_{surroundings} > 0 \quad (irreversible\ cycle)$$

State variables: Internal energy, U, entropy, S, volume, V, temperature, T, and pressure, P, are referred to as state variables. State variables are path-independent. The closed integral over a state variable is zero for any path of integration – e.g., $\oint dU = 0$.

Work and heat are not state variables, since in general $\oint dW \neq 0$ and $\oint dQ \neq 0$. It is interesting to note that $\oint dS = 0$, but $\oint TdS \neq 0$ in general, and, similarly, $\oint dV = 0$, but $\oint PdV \neq 0$ in general. Furthermore, although $\oint TdS \neq 0$ and $\oint PdV \neq 0$ in general, for a system with constant mole numbers, it is always true that $\oint (TdS - PdV) = 0$ because $\oint (TdS - PdV) = \oint (dQ - dW) = \oint dU$. That is, although both Q and W are path-dependent, the combination $dQ - dW$, which corresponds to the differential internal energy change, dU, is path-independent.

5.2 Heat Engines and Refrigerators

Heat engine: A heat engine is a device that utilizes heat in order to perform mechanical work. The general features of a heat engine are:
- Input: A system absorbs heat, $Q_{in} > 0$, from a high-temperature thermal reservoir.
- Desired output: The system uses this absorbed heat to perform useful mechanical work, $W_{out} > 0$.
- Exhaust (unwanted output): Heat absorbed by the system that is not converted to mechanical work is rejected to a low-temperature thermal reservoir, which is called the exhaust, $Q_{out} < 0$.

The system is not defined as the entire heat engine, but refers to the place where the heat is absorbed and which subsequently delivers mechanical work (to the work station). The high-temperature and low-temperature reservoirs and the recipient of the mechanical work are external to the system.

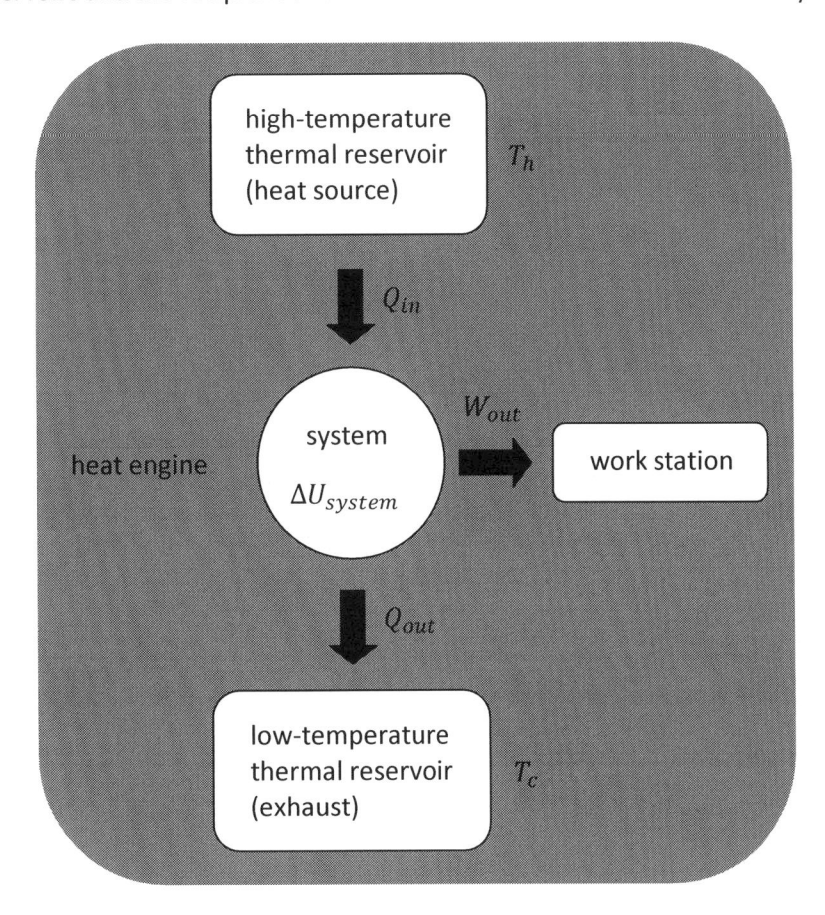

According to the first law of thermodynamics, for a heat engine with a system that has constant mole number, $\Delta U = Q - W$. For a complete cycle of a heat engine, $\Delta U = 0$ and $Q = W$, so a heat engine requires a continual supply of heat from the high-temperature reservoir in order to continually produce mechanical work (and therefore it continually produces exhaust). The continual supply of heat and production of mechanical work is achieved by repeating the thermodynamic cycle over and over.

During the input stage of each cycle, the system absorbs heat, which raises its internal energy: $Q \to \Delta U$. During the output stage of each cycle, the internal energy is converted partly into work, and the remainder (of the heat that was absorbed during the input stage) is rejected by the system as exhaust: $\Delta U \to Q - W$. It is in the output stage that work is produced, and the system does this work as a direct result of reducing its internal energy. The system has internal energy, and we want to use this internal energy to perform useful work. We must add heat to the system — from a high-temperature thermal reservoir (the heat source) — during each cycle in order to replenish the system's internal energy, in order that we may continually convert the system's internal energy into mechanical work. Thus, a heat engine converts internal energy into work by utilizing heat from a high-temperature reservoir.

A heat engine also inherently involves the second law of thermodynamics. According to the second law of thermodynamics, there is a cost that must be paid: Some of the heat absorbed by the system, which raised the internal energy of the system during the input stage, does not get converted into mechanical work. The unused heat that is rejected to a low-temperature thermal reservoir during each cycle is called the exhaust. The second law states that some amount of exhaust is inevitable. Put another way, all of the heat absorbed by the system cannot be converted into mechanical work.

During the input stage, heat is absorbed by the system from a high-temperature reservoir: $Q_{in} > 0$. The heat absorbed by the system raises the internal energy of the system: $\Delta U_{in} > 0$. In the output stage, the system reduces its internal energy: $\Delta U_{out} < 0$. This negative internal energy change leads to a positive amount of work done, $W_{out} > 0$, since $\Delta U_{out} = Q_{out} - W_{out}$. The fraction of internal energy that does not get converted into work is heat that is emitted (rejected) by the system in the form of exhaust: $Q_{out} < 0$.

For each complete cycle of a heat engine, $\Delta U_{net} = \Delta U_{in} + \Delta U_{out} = 0$, such that $\Delta U_{out} = -\Delta U_{in}$. A net amount of heat is added to the system during each cycle, $Q_{net} = Q_{in} + Q_{out} = Q_{in} - |Q_{out}|$, and a net amount of work is done by the system during each cycle, $W_{net} = W_{out}$. The net work done during each cycle is interpreted as the work output. However, it is quite possible (and probable) that work will be done on the system during one (or more) process(es) of the thermodynamic cycle, so that the net work is the sum of the work done by the system and the work done on the system: $W_{net} = W_{out} = W_{by} + W_{on} = W_{by} - |W_{on}|$. The absolute values correspond to $Q_{in} > 0$, $Q_{out} < 0$, $W_{by} > 0$, and $W_{on} < 0$. Conceptually, it can be easier to correctly navigate all of the signs by writing the equations with − signs and absolute values than to use + signs and remember that some of the quantities are actually negative.

The goal of a heat engine is to obtain as much useful mechanical work, $W_{out} > 0$, as possible for the given heat input, $Q_{in} > 0$. From the first law of thermodynamics, the most useful work we could hope to achieve during one cycle of a heat engine would be equal to Q_{in}, which would require zero exhaust, Q_{out}. However, according to the second law of thermodynamics, there must be some exhaust, $Q_{out} < 0$, which limits the percentage of mechanical work that can be performed: $W_{out} < Q_{in}$.[188]

---

[188] From conservation of energy, the most useful work we can hope to get is limited by Q_{in} — since you can't get more out than you put in. From the second law, $W_{out} < Q_{in}$ — i.e. you can't even get out what you put in. This does not mean that energy is lost, however. The fraction of Q_{in} that doesn't become W_{out} is rejected as exhaust, Q_{out}, and energy is conserved: $Q_{in} + Q_{out} = W_{out} < Q_{in}$ (since $Q_{out} < 0$). So you really do get out everything you put in, it's just that you don't want some of what you get out. Ultimately, the second law says that you can't get everything you want — you must also get some that you don't — since you can't reduce the overall entropy.

The heat input, $Q_{in} > 0$, must be supplied by a high-temperature reservoir and the heat output, $Q_{out} < 0$, must be delivered to a low-temperature reservoir, since heat naturally flows from high-temperature to low-temperature. That is, heat flows from a reservoir to the system during the input stage, $Q_{in} > 0$, and so must come from a source that is at a higher temperature, and heat flows from the system to a reservoir when the system rejects the unused heat (exhaust) during the output stage, $Q_{out} < 0$, and so must go to a reservoir that is at a lower temperature.

The purpose of drawing heat from and delivering heat to thermal reservoirs is that a thermal reservoir has such a large supply of thermal that it has a relatively constant temperature, despite the relatively (in comparison) amount of heat that may be extracted from or deposited into the thermal reservoir. That is, the temperatures of the high-temperature and low-temperature thermal reservoirs, T_h and T_c, may be treated as constants. Otherwise, if the high-temperature source were not a thermal reservoir, its temperature would drop over time – this means that less work would be done with each cycle, or an extra cost would have to be paid to maintain the temperature of the source. Therefore, the use of thermal reservoirs allows the maximum amount of work to be done by the heat engine. In principle, we could draw heat from any source – not necessarily a thermal reservoir – and could deposit exhaust anywhere – not necessarily to a thermal reservoir – but doing so wouldn't be nearly as practical.

Thermal reservoir: A heat source that offers a virtually limitless supply of thermal energy, or may receive a virtually limitless amount of thermal energy, with virtually no effect on the temperature of the heat source is said to be a thermal (or heat) reservoir. The atmosphere, for example, serves as a thermal reservoir (and can also serve as a pressure reservoir) operating at STP. Your house continually exchanges heat with the atmosphere, such that if you are not running a heater or air conditioner eventually your house will come to thermal equilibrium, in which case the air in your house will have the same temperature as the atmosphere (in a simplified picture that neglects any other factors). If you are running the air conditioner on a hot summer day, opening your front door will have a significant impact on the temperature of the air in your house (or, equivalently, how much your air conditioner will need to work to maintain the lower temperature of your room air). However, opening your door will have no noticeable effect on the temperature of the atmosphere, which acts as a thermal reservoir.

For a typical heat engine, the atmosphere serves as a convenient low-temperature thermal reservoir in which to deposit the exhaust (well, 'convenient' from the perspective of the company running the heat engine – too many heat engines depositing exhaust into the atmosphere may not be so 'convenient' for everyone else). A higher-temperature thermal reservoir is generally needed for the heat source, such as may be provided by a furnace (so we invest in the heat input, produced by the furnace, in order to produce useful mechanical work).

Heat source: A heat engine requires a high-temperature thermal reservoir to serve as a heat source. A typical heat engine uses a furnace as the heat source. The temperature of such a heat source, T_h, remains constant over the repeated thermodynamic cycles of the heat engine, regardless of how much heat is extracted from the heat source.

Exhaust: A heat engine requires a low-temperature thermal reservoir in which to deposit exhaust – the undesired heat output. The atmosphere often serves as the recipient of the exhaust. The temperature of the low-temperature thermal reservoir, T_c, remains constant over the repeated thermodynamic cycles of the heat engine, regardless of how much exhaust is delivered.

Heat engine cycle: Each cycle of a heat engine involves thermodynamic processes that draw heat, $Q_{in} > 0$, from a high-temperature reservoir in order to raise the internal energy of the system, and then reduce the internal energy of the system (restoring it to its original value) in order to produce useful work, $W_{out} > 0$, in addition to the undesired output (exhaust) of heat, $Q_{out} < 0$, into a low-temperature reservoir. The internal energy and entropy changes of the system are zero for each complete cycle of a heat engine: $\Delta U_{net} = 0$ and $\Delta S_{system} = 0$.

Work done by a heat engine: The net work done by a heat engine, which we label as W_{out}, equals the sum of the heat input and heat output (or, since the heat output is negative, the difference between the absolute values of the heat input and heat output) from the first law because $\Delta U_{net} = 0$:[189]

$$W_{out} \equiv W_{net} = Q_{net} - \Delta U_{net} = Q_{net} = Q_{in} + Q_{out} = Q_{in} - |Q_{out}|$$

If the cycle is carried out quasistatically, W_{out} will equal the area enclosed by the P-V curve for one cycle of the heat engine. Since $W_{out} > 0$ for a heat engine, the path must be clockwise – so that the work done during the expansion, which is positive, outweighs the work done during the contraction, which is negative. A reversible cycle of a heat engine appears as a clockwise path on a P-V diagram.

Efficiency of a heat engine: The ratio of the (net) work output, W_{out}, to the heat input, Q_{in}, which is inherently less than one according to the second law of thermodynamics, is naturally interpreted as the thermal efficiency, e, of once operating cycle of the heat engine:

$$e = \frac{W_{out}}{Q_{in}} = \frac{Q_{in} + Q_{out}}{Q_{in}} = 1 + \frac{Q_{out}}{Q_{in}} = 1 - \frac{|Q_{out}|}{Q_{in}}$$

We will explore the limited efficiency of heat engines and the relationship between the efficiency of a heat engine and the second law of thermodynamics in Sec. 5.4.

Entropy of a heat engine: Since entropy is path-independent, like internal energy, the change in the entropy of the system for one complete cycle of a heat engine is zero: $\Delta S_{system} = 0$. This does not mean that the entropy of the system is constant – just like the internal energy of a cycle is not constant throughout the cycle – but that the entropy of the system returns to its original value at the completion of the cycle.

If the heat engine cycle is reversible, $\Delta S_{surroundings} = 0$ for one complete cycle. The entropy of the surroundings includes the entropy of the thermal reservoirs; based on the direction of heat flow, we expect to have $\Delta S_{heat\ source} < 0$ and $\Delta S_{exhaust} > 0$. For a reversible cycle, $\Delta S_{overall} = 0$ even for a partial cycle. If instead the cycle is irreversible (which is the case even if just one process is irreversible), then $\Delta S_{surroundings} > 0$ for the complete cycle, such that $\Delta S_{overall} > 0$ for the complete cycle. In Sec. 5.4, we will learn that a reversible heat engine provides a greater maximum efficiency than does an irreversible heat engine.

---

[189] As usual, we assume that the system has constant mole numbers. We'll finally learn the correction term for the case of variable mole numbers in Chapter 6. However, most practical problems involve a system with constant n.

Conceptual Example. Most heat engines used for transportation use combustion engines. Combustion is an exothermic chemical reaction in which a fuel – such as gasoline – combines with oxygen, giving off a significant amount of energy in the formation of the products. For example, the combustion of propane, C_3H_8, releases energy in the formation of carbon dioxide and steam: $C_3H_8 + 5O_2 \rightarrow 3CO_2 + 4H_2O$.

    The steam engines that were once commonly employed to power trains and boats were external combustion engines – so-called because combustion occurs outside of the engine. Steam engines burned fuel – generally, coal or wood – in the heat source. The heat produced by burning the fuel was fed into the system, causing liquid water to evaporate into steam. The heat of transformation, $Q_{in} = ML$, that brings about this phase change increases the internal energy of the water molecules (which are much faster, on average, in steam than they are in liquid water). The increased internal energy that the molecules of steam have is then converted into mechanical work – just let the steam molecules push against something (e.g. a piston), causing the volume of the steam to expand (since increased volume corresponds to mechanical work done). The mechanical work done as the volume of the steam expands serves as the mechanism for transportation – e.g. the expanding piston may be connected to a wheel or paddlewheel so that its linear motion (as the volume expands and contracts) causes the wheel or paddlewheel to rotate. Since the engine cannot convert 100% of the heat input into useful work, it has an exhaust port that channels the unused heat through a pipe and into the atmosphere.

    Modern transportation is generally powered by an internal combustion engine. In this case, the combustion occurs inside the engine itself, which turns out to be much more efficient. The diesel engine is an internal combustion engine that powers automobiles. A diesel engine uses the Otto cycle, which we will consider in detail in Sec. 5.5.

Refrigerators and air conditioners: A room heater channels heat from a higher-temperature heat source to a lower-temperature room in order to warm the room, in which case heat flows in is natural direction. In contrast, refrigerators and air conditioners channel heat from a lower-temperature environment to a higher-temperature environment, which is against the direction of spontaneous heat flow. A refrigerator transfers heat from a lower-temperature food storage compartment to the higher-temperature room air in order to first (when the refrigerator is first plugged in, or when the thermostat is adjusted to a lower temperature) cool the food storage compartment and to later (once the thermostat's setting is reached) keep the food storage compartment cold. An air conditioner (which is a heat pump that is run in a cooling mode) similarly transfers heat from lower-temperature room air to the higher-temperature air outside of the house. Refrigerators and air conditioners are essentially heat engines that are run backwards.[190]

---

[190] It is conventional to say that a refrigerator or heat pump is a heat engine run in 'reverse.' However, the concept of reversibility already has a specific fundamental meaning, and relates to the overall entropy through the second law. The cycle of a heat engine is not necessarily reversible – that depends upon whether or not all of the thermodynamic process that comprise the cycle are reversible. However, whether or not it is reversible, it can still be run backwards – i.e. the processes can be carried out in the opposite order. If the cycle is reversible, carrying out the steps in the opposite order will lead to the reverse cycle, but if the cycle is irreversible, carrying out the steps in the opposite order will create an effect that is conceptually backwards, but will not technically be the 'reverse' of the original cycle (the overall entropy increases either way it is run). So 'backwards' avoids confusion.

Following is a description of what we mean when we say that a refrigerator or air conditioner is essentially a heat engine that is run backwards. When a heat engine runs forwards – i.e. normally – an amount of heat, $Q_{in} > 0$, is extracted from a thermal reservoir operating at a high temperature, T_h, and delivered to a system; this heat input ultimately leads to the production of work, $W_{out} > 0$, by the system, in addition to the undesired output of exhaust, $Q_{out} < 0$, to a thermal reservoir operating at a low temperature, T_c. When a heat engine runs backwards, an amount of heat, $Q_{in} > 0$, is removed from a thermal reservoir operating at a low temperature, T_c, and delivered to a system; this is achieved by having work done on the system, $W_{in} < 0$; this leads to the delivery heat, $Q_{out} < 0$, to a thermal reservoir operating at a high temperature, T_h. The removal of heat from a lower-temperature reservoir and delivery to a higher-temperature reservoir performs the essential effect of a refrigerator or air conditioner. In this case, the second law of thermodynamics requires work input – i.e. there is no such thing as a spontaneous refrigerator or air conditioner: You must do work to cause heat to flow against its natural tendency.

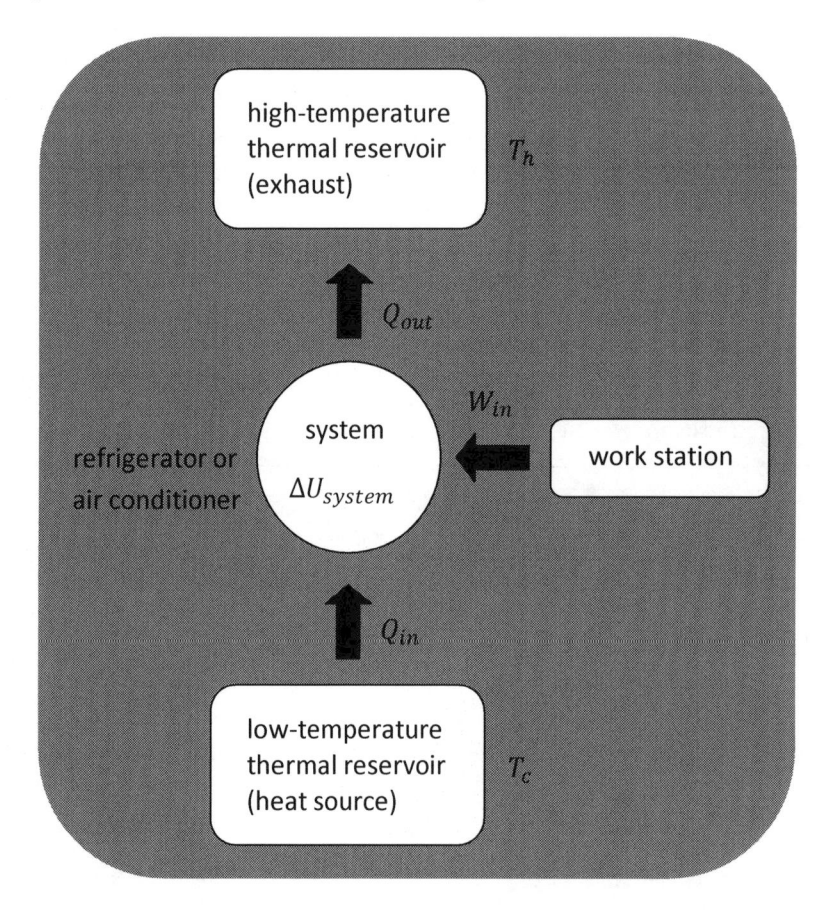

A refrigerator or air conditioner is a device in which mechanical work is performed in order to deliver heat from a lower-temperature source to a higher-temperature thermal reservoir. The general features of a refrigerator or air conditioner are:

- Desired input: A system absorbs heat, $Q_{in} > 0$, from a low-temperature thermal reservoir.[191]
- Output: The system delivers heat, $Q_{out} < 0$, to a high-temperature thermal reservoir.
- Undesired input: Mechanical work, $W_{in} < 0$, must be performed in order to extract heat from a lower-temperature thermal reservoir and deliver heat to a higher-temperature thermal reservoir.

A reversible cycle of a refrigerator or air conditioner appears as a counter-clockwise path in a P-V diagram, whereas the reversible cycle of a heat engine runs clockwise. In the case of a reversible heat engine, it can truly be reversed to make a refrigerator or air conditioner – simply change the direction of the arrow of the reversible path.

Heater: A heater is essentially no different from a refrigerator or air conditioner. A refrigerator or air conditioner extracts heat from a low-temperature thermal reservoir and deposits it into a high-temperature thermal reservoir, which is exactly the same thing that a heater does. Basically, the only difference between a room heater and a room air conditioner is that in the case of a room heater, the room is the low-temperature thermal reservoir and the atmosphere is the high-temperature thermal reservoir, whereas in the case of a room air conditioner, the room is the high-temperature thermal reservoir and the atmosphere is the low-temperature thermal reservoir (low and high being relative terms, of course). In principle, if we simply swap the two ports, we convert a room air conditioner to a heater, and vice-versa. Heaters, refrigerators, and air conditioners are different types of heat pumps.

Notice that a heat engine does not function as a heater. Rather, a heater is obtained by running the cycle of a heat engine backwards. A heat engine transfers heat from a high-temperature thermal reservoir to a low-temperature thermal reservoir in order to do useful work, whereas a heater does work to transfer heat from a low-temperature thermal reservoir to a high-temperature thermal reservoir. The heat engine makes use of the natural direction of heat flow (from the high-temperature thermal reservoir to the low-temperature one) in order to do mechanical work, while the heater does work to effectively cause heat to flow against its nature. Thus, heaters and heat engines are backwards in both conceptual design and function. On the other hand, a heater and an air conditioner (or refrigerator) involve exactly the same cycles – neither one is backwards compared to the other.

Important Distinction. It may be intuitive to want to think of a heat engine and a heater as being synonymous, and also to want to think of a heater as being backwards compared to an air conditioner or refrigerator – however, this would be incorrect. It is important to realize that a heater is actually a heat engine run backwards, and that a heater actually involves the exact same cycle as a refrigerator or air conditioner. What makes a room heater effectively different from a room air conditioner primarily boils down to whether the room functions as the low-temperature thermal reservoir from which heat is removed or the high-temperature thermal reservoir that receives heat – which corresponds to simply swapping the two ports.

---

[191] The food storage compartment of a refrigerator, or the air in a room where a heat pump (such as an air conditioner) operates, has constant temperature – like a low-temperature thermal reservoir – once the steady-state temperature has been established. When the refrigerator or heat pump is first activated (or just after the thermostat is set), the temperature of the food storage compartment, or the air in a room where a heat pump operates, will decrease – unlike a low-temperature thermal reservoir – until a steady state is reached.

Heat pumps: A device that pumps heat from a low-temperature thermal reservoir to a high-temperature thermal reservoir is called a heat pump. Heaters, refrigerators, and air conditioners (but not heat engines) are all heat pumps. A heat pump may run in heating mode – like a room heater – or in cooling mode – like a refrigerator or air conditioner. The difference is that in heating mode we are trying to generate the most $|Q_{out}|$ that we can for the given $|W_{in}|$ (or, equivalently, trying to use the least $|W_{in}|$ necessary to generate a given $|Q_{out}|$), since we are focused on trying to heat the higher-temperature thermal reservoir, while in cooling mode our goal is to extract the most Q_{in} that we can for the given $|W_{in}|$, since we are trying to remove heat from the lower-temperature thermal reservoir.

Coefficient of performance: A heat pump is more thermally effective if more heat, Q_{in}, can be extracted (in cooling mode) from the lower-temperature thermal reservoir – or more heat, $|Q_{out}|$, can be delivered (in heating mode) to the higher-temperature thermal reservoir – for a given amount of mechanical work, $|W_{in}|$. This definition of the thermal effectiveness of a heat pump is fundamentally different from the definition of the thermal efficiency of a heat engine run forwards, so we give it a different name – the coefficient of performance (COP). The COP of a heat pump is defined as:

$$\text{COP}_{cooling} \equiv -\frac{Q_{in}}{W_{in}} = \frac{-Q_{in}}{Q_{in} + Q_{out}} = \frac{-Q_{in}}{Q_{in} - |Q_{out}|} = \frac{1}{\frac{|Q_{out}|}{Q_{in}} - 1}$$

$$\text{COP}_{heating} \equiv \frac{Q_{out}}{W_{in}} = \frac{Q_{out}}{Q_{in} + Q_{out}} = \frac{1}{1 + \frac{Q_{in}}{Q_{out}}} = \frac{1}{1 - \frac{Q_{in}}{|Q_{out}|}}$$

The COP of a heat pump shows that there is a difference in the thermal effectiveness of a heat pump when it is used as a room heater rather than as a room air conditioner. Conceptually, the distinction arises because when designing a room air conditioner you wish to maximize Q_{in}, while the design of a room heater involves maximizing $|Q_{out}|$, for a given $|W_{in}|$.

Quite unlike the thermal efficiency of a heat engine, the COP of a heat pump can – and generally does – exceed 1. Therefore, it is very important not to confuse COP with thermal efficiency. The thermal efficiency, e, of a heat engine must be less than 100% because the second law of thermodynamics prevents a heat engine from delivering as much mechanical work, W_{out}, as heat, Q_{in}, is put in. In contrast, the fact that $\text{COP}_{cooling}$ may exceed 1 reflects that it is possible to extract more heat, Q_{in}, or deliver more heat, $|Q_{out}|$, than mechanical work, $|W_{in}|$, is put in. The second law of thermodynamics prohibits the exhaust, Q_{out}, of a heat engine from being zero, and also prevents the work input, W_{in}, of a heat pump from being zero. The requirement $|Q_{out}| > 0$ algebraically implies that $e < 100\%$, whereas algebraically $|W_{in}| > 0$ does not similarly restrict the COP. In the case of a heat pump, *some* mechanical work is required, yet more heat may be extracted than the amount of mechanical work put in.

Important Distinction. The thermal efficiency, $e = W_{out}/Q_{in}$, of a heat engine run forwards must be less than unity, $e < 1$, according to the second law of thermodynamics. It is important not to confuse this with the COP of a heat engine run backwards, which has a similar definition, $\text{COP}_{cooling} \equiv -Q_{in}/W_{in}$, but may be, and generally is, great than unity (> 1).

Flexible endpoints: Observe that the initial and final equilibrium states – which are identical equilibrium states – are flexible for a heat engine, refrigerator, or heat pump. That is, the order of the steps is not important (however, whether you go forwards or backwards makes a difference). For example, if one cycle of a heat engine consists of four processes – call them a, b, c, and d – the exact same effect is achieve whether you perform the processes starting with a, then doing b, following with c, and finishing with d, or if you perform them in the order b, c, d, and a. Either way, the net work and neat heat for the cycle are exactly the same.

5.3 The Carnot Cycle

Carnot cycle: The Carnot cycle is especially significant in thermodynamics because, as we will learn in the next section, the efficiency of the Carnot engine establishes an upper limit on the efficiency of any conceivable heat engine. The Carnot cycle consists of four reversible thermodynamic processes – two reversible isotherms and two reversible adiabats, one of each being a compression and one of each being an expansion.

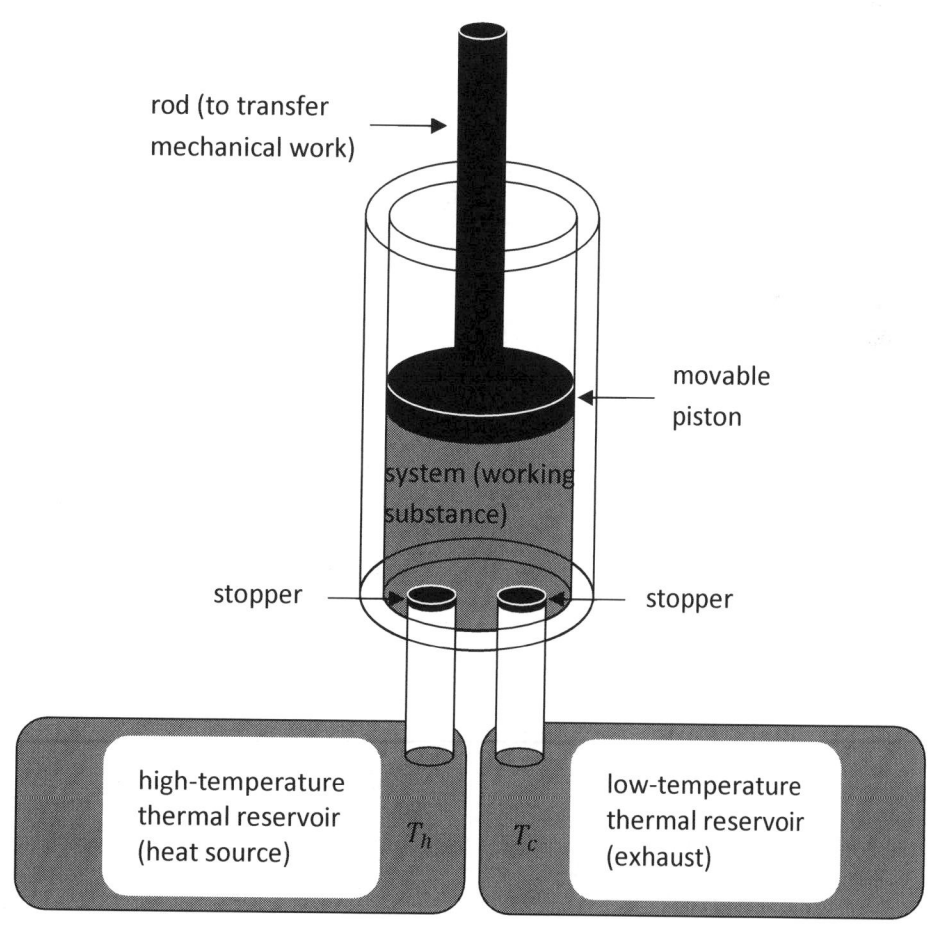

5 Heat Engines and Refrigerators

Carnot engine: The working substance – e.g. a gas – is placed inside a cylinder with perfectly insulating walls. The working substance will have constant mole number(s) throughout the process. There is a perfectly insulating movable piston at one end (which we choose to be the top, so that we may visualize quasistatic processes carried out with the technique of adding or removing grains of sand), which can slide without friction. The working substance serves as the system. The system can do work by moving the piston: The system does work when it expands, and work is done on the system when it contracts. The sliding piston delivers mechanical work to the region above it (the 'work station') – e.g. a rod may connect the piston to other machinery, such that the motion of the piston results in the rotation of some wheel, which does the desired useful work. The cylinder has two ports, which are covered with perfectly insulating stoppers (or stopcocks or equivalent means of sealing/unsealing the ports) when not in use; these ports connect to two thermal reservoirs operating at high and low temperatures (the low-temperature thermal reservoir may be the air, and the high-temperature thermal reservoir that supplies the heat may be furnace – the cost of running the furnace is what we pay to benefit from the work that is done).

The Carnot engine begins with the working substance in the initial equilibrium state, (P_a, V_a, T_a).[192] Each cycle of the Carnot engine involves carrying out the following four steps in order, with the fourth step returning the working substance to its original equilibrium state, (P_a, V_a, T_a):

- The stopper is removed from the port that connects the system to the high-temperature thermal reservoir. The system undergoes a reversible isothermal expansion (e.g. by slowly removing grains of sand from the movable piston) to a new volume, $V_b > V_a$. The system absorbs heat, Q_{ab}, from the high-temperature thermal reservoir, and does work, W_{ab}, as the system expands. The piston stops when the volume equals V_b, the stopper is put back in place, and the equilibrium state at the completion of the isothermal expansion is (P_b, V_b, T_a).
- With both stoppers in place, the system undergoes a reversible adiabatic expansion (e.g. by slowly removing grains of sand from the movable piston) to a new volume, $V_c > V_b$. This time there are no heat exchanges while the system does work, W_{bc}, as the system expands. The piston stops when the volume equals V_c and the equilibrium state at the completion of the adiabatic expansion is (P_c, V_c, T_c). The new temperature, T_c, is less than the previous temperature, T_a, as a result of the adiabatic expansion: $T_c < T_a$.[193]
- The stopper is removed from the port that connects the system to the low-temperature thermal reservoir. The system undergoes a reversible isothermal compression (e.g. by slowly adding grains of sand to the movable piston) to a new volume, $V_d < V_c$. The system releases its exhaust, Q_{cd}, to the low-temperature thermal reservoir, and work, W_{cd}, is done on the system as it is compressed. The piston stops when the volume equals V_d, the stopper is put back in place, and the equilibrium state at the completion of the isothermal compression is (P_d, V_d, T_c).

---

[192] Only two of these three variables are independent: For example, if the working substance is an ideal gas, we know that $P_a V_a = nRT_a$, such that (P_a, V_a) completely characterizes each equilibrium state. However, we will express each equilibrium state in the form (P_a, V_a, T_a) in order to show clearly which variables are changing at each stage of the cycle.

[193] The equation for an ideal gas, $PV = nRT$, can be combined with the equation for an adiabat for an ideal gas, $P_0 V_0^\gamma = PV^\gamma$, where the adiabatic index, γ, equals $\gamma = c_P/c_V > 1$, to show that $T_c < T_a$ if the working substance is an ideal gas (see the coming example). If a different working substance is used, the corresponding equation of state can be applied to see if $T_c < T_a$.

- With both stoppers in place, the system undergoes a reversible adiabatic compression (e.g. by slowly adding grains of sand to the movable piston) to its original volume, V_a. There are no heat exchanges while work, W_{da}, is done on the system as it is compressed. The piston stops when the volume equals V_a and the original equilibrium state, (P_a, V_a, T_a), is returned at the completion of the adiabatic compression.

State	Process	Work done by system	Heat absorbed by system
(P_a, V_a, T_a)			
	isothermal expansion	$W_{ab} > 0$	$Q_{ab} > 0$
(P_b, V_b, T_a)			
	adiabatic expansion	$W_{bc} > 0$	$Q_{bc} = 0$
(P_c, V_c, T_c)			
	isothermal compression	$W_{cd} < 0$	$Q_{cd} < 0$
(P_d, V_d, T_c)			
	adiabatic compression	$W_{da} < 0$	$Q_{da} = 0$
(P_a, V_a, T_a)			

Carnot P-V diagram: One cycle of a Carnot heat engine is illustrated below for an ideal gas. Conceptually, the Carnot cycle begins at point a.[194] The isothermal expansion, $a \to b$, is a hyperbola given by the equation, $P = P_a V_a / V$, which slopes down and to the right. The adiabatic expansion, $b \to c$, is a curve given by the equation, $P = P_b V_b^\gamma / V^\gamma$, which looks somewhat hyperbolic and also slopes down and to the right. The isothermal compression, $c \to d$, is a hyperbola given by the equation, $P = P_c V_c / V$, which slopes up and to the left. The adiabatic expansion, $d \to a$, is a curve given by the equation, $P = P_d V_d^\gamma / V^\gamma$, which looks somewhat hyperbolic and also slopes up and to the left, returning the system to its original state at point a. The net work, W_{out}, equals the area of the gray region enclosed by the P-V curve. The temperature remains constant over $a \to b$ and $c \to d$, where heat is added to and removed from the system, respectively.

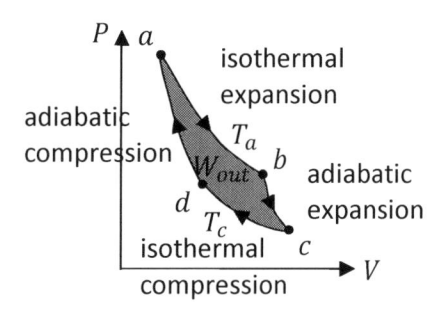

---

[194] Conceptually, it makes sense for the Carnot cycle to 'begin' by adding heat to the system from the high-temperature reservoir during the isothermal expansion and 'end' with the adiabatic compression that returns the working substance to its original state. Mathematically, however, a cycle is a cycle, and the same net work and heat result regardless of where you choose to 'start' (and therefore 'finish').

Carnot efficiency: The system absorbs heat, $Q_{in} = Q_{ab} > 0$, during the isothermal expansion and releases heat (exhaust), $Q_{out} = Q_{cd} < 0$, during the isothermal compression. The net heat equals $Q_{net} = Q_{in} + Q_{out} = Q_{ab} + Q_{cd} = Q_{ab} - |Q_{cd}|$. The system does work during both the isothermal and adiabatic expansions: $W_{by} = W_{ab} + W_{bc} > 0$. Work is done on the system during the isothermal and adiabatic compressions: $W_{on} = W_{cd} + W_{da} < 0$. The net work equals $W_{out} \equiv W_{net} = W_{by} + W_{on} = W_{ab} + W_{bc} + W_{cd} + W_{da} = W_{ab} + W_{bc} - |W_{cd}| - |W_{da}|$. For the complete cycle, $\Delta U_{net} = 0$, so that $W_{out} = Q_{net}$. The efficiency of the Carnot engine is therefore:

$$e_C = \frac{W_{out}}{Q_{in}} = \frac{W_{ab} + W_{bc} - |W_{cd}| - |W_{da}|}{Q_{ab}} = \frac{Q_{net}}{Q_{in}} = \frac{Q_{ab} - |Q_{cd}|}{Q_{ab}} = 1 - \frac{|Q_{cd}|}{Q_{ab}}$$

The first law and the Carnot engine: According to the first law of thermodynamics, $\Delta U = Q - W$. Since $\Delta U_{net} = 0$, the net work done equals the net heat: $W_{out} = Q_{net}$. For the adiabats, since $Q_{bc} = Q_{da} = 0$, the internal energy change equals the negative of the work done: $\Delta U_{bc} = -W_{bc} < 0$ and $\Delta U_{da} = -W_{da} = |W_{da}| > 0$. For the isotherms, $\Delta U_{ab} = Q_{ab} - W_{ab}$ and $\Delta U_{cd} = Q_{cd} - W_{cd} = -|Q_{cd}| + |W_{cd}|$.

Process	Work done by system	Heat absorbed by system	Internal energy change of system
$a \to b$	$W_{ab} > 0$	$Q_{ab} > 0$	$\Delta U_{ab} = Q_{ab} - W_{ab}$
$b \to c$	$W_{bc} > 0$	$Q_{bc} = 0$	$\Delta U_{bc} = -W_{bc} < 0$
$c \to d$	$W_{cd} < 0$	$Q_{cd} < 0$	$\Delta U_{cd} = Q_{cd} - W_{cd}$
$d \to a$	$W_{da} < 0$	$Q_{da} = 0$	$\Delta U_{da} = -W_{da} > 0$
cycle	$W_{out} = Q_{ab} - |Q_{cd}|$	$Q_{net} = Q_{ab} - |Q_{cd}|$	$\Delta U_{net} = 0$

The second law and the Carnot engine: Since the Carnot cycle is reversible, the entropy change for any of its processes can be found by integrating over the actual path: $\Delta S_{system} = \int_i^f đQ_{system}/T$. The overall entropy must be constant throughout the cycle: $\Delta S_{overall} = 0$ for any interval. This means that the entropy of the system and surroundings must be equal and opposite for any interval: $\Delta S_{system} = -\Delta S_{surroundings}$. Like the internal energy, the net entropy change of the system must be zero for the complete cycle, $\Delta S_{system}^{cycle} = 0$, which, since the cycle is reversible, requires the same for the surroundings, $\Delta S_{surroundings}^{cycle} = 0$. Since the adiabats are reversible, the entropy of the system (and therefore the surroundings, too) is constant over the course of each adiabat: $\Delta S_{system}^{bc} = \Delta S_{system}^{da} = 0$ and $\Delta S_{surroundings}^{bc} = \Delta S_{surroundings}^{da} = 0$. Heat is exchanged between the system and surroundings for the reversible isotherms. Since temperature is constant along an isotherm, $\Delta S_{system}^{ab} = Q_{ab}/T_a$, $\Delta S_{system}^{cd} = Q_{cd}/T_c$, $\Delta S_{surroundings}^{ab} = -Q_{ab}/T_a$, and $\Delta S_{surroundings}^{cd} = -Q_{cd}/T_c$.

The entropy change of the system for the cycle equals the sum of the entropy changes for each of the processes of the cycle:

$$\Delta S_{system}^{cycle} = Q_{ab}/T_a + Q_{cd}/T_c = Q_{ab}/T_a - |Q_{cd}|/T_c = 0$$

$$\frac{T_c}{T_a} = \frac{|Q_{cd}|}{Q_{ab}}$$

From these entropy changes, it directly follows (as we will see momentarily) that the efficiency of the Carnot engine must be less than 100%.

Process	Entropy change of system	Entropy change of surroundings	Entropy change overall
$a \to b$	$\Delta S^{ab}_{system} = \dfrac{Q_{ab}}{T_a}$	$\Delta S^{ab}_{surroundings} = -\dfrac{Q_{ab}}{T_a}$	$\Delta S^{ab}_{overall} = 0$
$b \to c$	$\Delta S^{bc}_{system} = 0$	$\Delta S^{bc}_{surroundings} = 0$	$\Delta S^{bc}_{overall} = 0$
$c \to d$	$\Delta S^{cd}_{system} = \dfrac{Q_{cd}}{T_c}$	$\Delta S^{cd}_{surroundings} = -\dfrac{Q_{cd}}{T_c}$	$\Delta S^{cd}_{overall} = 0$
$d \to a$	$\Delta S^{da}_{system} = 0$	$\Delta S^{da}_{surroundings} = 0$	$\Delta S^{da}_{overall} = 0$
cycle	$\Delta S^{cycle}_{system} = 0$	$\Delta S^{cycle}_{surroundings} = 0$	$\Delta S^{cycle}_{overall} = 0$

Carnot efficiency in terms of temperature: We just derived the following equation by computing the entropy changes for each process in the Carnot cycle:

$$\frac{T_c}{T_a} = \frac{|Q_{cd}|}{Q_{ab}}$$

Previously, we found that the efficiency of the Carnot engine is

$$e_C = 1 - \frac{|Q_{cd}|}{Q_{ab}}$$

Therefore, the efficiency of the Carnot engine can be expressed purely in terms of temperature:[195]

$$e_C = 1 - \frac{T_c}{T_a}$$

where $T_c < T_a$ because T_c corresponds to the low-temperature thermal reservoir (which receives the exhaust) and T_a corresponds to the high-temperature thermal reservoir (the heat source). Note that we derived this result without assuming any particular model for the working substance (e.g. we did not know if it was an ideal gas, a van der Waals fluid, or something else) – we derived the result just by analyzing the entropy changes.

Important Distinction. We derived the equation $e_C = 1 - T_c/T_h$, where T_c and T_h are the temperatures of the colder and warmer thermal reservoirs, respectively, specifically for a Carnot engine. It is a common mistake for students to use $e_C = 1 - T_c/T_h$ instead of $e = 1 - |Q_{out}|/Q_{in}$ for other engines.

---

[195] From this result, in principle, the Carnot engine could serve as a means of measuring absolute temperature. This notion has much theoretical and aesthetic significance. As a practical thermometer, the more approximately a real heat engine can be made reversible, the more precise will be the results.

Efficiency of a real Carnot-like engine: Any effort in lab to construct a Carnot engine suffers some inherent experimental sources of error: For one, it is impossible to reach the lofty goal of completely eliminating all frictional and dissipative forces. At best, a real Carnot-like engine has isothermal and adiabatic processes that are approximately, but not quite, reversible.[196] Since a real Carnot-like engine is irreversible, $\Delta S_{overall}^{cycle} > 0$. Even for an irreversible cycle, $\Delta S_{system}^{cycle} = 0$, since the entropy is path-independent, which requires that $\Delta S_{surroundings}^{cycle} > 0$.[197] Although $\Delta S_{system}^{cycle} = 0$, whether or not the cycle is reversible, the entropy change for each process is different for a real Carnot-like engine than it is for a Carnot engine because the integral $\Delta S_{system} = \int_i^f đQ_{system}/T$ only applies to reversible processes. Therefore, ΔS_{system} may be nonzero for the irreversible adiabats, and ΔS_{system} may differ from Q/T for the irreversible isotherms. What we know for sure is that the sum of the entropy changes for the system will be zero. Thus, our derivation for the efficiency of a Carnot engine in terms of temperature does not apply to a real Carnot-like engine. The Carnot efficiency is still very useful, as it establishes an upper limit – an experimental goal – as we prove in the next section.

Example. Derive equations for the work, heat exchange, internal energy change, and entropy change for the system for each process and the complete cycle for a Carnot engine using an ideal gas as the working substance, and show that these equations agree with the efficiency of a Carnot engine.

The work done by (or on) the system during the isotherms can be computed by expressing the pressure in terms of volume with the ideal gas law, $P_0 V_0 = PV$, and then integrating over volume:

$$W_{ab} = \int_{V=V_a}^{V_b} P dV = \int_{V=V_a}^{V_b} \frac{P_a V_a}{V} dV = P_a V_a \ln\left(\frac{V_b}{V_a}\right) = P_b V_b \ln\left(\frac{V_b}{V_a}\right)$$

$$W_{cd} = P_c V_c \ln\left(\frac{V_d}{V_c}\right) = P_d V_d \ln\left(\frac{V_d}{V_c}\right)$$

where we used the fact that PV is constant over the isotherm – so, e.g., $P_a V_a = P_b V_b$. The work done by (or on) the system along the adiabats can similarly be found using the equation for an adiabat for an ideal gas, $P_0 V_0^\gamma = PV^\gamma$:

---

[196] The term 'Carnot engine' refers to the theoretically ideal case where the adiabatic and isothermal processes are reversible. A physical engine that is constructed in lab to carry out the Carnot cycle is inherently irreversible, and so does not meet the definition set forth by the label 'Carnot engine.' Therefore, we denote the physical heat engine, with a irreversible version of the Carnot cycle, by 'real Carnot-like engine.'

[197] Are you wondering how the path-independence of the entropy can require $\Delta S_{system}^{cycle} = 0$, and at the same time allow $\Delta S_{surroundings}^{cycle} > 0$? (You should be – unless you've already reasoned this out.) The initial and final equilibrium states of the system are identical, which is why $\Delta S_{system}^{cycle} = 0$. The inequality, $\Delta S_{surroundings}^{cycle} > 0$, implies that the initial and final states of the *surroundings* – not the *system* – cannot be identical. We are, in fact, extracting heat from one thermal reservoir and depositing a fraction of this into another thermal reservoir, which does minutely (for all other purposes, negligibly) affect the states of the thermal reservoirs. So the explanation is that the system, but not the surroundings, completes a cycle.

$$W_{bc} = \int_{V=V_b}^{V_c} PdV = \int_{V=V_b}^{V_c} \frac{P_b V_b^\gamma}{V^\gamma} dV = P_b V_b^\gamma \int_{V=V_b}^{V_c} V^{-\gamma} dV = \frac{P_b V_b^\gamma}{1-\gamma}[V^{1-\gamma}]_{V=V_b}^{V_c}$$

$$W_{bc} = \frac{P_b V_b^\gamma}{1-\gamma} V_c^{1-\gamma} - \frac{P_b V_b^\gamma}{1-\gamma} V_b^{1-\gamma} = \frac{P_c V_c^\gamma}{1-\gamma} V_c^{1-\gamma} - \frac{P_b V_b^\gamma}{1-\gamma} V_b^{1-\gamma} = \frac{P_c V_c - P_b V_b}{1-\gamma}$$

$$W_{da} = \frac{P_a V_a - P_d V_d}{1-\gamma}$$

where we used the fact that PV^γ is constant over the adiabat – so, e.g., $P_b V_b^\gamma = P_c V_c^\gamma$. The equations for work done can also be expressed in temperature using the ideal gas law, $PV = nRT$:

$$W_{ab} = nRT_a \ln\left(\frac{V_b}{V_a}\right) \quad , \quad W_{cd} = nRT_c \ln\left(\frac{V_d}{V_c}\right)$$

$$W_{bc} = \frac{nR(T_c - T_a)}{1-\gamma} \quad , \quad W_{da} = \frac{nR(T_a - T_c)}{1-\gamma}$$

Observe that the work done along the adiabat $b \to c$ is equal and opposite to the work done along the adiabat $d \to a$: $W_{bc} = -W_{da}$. The net work is therefore equal to $W_{out} \equiv W_{net} = W_{ab} + W_{cd} = W_{ab} - |W_{cd}|$, which can be expressed in terms of pressure as

$$W_{out} = P_a V_a \ln\left(\frac{V_b}{V_a}\right) + P_c V_c \ln\left(\frac{V_d}{V_c}\right)$$

or in terms of temperature as

$$W_{out} = nRT_a \ln\left(\frac{V_b}{V_a}\right) + nRT_c \ln\left(\frac{V_d}{V_c}\right)$$

The internal energy of an ideal gas equals $U = nc_V T$. Therefore, the internal energy is constant over the isotherms:

$$\Delta U_{ab} = \Delta U_{cd} = 0$$

The internal energy changes for the adiabats are

$$\Delta U_{bc} = nc_V(T_c - T_a) \quad , \quad \Delta U_{da} = nc_V(T_a - T_c)$$

since $T_b = T_a$ and $T_d = T_c$. Observe that the internal energy for adiabat $b \to c$ is equal and opposite to the internal energy change for the adiabat $d \to a$: $\Delta U_{bc} = -\Delta U_{da}$. This is consistent with the similar statement regarding work, $W_{bc} = -W_{da}$, since ΔU is equal to $-W$ along an adiabat (as $Q = 0$). Note that the signs are, in fact, consistent, since $\gamma > 1$. The equation for ΔU_{bc} is indeed equal to $-W_{bc}$, and similarly for ΔU_{da} and $-W_{da}$, as can be seen with a little algebra starting with the definition of the adiabatic index, $\gamma = c_P/c_V$, and the relation, $c_P = c_V + R$. Eliminating c_P from these two expressions,

$$c_p = c_V \gamma = c_V + R$$
$$c_V = \frac{R}{\gamma - 1}$$

The internal energy changes can therefore be written in the form:

$$\Delta U_{bc} = \frac{nR(T_c - T_a)}{\gamma - 1} \quad , \quad \Delta U_{da} = \frac{nR(T_a - T_c)}{\gamma - 1}$$

This means that the internal energy changes for the adiabats can also be expressed in terms of pressure and volume (since these expressions are equal and opposite to the work done along the adiabats, which has already been expressed in terms of pressure and volume):

$$\Delta U_{bc} = \frac{P_c V_c - P_b V_b}{\gamma - 1} \quad , \quad \Delta U_{da} = \frac{P_a V_a - P_d V_d}{\gamma - 1}$$

Of course, the net internal energy change for the cycle must be zero — which is exactly what you get if you add them up:

$$\Delta U_{net} = 0$$

There is no heat exchanged along the adiabats:

$$Q_{bc} = Q_{da} = 0$$

The heat exchanged between the system and surroundings (namely, the two thermal reservoirs) along the isotherms equals the work done along the isotherms (since $\Delta U = 0$ along the isotherms):

$$Q_{ab} = P_a V_a \ln\left(\frac{V_b}{V_a}\right) = P_b V_b \ln\left(\frac{V_b}{V_a}\right) \quad , \quad Q_{cd} = P_c V_c \ln\left(\frac{V_d}{V_c}\right) = P_d V_d \ln\left(\frac{V_d}{V_c}\right)$$

In terms of temperature, this can be expressed as

$$Q_{ab} = nRT_a \ln\left(\frac{V_b}{V_a}\right) \quad , \quad Q_{cd} = nRT_c \ln\left(\frac{V_d}{V_c}\right)$$

The total heat absorbed by the system from the high-temperature thermal reservoir is $Q_{in} = Q_{ab}$ and the total heat released by the system into the low-temperature thermal reservoir is $Q_{out} = Q_{cd}$ (which is the exhaust). The net heat can be written in terms of pressure as

$$Q_{net} = P_a V_a \ln\left(\frac{V_b}{V_a}\right) + P_c V_c \ln\left(\frac{V_d}{V_c}\right)$$

or in terms of temperature as

$$Q_{net} = nRT_a \ln\left(\frac{V_b}{V_a}\right) + nRT_c \ln\left(\frac{V_d}{V_c}\right)$$

Since the Carnot engine is reversible, the entropy is constant along the two adiabats:

$$\Delta S^{bc}_{system} = \Delta S^{da}_{system} = 0$$

The temperature is constant along the reversible isotherms, such that the entropy change along the isotherms depends only upon the volumes:

$$\Delta S^{ab}_{system} = \int_a^b \frac{dQ_{system}}{T} = \frac{Q_{ab}}{T_a} = nR \ln\left(\frac{V_b}{V_a}\right)$$

$$\Delta S^{cd}_{system} = \int_c^d \frac{dQ_{system}}{T} = \frac{Q_{cd}}{T_c} = nR \ln\left(\frac{V_d}{V_c}\right)$$

The entropy change for the system must be zero for the cycle:

$$\Delta S^{cycle}_{system} = nR \ln\left(\frac{V_b}{V_a}\right) + nR \ln\left(\frac{V_d}{V_c}\right) = nR \ln\left(\frac{V_b V_d}{V_a V_c}\right) = 0$$

using the identities $\ln(xy) = \ln x + \ln y$ and $\ln(x/y) = \ln x - \ln y$. The argument of the logarithm must be unity in order for the logarithm to vanish, which requires that

$$V_b V_d = V_a V_c$$

The definition of the efficiency of a heat engine leads to

$$e_C = \frac{W_{out}}{Q_{in}} = \frac{nRT_a \ln\left(\frac{V_b}{V_a}\right) + nRT_c \ln\left(\frac{V_d}{V_c}\right)}{nRT_a \ln\left(\frac{V_b}{V_a}\right)} = 1 + \frac{T_c \ln\left(\frac{V_d}{V_c}\right)}{T_a \ln\left(\frac{V_b}{V_a}\right)} = 1 - \frac{T_c \ln\left(\frac{V_c}{V_d}\right)}{T_a \ln\left(\frac{V_b}{V_a}\right)}$$

where we used the identity $\ln(1/x) = -\ln x$. Since $V_b V_d = V_a V_c$, or $V_c/V_d = V_b/V_a$, the ratio of the logarithms is unity, reducing the efficiency to

$$e_C = 1 - \frac{T_c}{T_a}$$

which agrees with our previous result. Note that the ratio of volumes could be found without using the entropy equations – by combining the ideal gas law with the equations for the adiabats:

$$\left(\frac{V_c}{V_d}\right)^\gamma = \frac{P_b V_b^\gamma / P_c}{P_a V_a^\gamma / P_d} = \frac{V_b^\gamma}{V_a^\gamma} \frac{P_b P_d}{P_a P_c} = \frac{V_b^\gamma}{V_a^\gamma} \frac{1}{P_a P_c} \frac{P_a V_a P_c V_c}{V_b V_d} \Rightarrow \left(\frac{V_c}{V_d}\right)^{\gamma-1} = \left(\frac{V_b}{V_a}\right)^{\gamma-1} \Rightarrow \frac{V_c}{V_d} = \frac{V_b}{V_a}$$

5 Heat Engines and Refrigerators

Carnot heat pump: The Carnot engine can be transformed into a Carnot heat pump – which includes a refrigerator, air conditioner, and heater – simply by running the cycle in reverse.[198] Reverse means that expansions become compressions, heat absorbed becomes heat released, work done by the system becomes work done on the system, and vice-versa; it also means repeating the steps in reverse order. Each cycle of the Carnot refrigerator begins with the working substance in an initial equilibrium state, (P_a, V_a, T_a), and repeats the following four steps in order, with the fourth step returning the working substance to its original equilibrium state, (P_a, V_a, T_a):

- With both stoppers in place, the system undergoes a reversible adiabatic expansion to a new volume, $V_d > V_a$. There are no heat exchanges while work, $W_{ad} > 0$, is done by the system as it expands. The piston stops when the volume equals V_d and the equilibrium state at the completion of the adiabatic expansion is (P_d, V_d, T_c). The new temperature, T_c, is less than the previous temperature, T_a, as a result of the adiabatic expansion: $T_c < T_a$.
- The stopper is removed from the port that connects the system to the low-temperature thermal reservoir. The system undergoes a reversible isothermal expansion to a new volume, $V_c > V_d$. The system absorbs heat, $Q_{dc} > 0$, from the low-temperature thermal reservoir, and work, $W_{dc} > 0$, is done by the system as it expands. The piston stops when the volume equals V_c, the stopper is put back in place, and the equilibrium state at the completion of the isothermal expansion is (P_c, V_c, T_c).
- With both stoppers in place, the system undergoes a reversible adiabatic compression to a new volume, $V_b > V_c$. This time there are no heat exchanges while work, $W_{cb} < 0$, is done on the system as the system is compressed. The piston stops when the volume equals V_b and the equilibrium state at the completion of the adiabatic compression is (P_b, V_b, T_a), which returns the system to its original temperature, T_a.
- The stopper is removed from the port that connects the system to the high-temperature thermal reservoir. The system undergoes a reversible isothermal compression to its original volume, V_a. The system releases heat, $Q_{ba} < 0$, to the high-temperature thermal reservoir, and work, $W_{ba} < 0$, is done on the system as the system is compressed. The piston stops when the volume returns to its original value, V_b, the stopper is put back in place, and the equilibrium state at the completion of the isothermal expansion is (P_a, V_a, T_a).

State	Process	Work done by system	Heat absorbed by system
(P_a, V_a, T_a)			
	adiabatic expansion	$W_{da} > 0$	$Q_{da} = 0$
(P_d, V_d, T_c)			
	isothermal expansion	$W_{dc} > 0$	$Q_{dc} > 0$
(P_c, V_c, T_c)			
	adiabatic compression	$W_{cb} < 0$	$Q_{cb} = 0$
(P_b, V_b, T_a)			
	isothermal compression	$W_{ba} < 0$	$Q_{ba} < 0$
(P_a, V_a, T_a)			

---

[198] As the Carnot cycle is a reversible cycle, there is no distinction between 'in reverse' and 'backwards.'

The P-V diagram for a Carnot heat pump is exactly the same for a Carnot engine, except for traversing the path in a counterclockwise, rather than clockwise, route.

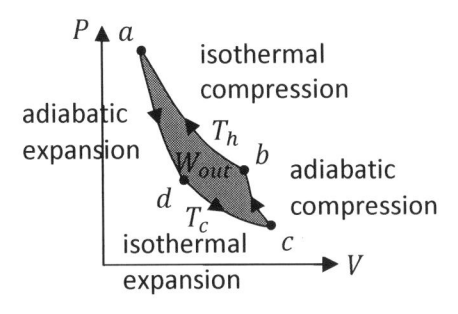

Carnot coefficient of performance: Since the Carnot cycle consists of exactly the same reversible processes and exactly the same equilibrium states[199] when run clockwise to function as a heat engine or when run counterclockwise to function as a heat pump, the same equations will apply, except for conceptual sign changes: The net work, W_{in}, will be negative (and we will think of it as input, W_{in}, rather than output, W_{out}); and what was an isothermal compression turns into an isothermal expansion, and vice-versa, causing Q_{ba} to be negative when Q_{ab} was positive, and Q_{dc} to be positive when Q_{cd} was negative (and we will interpret Q_{dc} as heat input, Q_{in}, when Q_{cd} was heat output, Q_{out}, and similarly Q_{ba} will now be interpreted as heat output, Q_{out}). These are minus signs all the way around, and so the equality of the ratio of the temperatures will again equal the ratio of the absolute value of the heat exchanges that we previously derived for the Carnot engine:

$$\frac{T_c}{T_a} = \frac{Q_{dc}}{|Q_{ba}|} = \frac{Q_{in}}{|Q_{out}|}$$

Recalling that $T_a > T_c$, we see that $|Q_{out}| > Q_{in}$. The coefficient of performance of a Carnot heat pump can be expressed in terms of temperature as:

$$\text{COP}_{cooling} \equiv -\frac{Q_{in}}{W_{in}} = \frac{1}{\frac{|Q_{out}|}{Q_{in}} - 1} = \frac{1}{\frac{T_a}{T_c} - 1} = \frac{T_c}{T_a - T_c}$$

$$\text{COP}_{heating} \equiv \frac{Q_{out}}{W_{in}} = \frac{1}{1 - \frac{Q_{in}}{|Q_{out}|}} = \frac{1}{1 - \frac{T_c}{T_a}} = \frac{T_a}{T_a - T_c}$$

Whereas the Carnot efficiency is restricted to be less than 1, the analogous restriction on the Carnot COP that derives from the second law of thermodynamics is merely that the COP is restricted to be greater than 0. Note that $\text{COP}_{heating} = 1 + \text{COP}_{cooling}$.

---

[199] The equilibrium states referred to are the four significant points where the four reversible processes meet. Since the processes must be carried out quasistatically, strictly speaking every point on the diagram corresponds (virtually) to an equilibrium state.

5.4 Maximum Efficiency of a Heat Engine

Natural processes: Many processes of the processes that are observed to occur in nature are only observed to proceed in one direction – which we call forward. Natural processes that are observed to occur both forwards and backwards are reversible, while those that are observed only to occur forwards are irreversible. The overall energy of an irreversible process would be conserved whether it occurred forwards or backwards, but the overall entropy of an irreversible process increases in the forward direction. The second law of thermodynamics thus determines whether or not a process can spontaneously occur in both directions or just in one direction.

An electron and positron can interact such that they annihilate and produce a pair of photons, and the reverse process – the pair-production of an electron and positron from two interacting photons – is also observed to occur in nature.[200] Pair annihilation is a reversible process, meaning that the overall entropy is conserved for the process.

When two objects of different temperature are placed in thermal contact, heat is observed to flow spontaneously from the higher-temperature object to the lower-temperature object, but not from the lower-temperature object to the higher-temperature object. For example, a cup of cold water left on a counter on a hot summer day is observed to get warmer, but is never observed to spontaneously get cooler (or on a more extreme notion, freeze). Similarly, a swing that is displaced from equilibrium and allowed to oscillate freely is observed to oscillate about equilibrium with smaller and smaller amplitude until coming to rest, but a swing that is at rest is never observed to spontaneously swing back and forth with increasing amplitude.[201] These processes are irreversible, and the overall entropy increases when they occur.

Observed and unobserved processes: The distinction between processes that are observed or not observed to occur both forwards and backwards has to do with reversibility and the second law of thermodynamics. This distinction can be cast in another form for processes that involve both mechanical work being done by or on the system and heat being exchanged between the system and its surroundings: Mechanical energy tends to be converted into internal energy; the reverse is not observed to occur spontaneously.

For example, a block that slides down an incline loses mechanical energy. This mechanical energy is lost as heat is exchanged between the block and the incline, and that heat exchange raises the internal energy of the block and incline. The lost mechanical energy of the block is macroscopically seen as the nonconservative work done by the friction force, while the increased internal energy raises the temperatures of the block and incline. Friction is always observed to cause a reduction in acceleration; it never contributes toward acceleration. The internal energy of the block never spontaneously exchanges heat with its surroundings in order to produce mechanical energy.

---

[200] A positron is the antiparticle to the electron. It has the same mass and spin as an electron, but all of its quantum numbers – such as electric charge – are reversed. A positron and electron do not always produce two photons when they interact – this is just one possible outcome. The outcomes are governed by probabilities.
[201] There may be some minute fluctuations from the motion of air molecules, or a hefty wind may tend to displace the swing in a particular direction, but the reverse of the process described is not observed to occur.

A freely oscillating swing similarly loses mechanical energy due to the combined effects of air resistance and the frictional torque exerted where the chains connect to the support bar. This mechanical energy is transferred to the air molecules, support bar, and swing in the form of heat, which raises their internal energies. The lost mechanical energy results in the diminishing amplitude, and the swing comes to rest when all of the mechanical energy has been spent. The internal energies never conspire together to cause the reverse process to occur, though.

A heat engine is ultimately a device that converts internal energy into mechanical energy, as the heat that is drawn from the high-temperature thermal reservoir and deposited into the system increases the internal energy of the system, and this increased internal energy of the system is used to do mechanical work. It is not possible to use 100% of this heat input to do mechanical work because this is not a natural process. We can drive the production of mechanical work from heat input, but with a cost: In the case of a heat engine, the cost is the production of less work output than heat input, with the difference in energy expelled to a low-temperature thermal reservoir as exhaust.

The second law of thermodynamics explains the distinction between reversibility and irreversibility in terms of the overall entropy, and thereby governs which processes may occur spontaneously. As a consequence, the second law of thermodynamics explains that mechanical energy may be spontaneously converted into internal energy, but not vice-versa. This is precisely the connection between the second law of thermodynamics and the limited efficiency of a heat engine.

Driven processes: It is possible to drive processes against the natural, spontaneous flow. For example, a heat pump can be used to cause heat to flow from low-temperature atmosphere on a cold winter day into a much higher-temperature living room. A girl can sit in a swing at rest in the equilibrium position and do mechanical work – leaning back and forth while pumping the air with her legs – in order to cause the amplitude of the swing to steadily increase. Heat from a furnace can be used to perform mechanical work, provided that some exhaust is channeled from the system to a low-temperature thermal reservoir. These driven processes do not occur naturally.

There is a cost to driving processes to occur in a direction that would otherwise result in a decrease in the overall entropy. That is, the second law of thermodynamics requires some driving mechanism in place in order that the overall entropy will be non-decreasing as the process evolves. In the case of a heat pump or driving a pendulum, the cost is mechanical work that must be provided as input: You must do mechanical work in order to drive an ordinarily irreversible process to occur in reverse. In the case of a heat engine, the cost is the production of some exhaust – you can't use all of the heat input to perform mechanical work. This is the reason that a heat engine cannot be perfectly efficient.

Efficiency and the first law of thermodynamics: For one cycle of a heat engine, the net internal energy change is zero, $\Delta U_{net} = 0$, so that the net work done, W_{out}, (the output work) is equal to the net heat, Q_{net}, exchanged with the surroundings: $W_{out} = Q_{net}$. As far as the first law of thermodynamics is concerned, the net heat, Q_{net}, may be entirely heat input, Q_{in}, from the high-temperature thermal reservoir; energy would still be conserved. Therefore, the first law of thermodynamics, $\Delta U = Q - W$, does not preclude the efficiency of a heat engine, $e = W_{out}/Q_{in}$, from equaling 100%. In general, conservation of energy does not preclude the natural conversion of internal energy into mechanical energy. It is only the second law of thermodynamics that places these prohibitions on the energy transformations that may occur spontaneously.

Coefficient of performance and the first law of thermodynamics: The net mechanical work, $W_{in} < 0$, that must be supplied to operate a heat pump (which is negative because this work is done on the system) equals the net heat, Q_{net}, exchanged between the system and the thermal reservoirs according to the first law of thermodynamics (since $\Delta U_{net} = 0$ for the cycle): $W_{in} = Q_{net} = Q_{in} + Q_{out}$. The net heat must be negative, $Q_{net} < 0$, since the net work is negative, meaning that the system loses a net amount of heat to the surroundings. The heat input, Q_{in}, is absorbed by the system from the low-temperature thermal reservoir, so $Q_{in} > 0$; the heat output, Q_{out}, is released by the system into the high-temperature thermal reservoir, so $Q_{out} < 0$. Therefore, conservation of energy requires that $|Q_{out}| > Q_{in}$ in order that $W_{in} < 0$. Note that the first law of thermodynamics does not impose that $W_{in} < 0$, but that this restriction is imposed by the second law of thermodynamics. Since the second law of thermodynamics imposes $W_{in} < 0$, it follows from the first law of thermodynamics that $|Q_{out}| > Q_{in}$. Furthermore, since we know which quantities must be negative, we can write $-|W_{in}| = Q_{in} - |Q_{out}|$, or, equivalently, $|W_{in}| = |Q_{out}| - Q_{in}$, which means that $|W_{in}| < |Q_{out}|$. This latter inequality demands that $\text{COP}_{heating} > 1$ because $\text{COP}_{heating} = Q_{out}/W_{in}$. Also, since Q_{in} can be virtually as large as $|Q_{out}|$, in principle, it follows that $\text{COP}_{cooling}$ may be > 1, since $\text{COP}_{cooling} = -Q_{in}/W_{in}$. In fact, it is common for $\text{COP}_{cooling}$ to exceed 1.

Efficiency and the second law of thermodynamics: The second law of thermodynamics expresses the observation that processes that are observed to occur in reverse have an overall entropy that is constant, while processes that occur only in one direction have an overall entropy that is monotonically increasing as time progresses: That is, the overall entropy must be non-decreasing.

Processes where internal energy would spontaneously be converted into mechanical energy are not observed to occur naturally, while processes where mechanical energy is converted into internal energy do occur spontaneously. Since processes where mechanical energy is converted into internal energy are observed to be irreversible, we conclude that they result in an increase in the overall entropy.

We can drive energy to be transformed in the opposite direction — i.e. from internal energy to mechanical energy — but with a cost. The cost is imposed by the second law of thermodynamics, which requires that by driving the process to occur backwards, the overall entropy must be non-decreasing.

In the case of a heat engine, the cost of converting internal energy to mechanical energy is that the system must expel some exhaust. The production of exhaust limits the efficiency of the heat engine. In this way, the second law of thermodynamics demands that the efficiency of a heat engine be less than 100%.

Coefficient of performance and the second law of thermodynamics: In the case of a heat pump, the cost of ultimately causing heat to flow from low temperature to high temperature is paid by the input of work. Heat will not flow spontaneously from low to high temperature, as this would lead to a decrease in the overall entropy. However, by paying the cost — i.e. by doing work on the system — heat can be made to flow from low to high temperature without a decrease in the overall entropy. In this way, the second law of thermodynamics demands that $W_{in} < 0$, and, as we have seen, this requires that $\text{COP}_{heating} > 1$ and allows for $\text{COP}_{cooling}$ to be > 1 (which is generally the case).

Kelvin-Planck form of the second law of thermodynamics: We have seen that the second law of thermodynamics comes in multiple forms. The notion that it is impossible to construct a heat engine in which the only effect of a cycle is for the system to absorb heat from a thermal reservoir and for the system to perform an equal amount of work is referred to as the Kelvin-Planck form of the second law of thermodynamics. This form of the second law of thermodynamics explains that it is impossible to have zero exhaust. Since the work output, W_{out}, must be less than the heat input, Q_{in}, the efficiency of a heat engine, $e = W_{out}/Q_{in}$, must be less than 100%. The Kelvin-Planck form of the second law of thermodynamics relates directly to the performance of a heat engine.

Clausius form of the second law of thermodynamics: The notion that it is impossible to construct a heat pump in which the only effect of a cycle is for the system to transfer heat from a low-temperature object to a high-temperature object without doing any mechanical work on the system is referred to as the Clausius form of the second law of thermodynamics. This form of the second law of thermodynamics explains that work must be done to drive a process to occur against its natural flow. Since $W_{in} < 0$, it follows that $\text{COP}_{heating} > 1$ and that $\text{COP}_{cooling}$ may be > 1 (and generally is the case). The Clausius form of the second law of thermodynamics relates directly to the performance of a heat pump.

Since a heat pump is a heat engine run backwards, and the same mathematics applies to a heat pump and a heat engine, except for some conceptual minus sign changes, the Kelvin-Planck and Clausius forms of the second law of thermodynamics are equivalent.

Carnot's theorem for isothermal heat exchanges: The relationship between the efficiency of a heat engine and whether or not its cycle is reversible is expressed by the two parts of Carnot's theorem. We presently restrict ourselves to Carnot's theorem for heat engines that employ high- and low-temperature thermal reservoirs to transfer heat to and from the working substance *isothermally*. We will later see that the results and application of Carnot's theorem are different if instead the temperature of the working substance changes as heat is exchanged with the thermal reservoirs. This distinction is very important – students who do not observe it often make critical mistakes when solving for the efficiency of a heat engine.[202]

The first part of Carnot's theorem states that *all* heat engines that operate on a *reversible* cycle between the same two temperatures have the *same* efficiency, provided that the heat exchanges are made isothermally. That is, all reversible heat engines are equally efficient if they operate between the same two thermal reservoirs with only isothermal heat exchanges. The first part of Carnot's theorem ultimately derives from the fact that the overall entropy is constant for reversible processes. The second part of Carnot's theorem states that it is *impossible* for a heat engine that operates on an *irreversible* cycle to have a *greater* efficiency than a heat engine that operates on a *reversible* cycle between the same two temperatures with only isothermal heat exchanges. That is, irreversible heat engines cannot be more efficient than reversible heat engines that make only isothermal heat exchanges if they operate between the same two thermal reservoirs. The second part of Carnot's theorem ultimately derives from the fact that the overall entropy increases for irreversible processes.

---

[202] There is an inclination of students to want to use the formula $e_C = 1 - T_c/T_h$, which is specific to the Carnot cycle, rather than $e = 1 - |Q_{out}|/Q_{in}$, which applies in general, when asked to compute the efficiency of a particular heat engine. The problem with this is that most other cycles do not involve isothermal heat exchanges.

> **Important Distinction.** Carnot's theorem does not state that all reversible heat engines operating between two temperatures have the same efficiency as the Carnot heat engine operating between the same two temperatures.[203] Rather, Carnot's theorem states that all reversible heat engines that exchange heat isothermally with two thermal reservoirs have the same efficiency as the Carnot heat engine using the same two thermal reservoirs. The distinction lies in how heat is exchanged. For example, in the following section, we will learn that the reversible Otto cycle is less efficient than the Carnot cycle for the same temperature range, which is due to the fact that the working substance changes temperature during the heat exchanges in the Otto cycle.

Proof of Carnot's theorem for isothermal heat exchanges: We can prove Carnot's theorem by considering a heat engine that is connected to a heat pump such that the work that is done by the heat engine is fed directly into the heat pump, where the same two thermal reservoirs are used by both the heat pump and the heat engine. We denote by T_c and T_h the temperatures of the low- and high-temperature thermal reservoirs, respectively (i.e. $T_h > T_c$). The heat engine's system absorbs heat, Q_e^{in}, from the high-temperature thermal reservoir and deposits its exhaust, Q_e^{out}, into the low-temperature thermal reservoir, while the heat pump's system absorbs heat, Q_p^{in}, from the low-temperature thermal reservoir and deposits heat, Q_p^{out}, into the high-temperature thermal reservoir. The heat engine outputs work, W_e^{out}, which is fed directly into the heat pump, W_p^{in}; assuming no loss of mechanical energy in this transfer, $W_e^{out} = -W_p^{in}$ or $W_e^{out} = |W_p^{in}|$. The net internal energy change for each system equals zero for one cycle: $\Delta U_e^{net} = \Delta U_p^{net} = 0$. The net work done by the heat engine must then equal the net heat exchanged between its system and the two thermal reservoirs, and similarly for the heat pump, according to the first law of thermodynamics:

$$Q_e^{in} - |Q_e^{out}| = Q_p^{in} - |Q_p^{out}|$$

The efficiency of the heat engine equals

$$e_e = \frac{W_e^{out}}{Q_e^{in}} = \frac{Q_e^{in} - |Q_e^{out}|}{Q_e^{in}}$$

We can imagine running the composite device backwards, turning the heat engine into a heat pump and the heat pump into a heat engine. This actually turns out to be a useful step in the proof of Carnot's theorem. It is therefore useful to apply the usual definition of efficiency to the heat pump, since when we consider running it backwards it will serve as a heat engine, rather than the COP:[204]

---

[203] Actually, sometimes Carnot's theorem is stated very much in these words. In that case, there is implicit understanding that 'operating between two temperatures' specifically refers to the use of two thermal reservoirs in order to exchange heat isothermally with the reservoirs. This wording, if not clarified, leads to confusion when, for example, the student comes across the Otto cycle – since the reversible Otto cycle is less efficient than the Carnot cycle (as we will see in the following section) with the same minimum and maximum temperatures.

[204] By 'usual,' we must interpret that W_e^{out} corresponds to $-W_p^{in}$; also, the heat associated with the higher-temperature thermal reservoir, Q_e^{in}, becomes $-Q_p^{out}$. This way, the 'efficiency' of the heat pump will correspond to the 'usual' definition of efficiency when it is run backwards as a heat engine.

$$e_p = \frac{W_p^{in}}{Q_p^{out}} = \frac{Q_p^{in} - |Q_p^{out}|}{|Q_p^{out}|}$$

Now we have the ingredients with which to construct the proof of Carnot's theorem. We begin by assuming that both the heat engine and the heat pump are reversible, corresponding to the first part of Carnot's theorem. We will prove the first part of Carnot's theorem using the following logic: We will assume that the first part of Carnot's theorem is not true, and then show that our assumption leads to a contradiction, thereby implying that the first part of Carnot's theorem is, in fact, correct.

Let us assume that the reversible heat engine is more efficient than the reversible heat pump:

$$e_e > e_p$$
$$\frac{Q_e^{in} - |Q_e^{out}|}{Q_e^{in}} > \frac{Q_p^{in} - |Q_p^{out}|}{|Q_p^{out}|}$$

Recalling that the first law of thermodynamics requires that $Q_e^{in} - |Q_e^{out}| = Q_p^{in} - |Q_p^{out}|$, the numerators of the inequality cancel:

$$\frac{1}{Q_e^{in}} > \frac{1}{|Q_p^{out}|}$$
$$|Q_p^{out}| > Q_e^{in}$$

Now we reach the point in the proof where the nature of the heat exchanges becomes important. Let us assume that the heat exchanges are made isothermally at the temperatures of the thermal reservoirs. In that case, Q_p^{out} is heat that is deposited by the reversible heat pump into the thermal reservoir operating at the higher temperature, T_h, and Q_e^{in} is heat that is withdrawn by the reversible heat engine from the same thermal reservoir operating at T_h. The inequality, $|Q_p^{out}| > Q_e^{in}$, says that a net amount of heat is being expelled into the higher-temperature thermal reservoir by the composite system. Since the net work of the composite system is zero and the net internal energy change is zero for each cycle, the same amount of net heat is being withdrawn from the lower-temperature thermal reservoir by the composite system. The Clausius form of the second law of thermodynamics states that this effect is impossible, which means that our original assumption, that $e_e > e_p$, leads to a contradiction. This proves that $e_e \leq e_p$.

The proof of the first part of Carnot's theorem can now be completed by actually reversing the composite system, such that the reversible heat engine serves as a reversible heat pump and the original reversible heat pump serves as a reversible heat engine. With e_e designating what was originally a reversible heat engine, but now is a reversible heat pump, and e_p similarly referring to what was originally a reversible heat pump, but now is a reversible heat engine, the exact same proof will show that $e_e \geq e_p$.

When the composite system is run forwards, we find that $e_e \leq e_p$. When it is reversed, we find that $e_e \geq e_p$. The only way that both statements can be true is if $e_e = e_p$. This proves that both the reversible heat engine and the reversible heat pump must be equally efficient. Note that our proof does not make any assumptions about the working substance that is used.

For the second part of the proof of Carnot's theorem, we assume that the heat engine is irreversible, but that the heat pump is reversible. We will again find that $e_e \leq e_p$. However, we will not be able to reverse the composite system, since the heat engine is irreversible. Therefore, the only conclusion we can draw is that the irreversible heat engine cannot be more efficient than the reversible heat pump (where the efficiency of the heat pump refers to the efficiency that it would have when functioning in reverse as a heat engine). This completes the proof of both parts of Carnot's theorem.

Observe that Carnot's theorem is independent of the nature of the working substance. Also, in our proof we assumed that the heat exchanges were made isothermally at the temperatures of the thermal reservoirs. We shall next consider the impact of Carnot's theorem for a heat engine in which the temperature of the working substance differs from the temperatures of the thermal reservoirs during the heat exchanges.

Carnot's theorem and variable-temperature heat exchanges: Let us now address the more general possibility that the cycle of a heat engine involves one or more processes in which the system changes temperature as it exchanges heat with its surroundings. For this more general heat engine, the working substance may absorb heat over a range of temperatures, $T_a < T < T_h$, and similarly may release heat to over a range of temperatures, $T_c < T < T_r$. Here, T_h and T_c are the maximum and minimum temperatures, respectively, of the working substance. These heat exchanges would be more efficient if heat were absorbed exclusively at T_h and released exclusively at T_c. We therefore draw the following conclusion: A heat engine with maximum temperature T_h and minimum temperature T_c cannot be more efficient than a Carnot engine operating between temperatures T_h and T_c. This explains why, for example, a reversible Otto cycle is less efficient than the Carnot cycle.

Application of Carnot's theorem: Carnot's theorem shows that the Carnot heat engine establishes an upper limit on the efficiency of any heat engine. Furthermore, the maximum efficiency depends only upon the temperatures of the two thermal reservoirs, and is independent of the nature of the working substance. A heat engine with maximum temperature T_h and minimum temperature T_c has an efficiency that is limited by:

$$e \leq 1 - \frac{T_c}{T_h}$$

In this way, the Carnot engine is very significant for all heat engines, even if they do not operate on a Carnot cycle.

Real heat engines: Reversible heat engines are very significant theoretically, as they are simpler to model mathematically and provide a practical upper limit to what can be achieved physically. With enough patience, processes can be carried out as quasistatically as desired, but frictional and dissipative forces cannot be completely eliminated. For this reason, all real heat engines inherently operate on irreversible cycles, and all real heat engines are observed to be less efficient than the Carnot cycle – in agreement with Carnot's theorem. For example, for an internal combustion engine with one thermal reservoir at room temperature and the other at which the gasoline burns, 1500 °F, the maximum theoretical efficiency is $e_{max} = 73\%$, while the best real internal combustion engines are limited, in practice, to about 50%, and commonly (in cars, for example) have efficiencies closer to 20%.

Real heat pumps: The COP of a heat pump is limited, in practice, to about 10, and is commonly closer to half this value or less (for refrigerators, for example). Since the COP of a real heat pump generally exceeds 1 by a considerable factor, a real heat pump is very thermally efficient compared to a gas furnace or an electric heater, which deliver less than 100% heat output compared to the (gas or electrical) energy input.

Carnot temperature: For a Carnot heat engine, we have previously derived that the ratio of the heat exchanges between the system and the two thermal reservoirs equals the corresponding ratio of the temperatures of the two thermal reservoirs:[205]

$$\frac{T_c}{T_h} = \frac{|Q_{out}|}{Q_{in}}$$

where T_c and T_h are the temperatures of the low- and high-temperature thermal reservoirs, respectively. This result, along with Carnot's theorem, shows that a heat engine can be utilized as a thermometer for absolute temperature. One of the two thermal reservoirs could be the atmosphere at STP, for which the temperature is known (273.16 K), and the other thermal reservoir could be the object of unknown temperature that we wish to measure. We could then carry out one cycle of a Carnot heat engine. By measuring the input and output heats exchanged between the system and thermal reservoirs along the isothermal processes, the above ratio could then be used to compute the absolute temperature of the desired thermal reservoir.

Efficiency and the third law of thermodynamics: The maximum possible efficiency that any heat engine can hope to attain is the Carnot efficiency,

$$e_{max} = 1 - \frac{T_c}{T_h}$$

where T_c and T_h are the temperatures of the low- and high-temperature thermal reservoirs, respectively. In the limit that the low-temperature thermal reservoir approaches zero, $T_c \to 0$, the Carnot efficiency approaches 100%. The second law of thermodynamics prohibits the efficiency of a heat engine from reaching 100%, which therefore implies that T_c cannot reach its limiting value of zero Kelvin. We thus see the relationship between the second law of thermodynamics, the efficiency of a heat engine, and the third law of thermodynamics.

Coefficient of performance and the third law of thermodynamics: Recall that the COP of a heat pump is defined for cooling and heating functions as:

$$\text{COP}_{cooling} \equiv -\frac{Q_{in}}{W_{in}} = \frac{T_c}{T_h - T_c}$$
$$\text{COP}_{heating} \equiv \frac{Q_{out}}{W_{in}} = \frac{T_h}{T_h - T_c}$$

---

[205] Recall that we derived this in general – without a specific working substance – by considering entropy changes.

in terms of the low, T_c, and high, T_h, temperatures of the two thermal reservoirs. In the limit that T_c approaches zero, $COP_{cooling} \to 0$ and $COP_{heating} \to 1$. However, T_c can never reach absolute zero (but it can get extremely close) according to the third law of thermodynamics, which implies that $COP_{cooling} > 0$ and $COP_{heating} > 1$.

5.5 Practical Heat Engines

Automobile engine: The standard internal combustion engine that powers automobiles is an irreversible heat engine that operates on a four-stroke combustion cycle. In this case, the working substance is air, and the system is the space above the piston (see the diagram below). A mixture of gasoline and air can be drawn into the system through an intake port or expelled from the system through an exhaust port. Rocker arms and springs allow a rotating camshaft to control the opening and closing of the intake and exhaust valves. A sparkplug makes it possible to ignite the mixture of gasoline and air. The raising and lowering of the piston, as work is done on or by the system, respectively, turns a crankshaft, which transfers the work output from the heat engine to the rotating tires of the car. The illustration below shows a single-cylinder gasoline engine, which is typical of lawnmowers, while an automobile engine usually has four, six, or eight cylinders.

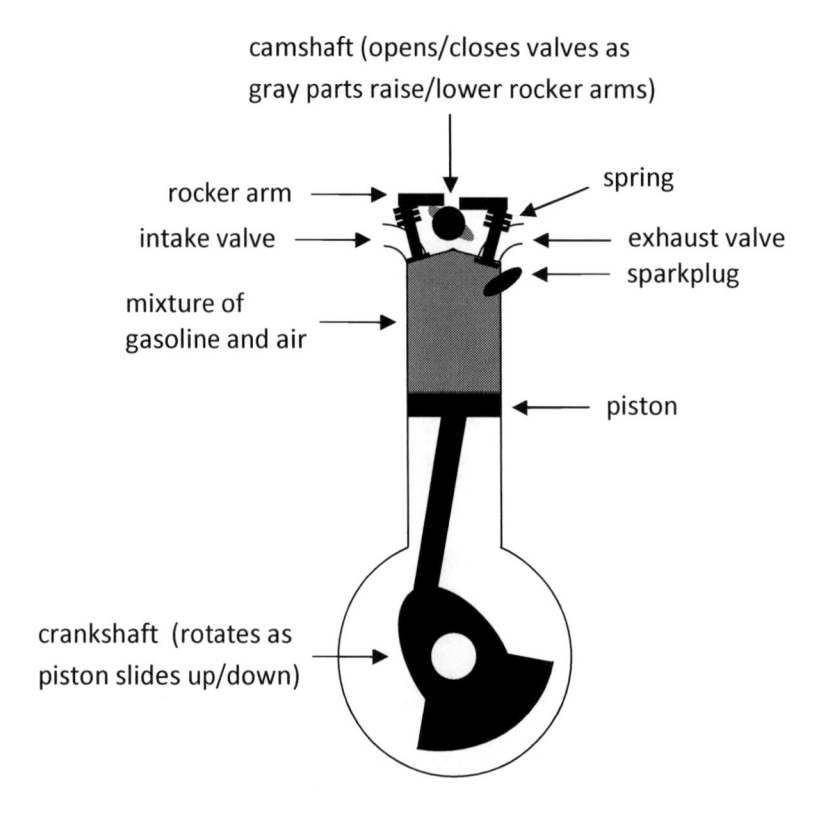

Air-standard Otto cycle: The standard combustion engine approximately follows the processes that comprise the (reversible) air-standard Otto cycle. The air-standard Otto cycle is characterized by four strokes and six processes. An engine stroke refers to a process in which the piston moves more than a negligible amount; two of the processes of the air-standard Otto cycle are isochoric, and therefore are not counted as strokes. The air-standard Otto cycle begins with air as the working substance in the initial equilibrium state (P_0, V_0, T_0):

- The intake valve is opened. A mixture of fuel and air enters the system. The piston is lowered in a reversible isobaric expansion to a new volume, $V_a > V_0$. Energy E_0 is added to the system by mass transfer – both in the form of chemical work (i.e. the work done by increasing the mole numbers of air and gasoline molecules) and the internal energy increase (especially, the potential of the mixture of air and gasoline for internal combustion). The intake valve is closed and the piston stops when the volume equals V_a. The equilibrium state of the air at the end of this process is (P_0, V_a, T_a). This reversible isobaric expansion is called the intake stroke.
- With both valves closed, the system undergoes a reversible adiabatic compression to its original volume, V_0. There are no heat exchanges while work, W_{ab}, is done on the system as it is compressed. The piston stops when the volume equals V_0. The equilibrium state of the air at the end of this process is (P_b, V_0, T_b). The new temperature, T_b, is greater than the previous temperature, T_a, as a result of the adiabatic compression: $T_b > T_a$. This reversible adiabatic compression is called the compression stroke.
- With both valves still closed, the spark plug is fired. A spark is emitted from the spark plug, which ignites the gasoline. Combustion occurs explosively. In a very short time interval, the pressure and temperature of the air increases significantly: $P_c > P_b$ and $T_c > T_b$. The volume is relatively constant during this very brief chemical reaction. In the (reversible) air-standard Otto cycle, we treat this as a reversible isochor. Therefore, no work is done on the system. The internal energy, ΔU_{bc}, of the air increases. Thus, the first law of thermodynamics implies that the air receives heat, Q_{bc}, from the gasoline during the process of combustion.[206] The equilibrium state of the air at the end of this process is (P_c, V_0, T_c). This process is called combustion, but is not counted as an engine stroke as the piston scarcely moves during this very brief process.
- With both valves still closed, the system undergoes a reversible adiabatic expansion to its previous volume, V_a. There are no heat exchanges while work, W_{cd}, is done by the system as it expands. The piston stops when the volume equals V_a. The equilibrium state of the air at the end of this process is (P_d, V_a, T_d). The new temperature, T_d, is lower than the previous temperature, T_c, as a result of the adiabatic expansion: $T_d < T_c$. This reversible adiabatic expansion is called the power stroke.

---

[206] Notice that the air is treated as the working substance, and so the gasoline is part of the air's surroundings, even though the gasoline and air are mixed together in the same region. The gasoline thus plays the role of the high-temperature thermal reservoir. Compare to the steam engine, where there is a separate high-temperature thermal reservoir. The gasoline has a great deal of internal energy in its chemical bonds, which is transferred to internal energy of the air molecules (namely, in the form of kinetic energy) during the process of combustion. We treat this decrease in the internal energy of the gasoline molecules as a transfer of heat to the air molecules, which increases the internal energy of the air molecules. This serves as the heat input, Q_{in}, for the Otto cycle.

- The exhaust valve is opened, connecting the system to the low-temperature thermal reservoir (the ambient atmosphere). In a very short timer interval, the pressure and temperature of the air decrease back to the values P_0 and T_a. The volume of the system is relatively constant during this very brief process. In the (reversible) air-standard Otto cycle, we treat this as a reversible isochor. Therefore, no work is done on the system. The internal energy, ΔU_{da}, of the air decreases as heat, Q_{da}, is transferred from the air to the low-temperature thermal reservoir (the atmosphere). The equilibrium state of the air at the end of this process is (P_0, V_a, T_a). This decompression is not counted as an engine stroke as the piston scarcely moves during this very brief process.
- While the exhaust valve remains open, the piston is raised in a reversible isobaric contraction to its original volume, V_0. Energy E_0 is removed to the system by mass transfer as gaseous residue is transferred to the atmosphere (as exhaust). The exhaust valve is closed and the piston stops when the volume equals V_0. The equilibrium state of the air at the end of this process is (P_0, V_0, T_0). Note that the initial and final pressure P_0 equals atmospheric pressure. This reversible isobaric compression is called the exhaust stroke.

State	Process	Work done by system	Heat absorbed by system
(P_0, V_0, T_0)			
	isobaric expansion	pre-cycle energy exchanges	
(P_0, V_a, T_a)			
	adiabatic compression	$W_{ab} < 0$	$Q_{ab} = 0$
(P_b, V_0, T_b)			
	isochoric pressurization	$W_{bc} = 0$	$Q_{bc} > 0$
(P_c, V_0, T_c)			
	adiabatic expansion	$W_{cd} > 0$	$Q_{cd} = 0$
(P_d, V_a, T_d)			
	isochoric depressurization	$W_{da} = 0$	$Q_{da} < 0$
(P_0, V_a, T_a)			
	isobaric contraction	post-cycle energy exchanges	
(P_0, V_0, T_0)			

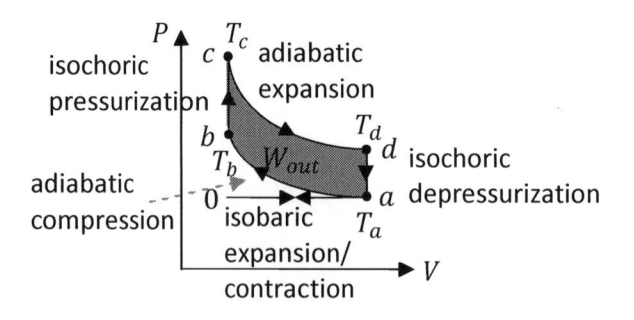

The second thru fifth steps of the air-standard Otto cycle enclose area on a P-V diagram. The last step retraces the same path as the first step in the opposite direction; the exhaust stroke effectively cancels the intake stroke. Steps two thru five comprise a complete cycle by themselves. For this reason, mathematically we need only treat these four steps (which do not correspond to the four strokes).[207]

Real air-standard Otto-like engines: The air-standard Otto engine involves a reversible cycle that models the internal combustion engine. A real internal combustion engine burns gasoline during the isochoric pressurization, in which the mole numbers of the gas change – which is a discrepancy between the simplistic Otto cycle and real combustion. A real internal combustion engine also has isochors that occur over a very short duration, whereas the Otto cycle's reversible isochors occur quasistatically. The adiabats of a real combustion engine also are noticeably not quasistatic. As a result, real internal combustion engines may have a thermal efficiency that is one-half or less that of the reversible air-standard Otto engine. Nonetheless, this is a fair sacrifice for the power that is delivered by real internal combustion engines compared to what would result if they were to operate much more quasistatically.

Efficiency of the air-standard Otto engine: No heat is exchanged along the adiabats: $Q_{ab} = Q_{cd} = 0$. The heat input and output thus equal the heat absorbed and released by the working substance during the isochors, which can be expressed in terms of the molar specific heat capacity at constant volume (since volume is constant along an isochor): $Q_{in} = Q_{bc} = nc_V(T_c - T_b) > 0$ and $Q_{out} = Q_{da} = nc_V(T_a - T_d) < 0$. The efficiency of the air-standard Otto engine therefore equals

$$e_O = 1 - \frac{|Q_{out}|}{Q_{in}} = 1 - \frac{T_d - T_a}{T_c - T_b}$$

The high and low temperatures of the cycle are T_c and T_a, respectively. The maximum possible efficiency according to Carnot's theorem is then

$$e_{max} = 1 - \frac{T_a}{T_c}$$

The air-standard Otto engine is less efficient than this: $e_O < 1 - T_a/T_c$. We will see this in a coming example for the case where the working substance is treated as an ideal gas.

Compression ratio: For an internal or external combustion engine, the ratio of the maximum volume, V_{max}, to the minimum volume, V_{min}, is called the compression ratio, r.[208] Conceptually, this ratio, $r = V_{max}/V_{min}$, expresses by what factor the working substance's volume is compressed. An internal or external combustion engine that has a higher compression ratio tends to be more thermally efficient because the oxygen molecules in the air and fuel molecules mix and react better in a tinier space.

---

[207] The pressures at points b and d – P_b and P_d, respectively – generally differ, even though they might look close in the sample P-V diagram for the air-standard Otto cycle on the previous page.
[208] The word 'compression' means to reduce the volume of the working substance through applied pressure – so compression generally implies both an increase in pressure and a reduction in volume. The compression ratio is a ratio of volumes, but not a ratio of pressures – there is another quantity called pressure ratio for this.

Number of engine cylinders: The net power obtained from an engine equals the number of engine cylinders, N_c, times the power delivered by each cylinder: $P_{net} = N_c P_c$. Thus, a 6-cylinder delivers 50% more power than a 4-cylinder engine.

Displacement volume: For an internal or external combustion engine, the difference between the maximum volume, V_{max}, and the minimum volume, V_{min}, is called the displacement volume, V_d. Conceptually, this difference, $V_d = V_{max} - V_{min}$, expresses how much volume the piston displaces.

Engine displacement: The engine displacement, V_e, equals the number of engine cylinders times the displacement volume of each cylinder: $V_e = N_c V_d$. The engine capacity refers to the engine displacement. For example, a 3-L V6 engine means that the engine displacement is $V_e = 3$ liters, which is spread out over 6 cylinders. The displacement volume of each cylinder is therefore $V_d = 0.5$ liters.

Cutoff ratio: For an internal or external combustion engine, the cutoff ratio, r_c, expresses how much the volume changes during the combustion process: $r_c = V_c/V_{c0}$, where V_{c0} and V_c are the initial and final volumes of the combustion reaction, respectively. The cutoff ratio, r_c, equals zero for the Otto cycle, but is significant for diesel engines.

Engine power: Since power is the instantaneous rate at which work is done, $P = dW/dt$, the power delivered by each cylinder of an internal or external combustion engine equals the net work output of one cycle times the frequency (which is one over the period) divided by 2 (since the crankshaft must complete two revolutions in order to deliver the four engine strokes): $P_{net} = N_c W_{out} f/2$. The frequency, f, is generally expressed in rpm (rotations per minute) for an engine, but must be converted to Hz (a Hertz being a cycle per second) – along with work expressed in Joules – in order to obtain power in Watts. A Watt can be converted to horsepower, $746 \text{ W} = 1.00 \text{ hp}$ – a more typical unit for engines. The symbol, P, represents both power and pressure, but the distinction should be clear from the context.

Engine torque: Ultimately, the engine provides torque to the wheels, and the torque and angular displacement of the wheels results in work done by the car. For linear motion, work done by a force that contributes toward the displacement of an object is given by the line integral: $W = \int_i^f \vec{F} \cdot d\vec{s}$. Since torque is the rotational analog for force and angular displacement is the rotational equivalent of displacement, the work done by a torque, τ, equals $W = \int_{\theta=\theta_0}^{\theta} \tau d\theta$. For a constant torque, $W = \tau\theta$. The net torque provided is related to the work output by $W_{out} = 4\pi \tau_{out}$. Recalling that the crankshaft completes two revolutions in order to deliver the four engine strokes, $\theta = 4\pi$.

Example. Derive an equation for the efficiency of the air-standard Otto engine in terms of the compression ratio assuming the working substance to be an ideal gas.

We previously derived an equation for the efficiency of the air-standard Otto engine in terms of the temperatures:

$$e_O = 1 - \frac{|Q_{out}|}{Q_{in}} = 1 - \frac{T_d - T_a}{T_c - T_b}$$

We need to express these temperatures in terms of the volumes. The equation for an ideal gas along an adiabat, $P_0 V_0^\gamma = P V^\gamma$, can be expressed in terms of temperature using the ideal gas law, $\frac{P_0 V_0}{T_0} = \frac{PV}{T}$:

$$P_0 V_a^\gamma = P_b V_0^\gamma \quad \Rightarrow \quad P_0 V_a^\gamma = \left(\frac{P_0 V_a T_b}{V_0 T_a}\right) V_0^\gamma \quad \Rightarrow \quad T_a V_a^{\gamma-1} = T_b V_0^{\gamma-1}$$

$$P_c V_0^\gamma = P_d V_a^\gamma \quad \Rightarrow \quad T_c V_0^{\gamma-1} = T_d V_a^{\gamma-1}$$

where $V_b = V_c = V_0$, $V_d = V_a$, and $P_a = P_0$. We can now express the efficiency in terms of the volumes:

$$e_O = 1 - \frac{\frac{T_c V_0^{\gamma-1}}{V_a^{\gamma-1}} - T_a}{T_c - \frac{T_a V_a^{\gamma-1}}{V_0^{\gamma-1}}} = 1 - \left(\frac{T_c V_0^{\gamma-1} - T_a V_a^{\gamma-1}}{T_c V_0^{\gamma-1} - T_a V_a^{\gamma-1}}\right)\frac{V_0^{\gamma-1}}{V_a^{\gamma-1}}$$

$$e_O = 1 - \frac{V_0^{\gamma-1}}{V_a^{\gamma-1}} = 1 - \left(\frac{1}{r}\right)^{\gamma-1}$$

where the compression ratio equals $r = V_a/V_0 > 1$. The air-standard Otto engine is more efficient for a higher compression ratio. For example, treating air as a diatomic ideal gas (which is appropriate, since air consists primarily of N_2 and O_2), for which $\gamma = 1.4$, a compression ratio of 5 provides an efficiency of 47%, whereas a compression ratio of 8 provides an efficiency of 56%.

Example. Treating the air as an ideal gas, show that the air-standard Otto engine is less efficient than a Carnot engine operating between the same minimum and maximum temperatures.

In the previous example, we found that the equations for the adiabats could be expressed as

$$T_a V_a^{\gamma-1} = T_b V_0^{\gamma-1}$$
$$T_c V_0^{\gamma-1} = T_d V_a^{\gamma-1}$$

These equations can be used to express the compression ratio as a ratio of temperatures:

$$\left(\frac{1}{r}\right)^{\gamma-1} = \frac{V_0^{\gamma-1}}{V_a^{\gamma-1}} = \frac{T_a}{T_b} = \frac{T_d}{T_c}$$

For an ideal gas, the efficiency of the air-standard Otto engine can alternately be written as

$$e_O = 1 - \frac{T_a}{T_b} = 1 - \frac{T_d}{T_c}$$

The lowest temperature in the air-standard Otto cycle is T_a and the highest temperature is T_c, so a Carnot engine operating between T_a and T_c would have an efficiency equal to

$$e_C = 1 - \frac{T_a}{T_c}$$

Since T_c is the highest temperature in the cycle, $T_c > T_b$, from which it follows that

$$\frac{T_a}{T_c} < \frac{T_a}{T_b}$$
$$1 - \frac{T_a}{T_b} < 1 - \frac{T_a}{T_c}$$
$$e_O < e_C$$

Diesel engine: The internal combustion engine that we have considered thus far is typical of a gasoline engine. We now consider a diesel engine, which is somewhat different. Like the gasoline engine, the diesel engine operates on a four-stroke combustion cycle, and the working substance is air. However, unlike the gasoline engine, fuel is not added to the diesel engine until after the compression stroke is completed. At the end of the compression stroke, fuel is injected into the system. No sparkplug is needed, as the mixture of fuel and air is at such a high temperature that combustion is spontaneous. This high temperature (ignition temperature) is achieved through a high compression ratio during the compression stroke. The high compression ratio and combustion temperature of the diesel engine provides for a higher efficiency compared to the gasoline engine. The diesel cycle is fundamentally different from the Otto cycle because the combustion process for a diesel engine is approximately isobaric, whereas the combustion process for the Otto engine is approximately isochoric.

Air-standard diesel cycle: Real diesel engines approximately follow the (reversible) air-standard diesel cycle. The air-standard diesel engine differs from the air-standard Otto engine in the following ways:
- Only air – not a combination of fuel and air – is drawn into the system during the intake stroke.
- The compression stroke occurs without fuel, and generally leads to a higher compression ratio.
- Fuel is injected into the system at the end of the compression stroke using a fuel injector.
- The high temperature achieved during the compression stroke provides for spontaneous combustion, so the air-standard diesel engine does not use a sparkplug.
- The combustion cycle is approximated by a reversible isobaric expansion – rather than an isochoric pressurization.

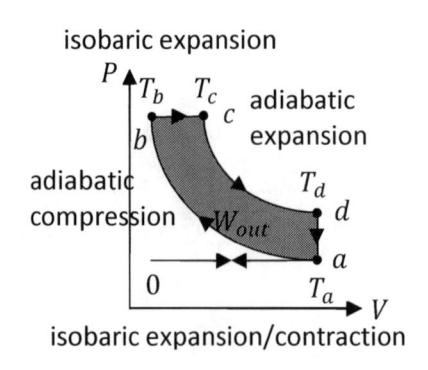

State	Process	Work done by system	Heat absorbed by system
(P_0, V_0, T_0)			
	isobaric expansion	pre-cycle energy exchanges	
(P_0, V_a, T_a)			
	adiabatic compression	$W_{ab} < 0$	$Q_{ab} = 0$
(P_b, V_0, T_b)			
	isobaric expansion	$W_{bc} > 0$	$Q_{bc} > 0$
(P_b, V_c, T_c)			
	adiabatic expansion	$W_{cd} > 0$	$Q_{cd} = 0$
(P_d, V_a, T_d)			
	isochoric depressurization	$W_{da} = 0$	$Q_{da} < 0$
(P_0, V_a, T_a)			
	isobaric contraction	post-cycle energy exchanges	
(P_0, V_0, T_0)			

Efficiency of the air-standard diesel engine: No heat is exchanged along the adiabats: $Q_{ab} = Q_{cd} = 0$. The heat input and output thus equal the heat absorbed and released by the working substance during the isobar and isochor, which can be expressed in terms of the molar specific heat capacity at constant pressure and volume, respectively: $Q_{in} = Q_{bc} = nc_P(T_c - T_b) > 0$ and $Q_{out} = Q_{da} = nc_V(T_a - T_d) < 0$. The efficiency of the air-standard diesel engine therefore equals

$$e_d = 1 - \frac{|Q_{out}|}{Q_{in}} = 1 - \frac{c_V}{c_P}\left(\frac{T_d - T_a}{T_c - T_b}\right) = 1 - \frac{1}{\gamma}\left(\frac{T_d - T_a}{T_c - T_b}\right)$$

since the adiabatic index equals $\gamma = c_P/c_V$. The air-standard diesel engine is generally more efficient than the air-standard Otto engine because, in practice, it generally features a much higher compression ratio and, as a consequence, a much higher combustion temperature.

Example. Derive an equation for the efficiency of the air-standard diesel engine in terms of the compression ratio, r, and cutoff ratio, r_c, assuming the working substance to be an ideal gas.

We just derived an equation for the efficiency of the air-standard diesel engine in terms of the temperatures:

$$e_d = 1 - \frac{|Q_{out}|}{Q_{in}} = 1 - \frac{1}{\gamma}\left(\frac{T_d - T_a}{T_c - T_b}\right)$$

Similar to the air-standard Otto cycle, the equation for an ideal gas along an adiabat, $P_0V_0^\gamma = PV^\gamma$, can be combined with the ideal gas law, $\frac{P_0V_0}{T_0} = \frac{PV}{T}$, in order to relate the temperatures to the volumes for the two adiabatic processes:

5 Heat Engines and Refrigerators

$$P_0 V_a^\gamma = P_b V_0^\gamma \implies P_0 V_a^\gamma = \left(\frac{P_0 V_a T_b}{V_0 T_a}\right) V_0^\gamma \implies T_a V_a^{\gamma-1} = T_b V_0^{\gamma-1}$$

$$P_b V_c^\gamma = P_d V_a^\gamma \implies T_c V_c^{\gamma-1} = T_d V_a^{\gamma-1}$$

where $V_b = V_0$, $V_d = V_a$, and $P_c = P_b$. Additionally, the ideal gas law, $\frac{P_0 V_0}{T_0} = \frac{PV}{T}$, can be combined with the definition of an isobar, $P_0 = P$, in order to relate temperature and volume for the isobar:

$$\frac{V_0}{T_b} = \frac{V_c}{T_c}$$

We now substitute this relations into the efficiency equation:

$$e_d = 1 - \frac{1}{\gamma}\left(\frac{\frac{T_c V_c^{\gamma-1}}{V_a^{\gamma-1}} - \frac{T_b V_0^{\gamma-1}}{V_a^{\gamma-1}}}{T_c - T_b}\right) = 1 - \frac{1}{\gamma}\left(\frac{\frac{V_c T_b V_c^{\gamma-1}}{V_0 V_a^{\gamma-1}} - \frac{T_b V_0^{\gamma-1}}{V_a^{\gamma-1}}}{\frac{V_c T_b}{V_0} - T_b}\right) = 1 - \frac{1}{\gamma}\left[\frac{r_c \left(\frac{r_c}{r}\right)^{\gamma-1} - \left(\frac{1}{r}\right)^{\gamma-1}}{r_c - 1}\right]$$

$$e_d = 1 - \frac{1}{\gamma}\left(\frac{1}{r}\right)^{\gamma-1}\left[\frac{r_c^\gamma - 1}{r_c - 1}\right]$$

where the compression ratio equals $r = V_a/V_0$, the cutoff ratio equals $r_c = V_c/V_0$, $r_c/r = V_c/V_a$, and $1 < r_c < r$.

It is interesting to note that the air-standard diesel engine is less efficient than the air-standard Otto engine for the *same* compression ratio. For example, treating air as a diatomic ideal gas with $\gamma = 1.4$, a compression ratio of 8 provides an air-standard Otto engine efficiency of 56%, while an analogous air-standard diesel engine with a cutoff ratio of 2 has an efficiency of 49%. However, the air-standard diesel engine typically features a much higher compression ratio than the air-standard Otto engine, which makes it more efficient overall. Treating air as a diatomic ideal gas, an air-standard diesel engine with a more typical compression ratio of 22 along with a cutoff ratio of 2 gives an efficiency of 66%, which is indeed higher than an air-standard Otto engine with a typical compression ratio of 8.

Gas turbine engine: The design of a gas turbine engine – which is used to power jets and ships, for example – is considerably different from the gasoline and diesel engines that we have considered thus far. The three main components of a gas turbine engine are a compressor, a combustion chamber, and an expansion turbine. The compressor is employed to compress a mixture of fuel and air adiabatically (like the gasoline and diesel engines). In the combustion chamber, the fuel is burned, heating the air. The combustion reaction occurs approximately isobarically (like the diesel engine, but unlike the gasoline engine). The expansion turbine provides an adiabatic expansion (again similar to the gasoline and diesel engines). The last step in the main cycle of the gas turbine engine involves an isobaric contraction (cf. the gasoline and diesel engines, which include an isochoric decompression). The processes of the gas turbine engine are closely approximated by the (reversible) Brayton cycle, where the adiabats of the real gas turbine engine are replaced with isentropes (reversible adiabats like those of the Carnot, Otto, and diesel cycles).

Brayton cycle: The gas turbine engine is modeled by the (reversible) Brayton cycle (aka Joule cycle), which involves two isentropes (reversible adiabats – like the other cycles that we have considered thus far) and two isobars. The Brayton cycle begins with air as the working substance in the initial equilibrium state (P_a, V_a, T_a), already mixed with fuel:[209]

- Using the compressor, the system undergoes an isentropic compression to a new volume, $V_b < V_a$. There are no heat exchanges while work, W_{ab}, is done on the system as it is compressed. The equilibrium state of the air at the end of this process is (P_b, V_b, T_b). The new temperature, T_b, is greater than the previous temperature, T_a, as a result of the adiabatic compression: $T_b > T_a$.
- Fuel is burned in the combustion chamber, adding heat, Q_{bc}, to the air. This combustion is treated as an isobaric expansion. The final volume of the isobaric expansion is $V_c > V_b$. The system does work, W_{bc}, during the expansion, and the internal energy, ΔU_{bc}, of the air increases. The equilibrium state of the air at the end of this process is (P_b, V_c, T_c). The new temperature, T_c, is greater than the previous temperature, T_b, as a result of the isobaric expansion: $T_c > T_b$.
- Using the expansion turbine, the system undergoes an isentropic expansion to its original pressure, P_a, and a new volume, $V_d > V_c$. There are no heat exchanges while work, W_{cd}, is done by the system as it expands. The equilibrium state of the air at the end of this process is (P_a, V_d, T_d). The new temperature, T_d, is smaller than the previous temperature, T_c, as a result of the adiabatic expansion: $T_d < T_c$.
- Gaseous residue is transferred to the atmosphere (as exhaust), and a fresh supply of fuel and air is gathered. The net effect is an isobaric contraction to the original equilibrium state (P_a, V_a, T_a). The system undergoes a net heat loss, Q_{da}, during this process. Work, W_{da}, is done on the system during the contraction and the internal energy, ΔU_{da}, of the air decreases until it is returned to its initial value.

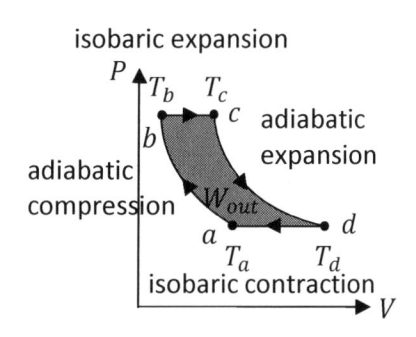

---

[209] The last step of the Brayton cycle includes the intake of new air and fuel, such that both air and fuel are already present in the first step. Compare this treatment of the intake and exhaust steps with the Otto and diesel cycles. The reason for the difference is that the isobaric processes associated with the intake and exhaust strokes occur outside of the closed cycle formed by the four middle processes of the Otto and diesel cycles, while they are inherently part of the Brayton cycle. Also, the Brayton cycle does not include the isochoric decompression of the Otto and diesel cycles, and so the intake and exhaust of the Brayton cycle are not complete reverses of one another in the P-V diagram of the Brayton cycle (like they are in the Otto and diesel cycles).

5 Heat Engines and Refrigerators

State	Process	Work done by system	Heat absorbed by system
(P_a, V_a, T_a)			
	isentropic compression	$W_{ab} < 0$	$Q_{ab} = 0$
(P_b, V_b, T_b)			
	isobaric expansion	$W_{bc} > 0$	$Q_{bc} > 0$
(P_b, V_c, T_c)			
	isentropic expansion	$W_{cd} > 0$	$Q_{cd} = 0$
(P_a, V_d, T_d)			
	isobaric contraction	$W_{da} < 0$	$Q_{da} < 0$
(P_a, V_a, T_a)			

Efficiency of the Brayton engine: No heat is exchanged along the adiabats: $Q_{ab} = Q_{cd} = 0$. The heat input and output thus equal the heat absorbed and released by the working substance during the isobars, which can be expressed in terms of the molar specific heat capacity at constant pressure (since pressure is constant for an isobar): $Q_{in} = Q_{bc} = nc_P(T_c - T_b) > 0$ and $Q_{out} = Q_{da} = nc_P(T_a - T_d) < 0$. The efficiency of the air-standard diesel engine therefore equals

$$e_B = 1 - \frac{|Q_{out}|}{Q_{in}} = 1 - \frac{T_d - T_a}{T_c - T_b}$$

Pressure ratio: The ratio of the maximum pressure, P_{max}, to the minimum pressure, P_{min}, is called the pressure ratio, r_P. Conceptually, this ratio, $r_P = P_{max}/P_{min}$, expresses by what factor the working substance's pressure is enhanced. A gas turbine engine that has a higher pressure ratio tends to be more thermally efficient.

Example. Derive an equation for the efficiency of the Brayton engine in terms of the pressure ratio assuming the working substance to be an ideal gas.

We previously derived an equation for the efficiency of the Brayton engine in terms of the temperatures:

$$e_O = 1 - \frac{|Q_{out}|}{Q_{in}} = 1 - \frac{T_d - T_a}{T_c - T_b}$$

We need to express these temperatures in terms of pressures. The equation for an ideal gas along an adiabat, $P_0 V_0^\gamma = PV^\gamma$, can be expressed in terms of temperature using the ideal gas law, $\frac{P_0 V_0}{T_0} = \frac{PV}{T}$:

$$P_a V_a^\gamma = P_b V_b^\gamma \implies P_a V_a^\gamma = P_b \left(\frac{P_a V_a T_b}{P_b T_a}\right)^\gamma \implies P_a^{1-\gamma} T_a^\gamma = P_b^{1-\gamma} T_b^\gamma$$

$$P_b V_c^\gamma = P_a V_d^\gamma \implies P_b^{1-\gamma} T_c^\gamma = P_a^{1-\gamma} T_d^\gamma$$

where $P_c = P_b$ and $P_d = P_a$. We can now express the efficiency in terms of the pressures:

$$e_B = 1 - \frac{T_c \left(\frac{P_b}{P_a}\right)^{\frac{1-\gamma}{\gamma}} - T_a}{T_c - T_a \left(\frac{P_a}{P_b}\right)^{\frac{1-\gamma}{\gamma}}} = 1 - \left[\frac{T_c(P_b)^{\frac{1-\gamma}{\gamma}} - T_a(P_a)^{\frac{1-\gamma}{\gamma}}}{T_c(P_b)^{\frac{1-\gamma}{\gamma}} - T_a(P_a)^{\frac{1-\gamma}{\gamma}}}\right]\left(\frac{P_b}{P_a}\right)^{\frac{1-\gamma}{\gamma}}$$

$$e_B = 1 - \left(\frac{P_b}{P_a}\right)^{\frac{1-\gamma}{\gamma}} = 1 - \left(\frac{P_a}{P_b}\right)^{\frac{\gamma-1}{\gamma}} = 1 - \left(\frac{1}{r_P}\right)^{\frac{\gamma-1}{\gamma}}$$

where the pressure ratio equals $r_P = P_b/P_a > 1$. The Brayton engine is more efficient for a higher pressure ratio. For example, treating air as a diatomic ideal gas with $\gamma = 1.4$, a pressure ratio of 20 provides an efficiency of 58%, whereas a pressure ratio of 40 provides an efficiency of 65%.

Endoreversible engines: The thermal efficiency of a heat engine provides a quantitative measure of what fraction of the heat input, Q_{in}, is converted into mechanical work output, W_{out}. A heat engine with a relatively low thermal efficiency wastes a large fraction of the heat input, Q_{in} – i.e. it produces more exhaust, $|Q_{out}|$, and less work output, W_{out}. A real heat engine will have a higher thermal efficiency – wasting less of its heat input, Q_{in}, which means producing less exhaust, $|Q_{out}|$, and more work output, W_{out} – if its cycle is more approximately reversible. Making a more thermally efficient heat engine thus means carrying out the processes very slowly – i.e. quasistatically – and reducing frictional and dissipative effects as much as possible.

However, while carrying out the processes very slowly makes the cycle of a real heat engine more approximately quasistatic and therefore more approximately reversible – and therefore makes a more efficient real heat engine – a heat engine that operates on a very slow time scale is not the most practical device. If you are using an internal combustion engine to power an automobile, for example, you don't want the mechanical work output to be performed so slowly that the wheels turn with very little angular acceleration. Thermal efficiency only reveals what fraction of the input heat is converted into mechanical work, but does not provide information about the rate at which the work is delivered.

Recall that power is the instantaneous rate at which the mechanical work is performed: $P_{out} = dW_{out}/dt$. The more rapidly the mechanical work is performed, the greater the power output of the heat engine. The problem with a quasistatic process is that it must be carried out slowly, which could be a long time to wait in order to obtain a desirable power output. From a practical perspective, it is desirable to sacrifice some thermal efficiency in order to significantly enhance power output.[210]

The endoreversible engine addresses the problem of significantly enhancing power output, with a modest sacrifice to thermal efficiency. Processes that involve a heat exchange between the system and one of the thermal reservoirs require significantly more time in order to be made approximately quasistatic compared to adiabatic processes. This can be understood as follows. Adiabatic processes need only last a reasonably long time compared to the relaxation times characteristic of the working substance, which tend to very short. In comparison, the large difference in temperature between the working substance and either of the thermal reservoirs requires a significant amount of time in order to allow only small deviations from thermal equilibrium.

---

[210] The design of a real heat engine should take several factors into account, including power output, thermal efficiency, cost, complexity of design, size, availability of resources, safety, health, and environmental effects.

From a practical point of view, then, we only really want to speed up processes that are not adiabatic. It is desirable to allow non-adiabatic processes to be noticeably irreversible in order to increase power output, and it is also desirable to include two adiabats in the cycle – since they can be reasonably reversible with negligible effect on the power output so as not to sacrifice thermal efficiency more than necessary. The endoreversible heat engine is designed to significantly improve power output through irreversible heat exchanges, with minimal sacrifice to the thermal efficiency by maintaining reversible adiabats. Specifically, a heat engine is endoreversible if it consists of two reversible adiabats and two irreversible processes where heat is exchanged between the system and two thermal reservoirs.

Endoreversible Carnot-like cycle: The power output of the Carnot engine is virtually zero because all of the processes are carried out quasistatically. The power output is significantly enhanced in the endoreversible Carnot-like engine, where the isothermal processes are carried out much more rapidly (and hence irreversibly). In the Carnot engine, the isothermal heat exchanges are carried out quasistatically through an infinitesimal temperature difference between the working substance and the thermal reservoir with which it is exchanging heat. The endoreversible Carnot-like engine produces much quicker heat exchanges by maintaining a finite temperature difference between the working substance and the thermal reservoir with which it is exchanging heat. In particular, the working substance has a constant temperature of T_{in} during the irreversible isothermal expansion, in which the system absorbs heat, Q_{in}, from the high-temperature thermal reservoir, which has a temperature of T_h, and the working substance has a constant temperature of T_{out} during the irreversible isothermal compression, in which the system releases heat, Q_{out}, to the low-temperature thermal reservoir, which has a temperature of T_c, where $T_c < T_{out} < T_{in} < T_h$.

The endoreversible Carnot-like engine begins and ends in the same equilibrium state, (P_a, V_a, T_{in}), and consists of the following steps:

- The system undergoes an irreversible isothermal expansion to a new volume, $V_b > V_a$. Heat, Q_{ab}, flows from the high-temperature thermal reservoir at temperature T_h to the system, which has a constant temperature of T_{in}. The system does work, W_{ab}, as the system expands. The equilibrium state at the completion of the irreversible isothermal expansion is (P_b, V_b, T_{in}).
- The system undergoes a reversible adiabatic expansion to a new volume, $V_c > V_b$. There are no heat exchanges. The system does work, W_{bc}, as the system expands. The equilibrium state at the completion of the reversible adiabatic expansion is (P_c, V_c, T_{out}). The temperature changes from T_{in} to $T_{out} < T_{in}$.
- The system undergoes an irreversible isothermal compression to a new volume, $V_d < V_c$. Heat (exhaust), Q_{cd}, flows from the system, which has a constant temperature of T_{out}, to the low-temperature thermal reservoir at temperature T_c. Work, W_{cd}, is done on the system as it is compressed. The equilibrium state at the completion of the irreversible isothermal compression is (P_d, V_d, T_{out}).
- The system undergoes a reversible adiabatic compression to its original volume, V_a. There are no heat exchanges. Work, W_{da}, is done on the system as it is compressed. The system is returned to its original equilibrium state, (P_a, V_a, T_{in}), at the completion of the reversible adiabatic compression.

Power delivered by the endoreversible Carnot-like engine: In Sec. 2.3, in an example we derived a formula for the steady-state rate of heat transfer by thermal conduction across the thickness L of a thermal conductor with uniform cross-sectional area A and uniform thermal conductivity k where the two end faces have different temperature and the in-between surfaces are thermally insulated:

$$P = kA\frac{\Delta T}{L}$$

It is convenient to define a thermal conductance, σ, as $\sigma \equiv kA/L$, such that the steady-state rate of heat transfer can be expressed compactly as

$$P = \sigma \Delta T$$

The rate of heat transfer is constant – i.e. $P = đQ/dt = Q/\Delta t$, where Δt is the time interval during wheat heat, Q, is transferred – in the steady state:

$$\frac{Q}{\Delta t} = \sigma \Delta T$$

In the endoreversible Carnot-like engine, during the isothermal expansion, heat $Q_{ab} > 0$ flows from the high-temperature thermal reservoir at temperature T_h to the working substance at temperature T_{in} for a time interval Δt_h across a conducting wall with thermal conductivity σ_h:

$$Q_{ab} = \sigma_h \Delta t_h (T_h - T_{in})$$

Similarly, during the isothermal compression, heat $Q_{cd} < 0$ flows from the working substance at temperature T_{out} to the low-temperature thermal reservoir at temperature T_c for a time interval Δt_c across a conducting wall with thermal conductivity σ_c:

$$Q_{cd} = \sigma_c \Delta t_c (T_c - T_{out})$$

According to the first law of thermodynamics, the net heat, $Q_{ab} + Q_{cd} = Q_{ab} - |Q_{cd}|$, exchanged between the system and the two thermal reservoirs equals the net work done, W_{out}, by the system (since the internal energy change is zero for the cycle):

$$Q_{ab} + Q_{cd} = W_{out}$$
$$W_{out} = \sigma_h \Delta t_h (T_h - T_{in}) + \sigma_c \Delta t_c (T_c - T_{out})$$

Neglecting the time of the reversible adiabatic processes compared to the irreversible isothermal heat exchanges – since the adiabatic processes need only have a duration that is long compared to the relatively small relaxation times characteristic of the system in order to be reasonably quasistatic – the net power delivered by the system equals:

$$P_{net} = \frac{W_{out}}{\Delta t_h + \Delta t_c} = \frac{\sigma_h \Delta t_h (T_h - T_{in}) + \sigma_c \Delta t_c (T_c - T_{out})}{\Delta t_h + \Delta t_c}$$

$$P_{net} = \frac{\sigma_h (T_h - T_{in}) + \sigma_c \frac{\Delta t_c}{\Delta t_h}(T_c - T_{out})}{1 + \frac{\Delta t_c}{\Delta t_h}}$$

The ratio of the time intervals can be expressed in terms of the ratio of the heat exchanges by dividing the two heat exchange equations:

$$\frac{\Delta t_c}{\Delta t_h} = \frac{Q_{cd}\sigma_h(T_h - T_{in})}{Q_{ab}\sigma_c(T_c - T_{out})}$$

The net power delivered can then be expressed in terms of the heat exchanges:

$$P_{net} = \frac{\sigma_h(T_h - T_{in}) + \sigma_h \frac{Q_{cd}}{Q_{ab}}(T_h - T_{in})}{1 + \frac{Q_{cd}\sigma_h(T_h - T_{in})}{Q_{ab}\sigma_c(T_c - T_{out})}} = \frac{\sigma_h(T_h - T_{in})\left(1 + \frac{Q_{cd}}{Q_{ab}}\right)}{1 + \frac{Q_{cd}\sigma_h(T_h - T_{in})}{Q_{ab}\sigma_c(T_c - T_{out})}}$$

$$P_{net} = \frac{1 + \frac{Q_{cd}}{Q_{ab}}}{\frac{1}{\sigma_h(T_h - T_{in})} + \frac{Q_{cd}}{Q_{ab}\sigma_c(T_c - T_{out})}}$$

The efficiency of the endoreversible Carnot-like engine is limited by Carnot's theorem to

$$e_C^{er} = 1 - \frac{|Q_{cd}|}{Q_{ab}} \leq 1 - \frac{T_{out}}{T_{in}}$$

which is less than the efficiency of a Carnot engine, $e_C = 1 - T_c/T_h$, operating between the same two thermal reservoirs: $e_C^{er} < e_C$. This follows because $T_{out}/T_{in} < T_c/T_h$. For given values of the temperatures, the maximum possible efficiency of the endoreversible Carnot-like engine is achieved when $-Q_{cd}/Q_{ab} = T_{out}/T_{in}$. In this case, the net power delivered equals

$$P_{net} = \frac{T_{in} - T_{out}}{\frac{T_{in}}{\sigma_h(T_h - T_{in})} + \frac{T_{out}}{\sigma_c(T_{out} - T_c)}}$$

Our goal is to determine which values of T_{in} and T_{out} maximize the net power delivered by the engine, and also to see what impact this has on the thermal efficiency of the engine.

We find the values of T_{in} and T_{out} that maximize the net power by setting a partial derivative of P_{net} with respect to T_{in} equal to zero, while holding the independent variable T_{out} constant, and also setting a partial derivative of P_{net} with respect to T_{out} equal to zero, while holding the independent variable T_{in} constant:

$$\frac{\partial P_{net}}{\partial T_{in}} = 0$$

$$\frac{\partial}{\partial T_{in}}\left[\frac{T_{in} - T_{out}}{\frac{T_{in}}{\sigma_h(T_h - T_{in})} + \frac{T_{out}}{\sigma_c(T_{out} - T_c)}}\right] = 0$$

$$\frac{1}{\frac{T_{in}}{\sigma_h(T_h - T_{in})} + \frac{T_{out}}{\sigma_c(T_{out} - T_c)}} - \frac{(T_{in} - T_{out})\left[\frac{1}{\sigma_h(T_h - T_{in})} + \frac{T_{in}}{\sigma_h(T_h - T_{in})^2}\right]}{\left[\frac{T_{in}}{\sigma_h(T_h - T_{in})} + \frac{T_{out}}{\sigma_c(T_{out} - T_c)}\right]^2} = 0$$

$$\frac{T_{in}}{\sigma_h(T_h - T_{in})} + \frac{T_{out}}{\sigma_c(T_{out} - T_c)} - \frac{T_h(T_{in} - T_{out})}{\sigma_h(T_h - T_{in})^2} = 0$$

$$\frac{T_{in}^2 - T_{out}T_h}{\sigma_h(T_h - T_{in})^2} = \frac{T_{out}}{\sigma_c(T_{out} - T_c)}$$

$$\frac{\partial P_{net}}{\partial T_{out}} = 0$$

$$\frac{\partial}{\partial T_{out}}\left[\frac{T_{in} - T_{out}}{\frac{T_{in}}{\sigma_h(T_h - T_{in})} + \frac{T_{out}}{\sigma_c(T_{out} - T_c)}}\right] = 0$$

$$\frac{-1}{\frac{T_{in}}{\sigma_h(T_h - T_{in})} + \frac{T_{out}}{\sigma_c(T_{out} - T_c)}} - \frac{(T_{in} - T_{out})\left[\frac{1}{\sigma_c(T_{out} - T_c)} + \frac{T_{out}}{\sigma_c(T_{out} - T_c)^2}\right]}{\left[\frac{T_{in}}{\sigma_h(T_h - T_{in})} + \frac{T_{out}}{\sigma_c(T_{out} - T_c)}\right]^2} = 0$$

$$-\frac{T_{in}}{\sigma_h(T_h - T_{in})} - \frac{T_{out}}{\sigma_c(T_{out} - T_c)} - \frac{T_c(T_{in} - T_{out})}{\sigma_c(T_{out} - T_c)^2} = 0$$

$$\frac{-T_{out}^2 + T_{in}T_c}{\sigma_c(T_{out} - T_c)^2} = \frac{T_{in}}{\sigma_h(T_h - T_{in})}$$

There are two equations and two unknown parameters, T_{in} and T_{out}. We proceed by eliminating an expression that is common to both equations:

$$\frac{\sigma_h(T_h - T_{in})}{\sigma_c(T_{out} - T_c)} = \frac{T_{in}^2 - T_{out}T_h}{T_{out}(T_h - T_{in})} = \frac{T_{in}(T_{out} - T_c)}{-T_{out}^2 + T_{in}T_c}$$

$$T_{in}^3 T_c + T_{out}^3 T_h = T_{in}T_{out}^2 T_h + T_{in}^2 T_{out} T_c$$

$$T_{in}^2 T_c(T_{in} - T_{out}) = T_{out}^2 T_h(T_{in} - T_{out})$$

$$\frac{T_{out}}{T_{in}} = \sqrt{\frac{T_c}{T_h}}$$

We have now found the relationship between the two parameters, T_{in} and T_{out}, which maximizes the pressure. We can now solve for T_{in} in terms of the constants by substituting this relation into the first equation on the top line of the previous group of equations:

5 Heat Engines and Refrigerators

$$\frac{\sigma_h(T_h - T_{in})}{\sigma_c(T_{out} - T_c)} = \frac{T_{in}^2 - T_{out}T_h}{T_{out}(T_h - T_{in})}$$

$$\sigma_h(T_h - T_{in})^2 = \left(-\sqrt{\frac{T_h}{T_c}}T_{in} + T_h\right)\sigma_c\left(T_c - \sqrt{\frac{T_c}{T_h}}T_{in}\right)$$

$$\sigma_h(T_h - T_{in})^2 = \sigma_c\left(T_hT_c + T_{in}^2 - 2T_{in}\sqrt{T_hT_c}\right)$$

$$\sigma_h(T_h - T_{in})^2 = \sigma_c\left(T_{in} - \sqrt{T_hT_c}\right)^2$$

$$T_h\sqrt{\sigma_h} - T_{in}\sqrt{\sigma_h} = T_{in}\sqrt{\sigma_c} - \sqrt{\sigma_c T_h T_c}$$

$$T_{in} = \frac{T_h\sqrt{\sigma_h} + \sqrt{\sigma_c T_h T_c}}{\sqrt{\sigma_h} + \sqrt{\sigma_c}}$$

from which it follows that

$$T_{out} = \frac{\sqrt{\sigma_h T_h T_c} + T_c\sqrt{\sigma_c}}{\sqrt{\sigma_h} + \sqrt{\sigma_c}}$$

These equations for T_{in} and T_{out} are the values of T_{in} and T_{out}, in terms of the constants, which maximize the power delivered by the endoreversible Carnot-like engine. The maximum power delivered by the engine can be found by substituting these results into the equation for power:

$$P_{net} = \frac{\dfrac{T_h\sqrt{\sigma_h} + \sqrt{\sigma_c T_h T_c}}{\sqrt{\sigma_h} + \sqrt{\sigma_c}} - \dfrac{\sqrt{\sigma_h T_h T_c} + T_c\sqrt{\sigma_c}}{\sqrt{\sigma_h} + \sqrt{\sigma_c}}}{\dfrac{\dfrac{T_h\sqrt{\sigma_h} + \sqrt{\sigma_c T_h T_c}}{\sqrt{\sigma_h} + \sqrt{\sigma_c}}}{\sigma_h\left(T_h - \dfrac{T_h\sqrt{\sigma_h} + \sqrt{\sigma_c T_h T_c}}{\sqrt{\sigma_h} + \sqrt{\sigma_c}}\right)} + \dfrac{\dfrac{\sqrt{\sigma_h T_h T_c} + T_c\sqrt{\sigma_c}}{\sqrt{\sigma_h} + \sqrt{\sigma_c}}}{\sigma_c\left(\dfrac{\sqrt{\sigma_h T_h T_c} + T_c\sqrt{\sigma_c}}{\sqrt{\sigma_h} + \sqrt{\sigma_c}} - T_c\right)}}$$

$$P_{net} = \frac{\dfrac{(\sqrt{\sigma_h T_h} + \sqrt{\sigma_c T_c})(\sqrt{T_h} - \sqrt{T_c})}{\sqrt{\sigma_h} + \sqrt{\sigma_c}}}{\dfrac{\sqrt{\sigma_h T_h} + \sqrt{\sigma_c T_c}}{\sigma_h\sqrt{\sigma_c}(\sqrt{T_h} - \sqrt{T_c})} + \dfrac{\sqrt{\sigma_h T_h} + \sqrt{\sigma_c T_c}}{\sigma_c\sqrt{\sigma_h}(\sqrt{T_h} - \sqrt{T_c})}}$$

$$P_{net} = \frac{\dfrac{(\sqrt{\sigma_h T_h} + \sqrt{\sigma_c T_c})(\sqrt{T_h} - \sqrt{T_c})}{\sqrt{\sigma_h} + \sqrt{\sigma_c}}}{\dfrac{(\sqrt{\sigma_h} + \sqrt{\sigma_c})(\sqrt{\sigma_h T_h} + \sqrt{\sigma_c T_c})}{\sigma_h\sigma_c(\sqrt{T_h} - \sqrt{T_c})}}$$

$$P_{net} = \sigma_h\sigma_c\left(\frac{\sqrt{T_h} - \sqrt{T_c}}{\sqrt{\sigma_h} + \sqrt{\sigma_c}}\right)^2$$

Efficiency of the endoreversible Carnot-like engine: We previously found the maximum efficiency of the endoreversible Carnot-like engine to be

$$e^{er}_{C,max} = 1 - \frac{T_{out}}{T_{in}}$$

For the combination of parameters, T_{in} and T_{out}, the power delivered by the engine is maximized and the corresponding maximum thermal efficiency is

$$e^{er}_{C,max} = 1 - \sqrt{\frac{T_c}{T_h}}$$

where we have employed the following result, which we derived previously:

$$\frac{T_{out}}{T_{in}} = \sqrt{\frac{T_c}{T_h}}$$

Conceptual Questions

The selection of conceptual questions is intended to enhance the conceptual understanding of students who spend time reasoning through them. You will receive the most benefit if you first try it yourself, then consult the hints, and finally check your answer after reasoning through it again. The hints and answers can be found, separately, toward the back of the book.

1. Describe a theoretical technique by which an isobaric process can be carried out quasistatically.

2. A block of ice left outside won't spontaneously melt on a freezing cold winter day. Can it be driven to do so? If not, explain. If so, describe how and explain whether or not this contradicts the second law of thermodynamics.

3. A cup of water freezes when placed in a perfectly insulating ice chest. What happens to the entropy of the water in the cup? Is this consistent with the second law of thermodynamics? Explain.

4. An electric fan is turned on in a perfectly insulated room. Does the temperature of the room decrease, increase, or remain unchanged? Explain. Compare this to opening the door of a refrigerator and the effect that this has on the temperature of a perfectly insulated kitchen. If the temperature of the room does not decrease, explain how a fan works to make you feel cooler.

5. Does a car burn more fuel on a cold winter day or a hot summer day? Explain.

6. A wooden crate is pushed 10 m to the north along a horizontal sidewalk and then pushed 10 m to the south, returning it to its original position. Is this process reversible? Explain.

7. Make a sketch each of the following processes in a T-S diagram (assuming the process to be reversible): an isentrope, an isotherm, an adiabat, an isobar, and an isochor.

8. Make a sketch of one cycle for each of the following heat engines in a T-S diagram: a Carnot engine, an air-standard Otto engine, an air-standard diesel engine, and a Brayton engine.

Practice Problems

The selection of practice problems primarily consists of problems that offer practice carrying out the main problem-solving strategies or involve instructive applications of the concepts (or both). You will receive the most benefit if you first try it yourself, then consult the hints, and finally check your answer after working it out again. The hints to all of the problems and the answers to selected problems can be found, separately, toward the back of the book.

Entropy

1. What is the change in entropy of the substance for a 250-g ice cube that slowly melts at 0 °C? The latent heat of fusion for water is 3.3×10^5 J/kg.

2. What is the change in entropy for 250-g of water that is quasistatically cooled from 80 °C to 20 °C? The specific heat capacity of water is 4.186 kJ/kg/°C.

3. An ideal gas expands along a reversible isobar until its volume doubles. (A) Derive an equation for the entropy change of the ideal gas in terms of its mole number and appropriate constants only. (B) Derive the same equation by integrating along the following reversible path: Connect the same initial and final states with a combination of a reversible isotherm and a reversible adiabat.

4. A monatomic ideal gas expands according to the equation $P^a V^b = const.$, where a and b are constants. Derive an equation for the entropy change in terms of its mole number and its initial and final volume (and appropriate constants) only.

5. One mole of an ideal gas freely expands to twice its initial volume. After the expansion, each molecule can be in one of two states: It can reside in the original volume, or outside of it. Compute the entropy change microscopically using this two-state picture and show that it agrees with the entropy change in the macroscopic picture.

Heat engines

6. One cycle of a heat engine consists of two reversible isobars and two reversible isochors. The working substance is an ideal gas. (A) Make a sketch of the P-V diagram for this cycle. (B) Derive an equation for the efficiency of this heat engine in terms of the initial and final volumes. (C) Compare this to the efficiency of a Carnot engine operating between the same minimum and maximum temperatures. (D) Derive equations for the coefficient of performance in cooling mode and heating mode for this heat engine if it is used as a heat pump.

7. One cycle of a heat engine consists of a reversible isobar, a reversible isotherm, and a reversible isochor. The working substance is an ideal gas. (A) Make a sketch of the P-V diagram for this cycle. (B) Derive an equation for the efficiency of this heat engine in terms of the initial and final volumes. (C) Compare this to the efficiency of a Carnot engine operating between the same minimum and maximum temperatures.

8. One cycle of a Stirling heat engine consists of two reversible isotherms and two reversible isochors. Assume that the working substance is an ideal gas. (A) Make a sketch of the P-V diagram for this cycle. (B) Derive an equation for the efficiency of this heat engine in terms of the high and low temperatures and the compression ratio.

9. Treating the air as an ideal gas, show that the air-standard diesel engine is less efficient than a Carnot engine operating between the same minimum and maximum temperatures.

10. Treating the air as an ideal gas, show that the Brayton engine is less efficient than a Carnot engine operating between the same minimum and maximum temperatures.

11. Treating the air as an ideal gas, show that the endoreversible Carnot-like engine is less efficient than a Carnot engine operating between the same minimum and maximum temperatures.

6 An Introduction to Thermodynamics

6.1 Formulation of Thermodynamics

Macroscopic coordinates: Measurements of macroscopic matter have scales of length and time that are enormous compared to the scales of individual molecules.[211] Macroscopic measurements involve averages over an enormous number of atomic coordinates, by which only a small number of relatively static states are observed compared to the incredibly large number of microstates – i.e. the various combinations of individual atomic coordinates. Examples of macroscopic thermodynamic coordinates include temperature, volume, pressure, mole number, internal energy, and entropy.

Homogeneous: A substance that has uniform density throughout is homogenous.

Isotropic: A substance that has physical characteristics that are the same in all directions is isotropic.

Important Distinction. Homogeneous means uniform throughout, while the similar term isotropic means the same in all directions. It is possible for a substance to be homogeneous, but not isotropic, or to be isotropic, but not homogenous. For example, the earth has a geological structure that is roughly isotropic, but not at all homogenous: It is isotropic to the extent that its layers have approximately spherical symmetry, but it is not homogeneous because it consists of multiple layers of different densities. A magnet has properties that are roughly homogeneous, but not at all isotropic: It can be homogenous to the extent that its molecules are evenly distributed throughout, but it is not isotropic because atomically there is a preferred direction of angular momentum.

Extensive parameters: For a homogenous system in equilibrium, the extensive thermodynamic parameters are those macroscopic coordinates that tend to be additive – i.e. the value for the whole equals the sum of the values for its parts. Examples of extensive parameters include volume, mole number, mass, internal energy, entropy, and heat capacity. If you divide such a system in half, each half will have half the volume, mole number, mass, and any other extensive parameter.

Intensive parameters: For a homogeneous system in equilibrium, the intensive thermodynamic parameters are those macroscopic coordinates that tend to be the same for any part as they are for the whole. Examples of intensive parameters include density, temperature, pressure, specific heat capacity, and molar quantities (e.g. molar volume or internal energy per mole). If you divide such a system in half, each half will have the same density, temperature, pressure, and any other intensive parameter.

---

[211] An atom has a size of ~1 Å = 10^{-10} m and vibrates with a period of ~1 fs = 10^{-15} s.

> **Important Distinction.** Thermodynamic coordinates can be classified as extensive or intensive parameters. For homogeneous systems in equilibrium, extensive parameters are additive, while intensive parameters are the same throughout. For example, any part of the system will have the same temperature, so temperature is intensive, whereas any part of the system will have proportionately less mass than the whole system, so mass is extensive. Subsystems of homogeneous systems in equilibrium will have the same value for intensive parameters – e.g. $T_1 = T_2 = \cdots = T_N = T$ – while for extensive parameters, their values will add up to the value for the system – e.g. $m_1 + m_2 + \cdots + m_N = m$. As another example, consider heat capacity and specific heat capacity: Specific heat capacity is a property of the material, and so is intensive, whereas heat capacity varies depending upon the mass of the substance, and so is extensive (recall that specific heat capacity is heat capacity per unit mass).[212]

Exact differentials: If an integral over a differential element is path-independent – i.e. the numerical value of the integral depends only upon the endpoints – the differential element is termed an exact differential. The internal energy (dU) and entropy (dS) differentials are examples of exact differentials: The finite internal energy or entropy found by integrating these differentials, $\int_i^f dU$ or $\int_i^f dS$, is independent of the thermodynamic process, and depends only on the initial and final states.

Inexact differentials: If an integral over a differential element is, in general, path-dependent – i.e. the numerical value of the integral varies, in general, depending upon the path of integration – the differential element is termed an inexact differential. The work $(đW)$ and heat $(đQ)$ differentials are examples of inexact differentials: Beginning and ending in the same thermodynamic state, the work done and heat exchanged generally depend upon which thermodynamic process is involved.

Simple systems: An ideal system that is homogeneous and isotropic, has negligible surface effects, and is free of gravitational, electrical, and magnetic interactions defines a simple system. Such a system primarily features thermodynamic properties.[213]

Equilibrium: Macroscopic systems have a tendency to progress toward equilibrium states. For example, if you shake a thermos bottle of coffee vigorously for a while, once you stop the coffee will be in motion for a while, and will be a little warmer than it was prior to the shaking, but as time progresses, the motion will dampen out and eventually the system will reach equilibrium. The equilibrium states are independent of any external influences which have been removed, but do depend on internal factors. Once attained, the system has a natural tendency to remain in equilibrium.[214] The equilibrium states of simple systems can be completely described, in macroscopic terms, by $r + 2$ independent extensive parameters: the internal energy (U), the volume (V), and the mole numbers of the r chemical components $(n_1, n_2, \cdots, n_r)$.

---

[212] In this case, however, it is really the inverse heat capacities that add up: E.g. $C_{V1}^{-1} + C_{V2}^{-1} + \cdots + C_{VN}^{-1} = C_V^{-1}$.

[213] The remaining mechanical property is that the volume of the system may be changed – e.g. by moving a piston – which manifests itself through the work done by the system.

[214] This is a sense of thermodynamic inertia. That is, macroscopic systems have a natural tendency to be in equilibrium. A net external influence is required to noticeably disturb a system from equilibrium. If the net external influence becomes zero, the system tends to return to equilibrium.

Fundamental relation: A thermodynamic system in equilibrium is fully described by the fundamental relation. For a simple system, the fundamental relation expresses the entropy in terms of the internal energy, volume, and mole numbers for the r chemical components: $S = S(U, V, n_1, n_2, \cdots, n_r)$. As a relationship among extensive parameters, the fundamental relation must satisfy the additivity property. Also, with thermodynamic extremum principles in mind, the derivative of the entropy must exist and the partial derivative of the entropy with respect to the internal energy must be positive:[215]

$$\left(\frac{\partial S}{\partial U}\right)_{V, n_1, n_2, \cdots, n_r} > 0$$

For a simple system, the fundamental relation can alternatively be inverted to express the internal energy as a function of the entropy, volume, and mole numbers: $U = U(S, V, n_1, n_2, \cdots, n_r)$. This form of the fundamental relation must similarly satisfy the additivity property. Additionally, the internal energy must be a single-valued function of the extensive parameters; and, if the partial derivative of the internal energy with respect to the entropy is zero, the entropy of the system must also be zero:

$$\left(\frac{\partial U}{\partial S}\right)_{V, n_1, n_2, \cdots, n_r} = 0 \quad \Rightarrow \quad S = 0$$

Additivity: Extensive parameters, such as volume, are additive: That is, if the system is divided up into k subsystems, the sum of the volumes (or other extensive parameter) of each of the subsystems equals the volume (or other extensive parameter) of the system: $V = \sum_{i=1}^{k} V_i$. The fundamental relations for each of the subsystems of a simple system has the form, $S_i = S_i(U_i, V_i, n_{i1}, n_{i2}, \cdots, n_{ir})$. The sum of these fundamental relations of the subsystems must equal the fundamental relation for the system as a whole – since the entropy is extensive: $S = \sum_{i=1}^{k} S_i$. Here is another way to look at it: The extensive parameters for any subsystem are a fraction, λ, of the extensive parameters for the system as a whole:

$$S_i = \lambda S \;,\; U_i = \lambda U \;,\; V_i = \lambda V \;,\; n_{i1} = \lambda n_1 \;,\; n_{i2} = \lambda n_2 \;,\cdots,\; n_{ir} = \lambda n_r$$

This implies that

$$S(\lambda U, \lambda V, \lambda n_1, \lambda n_2, \cdots, \lambda n_r) = \lambda S(U, V, n_1, n_2, \cdots, n_r)$$

This is the additivity property, which can also be inverted from the entropy representation to the energy representation as

$$U(\lambda S, \lambda V, \lambda n_1, \lambda n_2, \cdots, \lambda n_r) = \lambda U(S, V, n_1, n_2, \cdots, n_r)$$

Such a function is said to be a homogeneous, first-order function of its arguments.

---

[215] This means that the entropy must be a continuous, smooth function of the extensive parameters, and must be a monotonically increasing function of the internal energy for any fixed values of the volume and mole numbers. The subscripts on the partial derivative list which extensive parameters are being held constant.

> **Examples.** The function $z(x,y) = \sqrt{xy}$ is a homogeneous, first-order function of x and y because $z(\lambda x, \lambda y) = \sqrt{\lambda x \lambda y} = \lambda \sqrt{xy} = \lambda z(x,y)$. Inverting this relationship to think of x as a function of y and z, i.e. $x(y,z) = z^2/y$, the function $x(y,z)$ is also a homogeneous, first-order function of its arguments since $x(\lambda y, \lambda z) = (\lambda z)^2/\lambda y = \lambda z^2/y = \lambda x(y,z)$. In contrast, the function $w(x,y) = x^2 y^2$ is not a homogeneous, first-order function of x and y because $w(\lambda x, \lambda y) = (\lambda x)^2 (\lambda y)^2 = \lambda^4 x^2 y^2 = \lambda^4 w(x,y) \neq \lambda w(x,y)$.

Molar scaling: By setting the scale factor, λ, equal to $1/n$, the fundamental relation is scaled down (or up, as the case may be) to a single mole. For a simple system,

$$ns(u, v, x_1, x_2, \cdots, x_r) = S(U, V, n_1, n_2, \cdots, n_r)$$

where $s = S/n$ is the entropy per mole, $u = U/n$ is the internal energy per mole, $v = V/n$ is the molar volume, $n = \sum_{i=1}^{r} n_i$ is the total number of moles of the r chemical components, and $x_i = n_i/n$ is the mole fraction for component $i \in (1, 2, \cdots, r)$. Note that only $r + 1$ of these arguments are independent since $\sum_{i=1}^{r} x_i = 1$. These molar quantities are intensive parameters – the result of dividing two extensive parameters. For a simple system with a single component, molar scaling reduces to

$$ns(u, v) = S(U, V, n)$$

Extremum principles: For a fixed value of the internal energy, equilibrium is attained when the free internal parameters maximize the total entropy. Alternatively, for a fixed value of the entropy, equilibrium is attained when the free internal parameters minimize the internal energy. These extremum principles relate to the second law of thermodynamics.

To see that these two extremum principles are equivalent, imagine that a system has reached equilibrium, such that its entropy is maximum for its internal energy. Now if this internal energy were not minimum for the given entropy, this internal energy could be reduced by using it to do useful work, and then returned in the form of heat. However, this results in an immediate contradiction because $đQ = TdS$: That is, the heat added back to the system to return its internal energy back to its original value results in an increase of entropy – which was previously assumed to be maximum. Thus, the original equilibrium state must already have minimum internal energy – i.e. the internal energy cannot be decreased without some external influences that would disturb equilibrium.

Energy representation: The internal energy is expressed as a single-valued function of the entropy, volume, and mole numbers for a simple system in equilibrium in the energy representation as $U = U(S, V, n_1, n_2, \cdots, n_r)$. This fundamental relation can be expressed in terms of differentials as:[216]

$$dU = \left(\frac{\partial U}{\partial S}\right)_{V, n_1, n_2, \cdots, n_r} dS + \left(\frac{\partial U}{\partial V}\right)_{S, n_1, n_2, \cdots, n_r} dV + \sum_{i=1}^{r} \left(\frac{\partial U}{\partial n_i}\right)_{S, V, n_{j \neq i}} dn_i$$

---

[216] These differentials represent an infinitesimal departure from the present equilibrium state; hence the heat flux and work identifications that follow are termed quasistatic. For a brief review of the total differential, see Sec. 1.2.

These partial derivatives are intensive parameters with a natural interpretation:

$$T \equiv \left(\frac{\partial U}{\partial S}\right)_{V,n_1,n_2,\cdots,n_r} \quad , \quad P \equiv -\left(\frac{\partial U}{\partial V}\right)_{S,n_1,n_2,\cdots,n_r} \quad , \quad \mu_i \equiv \left(\frac{\partial U}{\partial n_i}\right)_{S,V,n_{j \neq i}}$$

where μ_i is the chemical potential of component i. In terms of the temperature, pressure, and chemical potentials, the internal energy is

$$dU = TdS - PdV + \sum_{i=1}^{r} \mu_i dn_i$$

The first two terms are recognized as the quasistatic heat flux ($đQ = TdS$) and quasistatic work ($đW = PdV$); the sum represents quasistatic chemical work $đW_c = \sum_{i=1}^{r} \mu_i dn_i$. Thus, we see that this mathematically eloquent framework has reproduced the first law of thermodynamics – and, more, it tells us what to do if the mole numbers are changing:

$$dU = đQ - đW + đW_c$$

For a simple system with a single component, in molar form this becomes

$$du = Tds - Pdv$$

Entropy representation: The entropy is expressed as a function of the entropy, volume, and mole numbers for a simple system in equilibrium in the entropy representation: $S = S(U, V, n_1, n_2, \cdots, n_r)$. In differential form,

$$dS = \left(\frac{\partial S}{\partial U}\right)_{V,n_1,n_2,\cdots,n_r} dU + \left(\frac{\partial S}{\partial V}\right)_{U,n_1,n_2,\cdots,n_r} dV + \sum_{i=1}^{r} \left(\frac{\partial S}{\partial n_i}\right)_{U,V,n_{j \neq i}} dn_i$$

In terms of the temperature, pressure, and chemical potentials, this becomes

$$dS = \frac{dU}{T} + \frac{P}{T}dV - \sum_{i=1}^{r} \frac{\mu_i}{T} dn_i$$

Important Distinction. You must make a choice between working with the energy representation or the entropy representation. Very often, one representation will be more convenient than the other for a particular problem. It is a common mistake for beginning students to try to mix and match ideas and equations from one relation with the other, which leads to problems, for example, when applying the extremum postulates. Choose one representation and stick with it, and ignore the other representation while applying the one. In the energy representation, energy is a dependent variable, while entropy is independent, and vice-versa in the entropy representation. This dependence cannot be swapped.

Thermodynamic definitions: Conceptually, thermodynamics begins with the following extensive parameters for a simple system in equilibrium: volume, mole numbers, and internal energy. The volume and mole numbers are basic notions, which are easy to observe macroscopically. The internal energy can be measured to change, macroscopically, and this change represents energy added or subtracted microscopically. The total energy of the system and surroundings obeys a fundamental conservation law. Also, the energy transformations can be controlled: That is, thermodynamic processes can be carried out with constant mole numbers, which are either adiabatic or isochoric – in these extremes, the change in internal energy equals the work done or heat exchanged, respectively.

For a simple system in equilibrium, these extensive parameters – internal energy, volume, and mole numbers – provide a full macroscopic description of its thermodynamic properties. One more extensive parameter, the entropy, is postulated to exist for equilibrium states, and to have properties that allow it to mathematically determine which equilibrium state a system will reach for a given internal energy. Specifically, the second law of thermodynamics puts a constraint on the partial derivative of entropy with respect to energy, and also on the partial derivative of internal energy with respect to entropy, as these partial derivatives relate to the entropy maximum and energy minimum postulates.

The total differential of the fundamental relation in the energy or entropy representation then leads to mathematical definitions of intensive parameters, two of which turn out to agree with our intuitive notions of temperature and pressure. That is, temperature, pressure, and chemical potential are defined by the following equations:

$$T \equiv \left(\frac{\partial U}{\partial S}\right)_{V,n_1,n_2,\cdots,n_r} \quad , \quad P \equiv -\left(\frac{\partial U}{\partial V}\right)_{S,n_1,n_2,\cdots,n_r} \quad , \quad \mu_i \equiv \left(\frac{\partial U}{\partial n_i}\right)_{S,V,n_{j\neq i}}$$

The conceptual interpretations of temperature and pressure follow from this, but do not define these physical quantities. We will see, for example, that these definitions agree with our expectations for equilibrium, and that the definition of temperature is consistent with the third law. These intensive parameters can thus be obtained if the fundamental relation is known.

After defining these intensive parameters, the total differential is seen to be a generalization of the first law of thermodynamics to the case where the mole numbers may not be constant. One term is interpreted as the quasistatic work: $đW = PdV$. Conceptually, this agrees with our notion that a system can do work by changing its volume, and mathematically it agrees with the general physics definition of work as $W = \int_i^f \vec{F} \cdot d\vec{s}$. The quasistatic heat flux is then defined as $đQ = dU + đW$ for the case of constant mole numbers. Once again, this definition is mathematical, and agrees with our intuitive sense of heat flux. Conceptually, heat is a means by which a system can transfer its macroscopically hidden energy. Comparison of equations shows that the quasistatic heat flux is also related to the entropy by $đQ = TdS$. That is, an exchange of heat is associated with a change in entropy. As the entropy was postulated to exist for equilibrium states, the integral $Q = \int_i^f TdS$ has endpoints at equilibrium. The central problem of thermodynamics is to begin in one equilibrium state, disturb equilibrium, and determine the equilibrium state that results. A new form of work, quasistatic chemical work, $đW_c = \sum_{i=1}^r \mu_i dn_i$, is also defined mathematically; conceptually, it represents energy changes associated with changing mole numbers.

Conjugate variables: In the differential form of the fundamental relation in the energy representation, the intensive parameters are associated with extensive parameters: T appears as the coefficient of dS, $-P$ appears as the coefficient of dV, and μ_i appears as the coefficient of dn_i. These pairs of intensive and extensive parameters — T with S, P with V, and μ_i with n_i — are called conjugate variables. The terms TdS, PdV, and $\mu_i dn_i$ are all forms of energy transfer — either heat or work.

The Nernst postulate: The third law of thermodynamics is also known as the Nernst postulate. One of the postulates of thermodynamics regarding the fundamental relation is that the entropy be a monotonically increasing function of the internal energy of the system, or, equivalently, that

$$\left(\frac{\partial S}{\partial U}\right)_{V, n_1, n_2, \cdots, n_r} > 0$$

This partial derivative is the inverse temperature, $1/T$. Thus, the absolute temperature must be a positive quantity, with a clear minimum that can be approached, yet not attained. Another postulate was that

$$\left(\frac{\partial U}{\partial S}\right)_{V, n_1, n_2, \cdots, n_r} = 0 \Rightarrow S = 0$$

In terms of temperature, this means that as absolute zero is approached, the entropy of the system also approaches zero — a limit that can be approached, but never quite attained.

Equations of state: As partial derivatives of extensive variables, the temperature, pressure, and chemical potentials are functions of the extensive variables: Specifically, in the energy representation for a simple system in equilibrium, $T = T(S, V, n_1, n_2, \cdots, n_r)$, $P = P(S, V, n_1, n_2, \cdots, n_r)$, and $\mu_i = \mu_i(S, V, n_1, n_2, \cdots, n_r)$. These relations are called equations of state. The equations of state are homogeneous, zero-order functions of the extensive parameters because they are partial derivatives of homogeneous, first-order functions: This means that the intensive parameters are independent of any scaling of the extensive parameters. For example, $P(\lambda S, \lambda V, \lambda n_1, \lambda n_2, \cdots, \lambda n_r) = P(S, V, n_1, n_2, \cdots, n_r)$. All of the equations of state together are equivalent to the fundamental relation in terms of thermodynamic content.

Examples. The function $z(x, y) = x/y$ is a homogeneous, zero-order function of x and y because $z(\lambda x, \lambda y) = \lambda x / \lambda y = x/y = z(x, y)$. However, inverting this relationship to think of x as a function of y and z, i.e. $x(y, z) = yz$, the function $x(y, z)$ is not a homogeneous, zero-order function of its arguments since $x(\lambda y, \lambda z) = \lambda y \lambda z = \lambda^2 yz = \lambda^2 x(y, z)$.

The extensive parameters are homogeneous, first-order functions of the other extensive parameters, and the intensive parameters are homogeneous, zero-order functions of the extensive parameters. Observe that in both cases the functions are expressed in terms of extensive parameters: Don't get carried away and try scaling with intensive parameters as arguments — as scaling only makes sense for extensive parameters because they are additive.

Example. Show that the relation, $S = \sqrt{n(aV + bU)}$, where a and b are positive constants, is a viable fundamental relation for a simple system in equilibrium, and determine its equations of state.

The fundamental relation must satisfy the postulates of thermodynamics. For one, it must satisfy the additivity property: $S(\lambda U, \lambda V, \lambda n) = \sqrt{\lambda n(\lambda aV + \lambda bU)} = \lambda\sqrt{n(aV + bU)} = \lambda S(U, V, n)$. Thus, this is a first-order, homogeneous equation. Secondly, inverting the fundamental equation to solve for the internal energy, the internal energy must be a single-valued function:

$$U = \frac{S^2}{bn} - \frac{aV}{b}$$

Thirdly, the partial derivative

$$\left(\frac{\partial S}{\partial U}\right)_{V,N} = \frac{bn}{2\sqrt{n(aV + bU)}}$$

is positive, as required. Also, in the limit that the related partial derivative

$$\left(\frac{\partial U}{\partial S}\right)_{V,N} = \frac{2\sqrt{n(aV + bU)}}{bn} = \frac{2S}{bn}$$

approaches zero, the entropy similarly approaches zero (since the alternative, i.e. for n to become infinite, it not plausible).

The equations of state are easily found through the partial derivatives:

$$T(U, V, n) = \left(\frac{\partial U}{\partial S}\right)_{V,n} = \frac{2\sqrt{n(aV + bU)}}{bn}$$

$$P(U, V, n) = -\left(\frac{\partial U}{\partial V}\right)_{S,n} = \frac{a}{b}$$

$$\mu(U, V, n) = \left(\frac{\partial U}{\partial n}\right)_{S,V} = -\frac{aV + bU}{bn}$$

Observe that each equation of state is a homogeneous, first-order equation of the extensive parameters. We worked in the entropy representation, treating internal energy, volume, and mole number as the independent variables, but could have just as easily worked in the energy representation, instead expressing the equations of state in terms of entropy, volume, and mole number.

Relating intensive parameters: The fundamental relation involves $r + 3$ extensive parameters for r chemical components. Including the $r + 2$ equations of state, there are $r + 3$ equations with $r + 3$ extensive parameters. In principle, the $r + 3$ extensive parameters can be eliminated, yielding an equation exclusively among the intensive parameters. Following this prescription, the temperature, for example, can be expressed as a function of the other intensive parameters: $T = T(P, \mu_1, \mu_2, \cdots, \mu_r)$.

Example. For the system with fundamental relation, $S = \sqrt{n(aV + bU)}$, where a and b are positive constants, derive a relation among the intensive parameters.

We had derived the equations of state in the previous example:

$$T = \frac{2\sqrt{n(aV + bU)}}{bn} = \frac{2S}{bn} \quad , \quad P = \frac{a}{b} \quad , \quad \mu = -\frac{aV + bU}{bn} = -\frac{S^2}{bn^2}$$

As it turns out, for this simple example, the pressure is constant, completely independent of the extensive (or intensive) parameters. The combination S/n is easily eliminated:

$$\frac{S^2}{n^2} = -b\mu = \frac{b^2 T^2}{4}$$

Thus, temperature is a function of the chemical potential only: $T(\mu) = -2\sqrt{\mu}/b$.[217]

Important Distinction. When we say that the fundamental relation, $U(S, V, n_1, n_2, \cdots, n_r)$, in the energy representation (or the equivalent relation in the entropy representation) applies to a simple system in equilibrium, we do not have in mind a particular equilibrium state. Rather, the fundamental relation applies to equilibrium states. The principal thermodynamic problem is to begin with a system (which may consist of subsystems) in one particular equilibrium state, disturb equilibrium in some way, and predict which equilibrium state will eventually be reached. Conservation of energy (the first law) and the extremum principles aid in this latter part.

In the previous example, we found that the temperature was a function of the chemical potential – or, in terms of extensive parameters, a function of the entropy and mole number. If the system is in thermal equilibrium, we expect the temperature to remain constant. Thus, this relationship determines how the temperature and other macroscopic coordinates are related at equilibrium – the value of the temperature attained at equilibrium is related to the chemical potential (and extensive parameters) attained. More than one thermal equilibrium state is possible – the exact circumstances determine which thermal equilibrium state will result.

Important Distinction. We generally have quasistatic thermodynamic processes in mind, as we derived the first law in terms of quasistatic heat flux and quasistatic work done (including chemical work) by making differential departures from equilibrium values. However, we can consider irreversible processes passing through a series of non-equilibrium states for path-independent quantities, such as internal energy and entropy, by finding an equivalent reversible (and hence quasistatic) thermodynamic process connecting the same initial and final equilibrium states. In this case, however, the path-dependent quantities, such as heat and work, will generally be different for the reversible path compared to the actual irreversible path; yet the combination of heat and work (including chemical work) will reproduce the same internal energy for either of these processes.

---

[217] Since T^2 must be positive, the chemical potential, in this case, must be negative.

The Euler equation: For a simple system in equilibrium, in the energy representation, the additivity property for the fundamental relation can be expressed in terms of a scaling factor, λ, as:

$$U(\lambda S, \lambda V, \lambda n_1, \lambda n_2, \cdots, \lambda n_r) = \lambda U(S, V, n_1, n_2, \cdots, n_r)$$

Taking a partial derivative with respect to the scaling factor on both sides of the equation,

$$\frac{\partial U(\lambda S, \lambda V, \lambda n_1, \lambda n_2, \cdots, \lambda n_r)}{\partial \lambda} = U(S, V, n_1, n_2, \cdots, n_r)$$

The partial derivative on the left-hand side of the equation can be applied via the chain rule:

$$U(S,V,n_1,n_2,\cdots,n_r) = \frac{\partial U(\lambda S, \lambda V, \lambda n_1, \lambda n_2, \cdots, \lambda n_r)}{\partial(\lambda S)}\frac{\partial(\lambda S)}{\partial\lambda} + \frac{\partial U(\lambda S, \lambda V, \lambda n_1, \lambda n_2, \cdots, \lambda n_r)}{\partial(\lambda V)}\frac{\partial(\lambda V)}{\partial\lambda}$$
$$+ \sum_{i=1}^{r} \frac{\partial U(\lambda S, \lambda V, \lambda n_1, \lambda n_2, \cdots, \lambda n_r)}{\partial(\lambda n_i)}\frac{\partial(\lambda n_i)}{\partial\lambda}$$
$$= S\frac{\partial U(\lambda S, \lambda V, \lambda n_1, \lambda n_2, \cdots, \lambda n_r)}{\partial(\lambda S)} + V\frac{\partial U(\lambda S, \lambda V, \lambda n_1, \lambda n_2, \cdots, \lambda n_r)}{\partial(\lambda V)}$$
$$+ \sum_{i=1}^{r} n_i \frac{\partial U(\lambda S, \lambda V, \lambda n_1, \lambda n_2, \cdots, \lambda n_r)}{\partial(\lambda n_i)}$$

Setting λ equal to one,

$$U = S\left(\frac{\partial U}{\partial S}\right)_{V,n_1,n_2,\cdots,n_r} + V\left(\frac{\partial U}{\partial V}\right)_{S,n_1,n_2,\cdots,n_r} + \sum_{i=1}^{r} n_i \left(\frac{\partial U}{\partial n_i}\right)_{S,V,n_{j\neq i}}$$

$$U = TS - PV + \sum_{i=1}^{r} \mu_i n_i$$

This last equation, which expresses a relationship between the extensive and intensive parameters, is known as the Euler equation. If all of the equations of state are known, they can be substituted into the Euler equation to give the fundamental relation. We thus conclude that the fundamental relation is equivalent to the complete set of equations of state.

Example. The equations of state for a system are: $T = a(U^2/Vn)^{1/3}$, $P = U/V$, and $\mu = -U/n$. Determine the fundamental relation for this system.
Substituting these equations of state for the intensive parameters into Euler's equation,

$$U = TS - PV + \mu n = aS\left(\frac{U^2}{Vn}\right)^{1/3} - U - U$$
$$S = 3(UVn)^{1/3}/a$$

The Gibbs-Duhem relation: Implicitly differentiating Euler's equation,

$$dU = TdS + SdT - PdV - VdP + \sum_{i=1}^{r}(\mu_i dn_i + n_i d\mu_i)$$

This result is the first law of thermodynamics in addition to extra terms. The extra terms must, therefore, sum to zero:

$$SdT - VdP + \sum_{i=1}^{r} n_i d\mu_i = 0$$

This equation is called the Gibbs-Duhem relation. Whereas the first law involves intensive coefficients of extensive differentials, the Gibbs-Duehm relation involves extensive coefficients of intensive differentials. For a system with a single chemical component, this can be expressed in molar form as

$$sdT - vdP + d\mu = 0$$

The Gibbs-Duhem relation can be used to obtain the equation of state for $\mu(s, v)$ for a single-component system if the other two equations of state are already known. However, this comes at a cost: Since this is a differential equation, the resulting equation of state will involve a constant of integration. Knowing two equations of state and integrating to obtain the third is therefore not equivalent to the fundamental relation; the difference is the undetermined constant of integration.

Example. Given $T = 4S/\sqrt{aVn}$ and $P = S^2/\sqrt{aV^3n}$, determine the third equation of state.

The Gibbs-Duhem relation can be integrated to determine the third equation of state, aside from an undetermined constant of integration. It is convenient to first write the equations of state in molar form, as this effectively reduces the number of variables involved:

$$T = \frac{4s}{\sqrt{av}}, \quad P = \frac{s^2}{\sqrt{av^3}}$$

The Gibbs-Duhem relation involves the intensive parameters in differential form. Thus, it is necessary to implicitly differentiate these equations of state:

$$dT = \frac{4ds}{\sqrt{av}} - \frac{2sdv}{\sqrt{av^3}}, \quad dP = \frac{2sds}{\sqrt{av^3}} - \frac{3s^2 dv}{2\sqrt{av^5}}$$

Substituting these differentials into the molar form of the Gibbs-Duhem relation:

$$sdT - vdP + d\mu = \frac{4sds}{\sqrt{av}} - \frac{2s^2 dv}{\sqrt{av^3}} - \frac{2sds}{\sqrt{av}} + \frac{3s^2 dv}{2\sqrt{av^3}} + d\mu = \frac{2sds}{\sqrt{av}} - \frac{s^2 dv}{2\sqrt{av^3}} + d\mu = 0$$

This last equation cannot be integrated term by term, and it is not separable because there are three differentials. However, observe that $2sds$ is the implicit derivative of s^2, while $-dv/2v^{3/2}$ is the implicit derivative of $1/v^{1/2}$. That is, these two terms look like the result of a product rule. We can exploit this by defining a new variable, $z \equiv s^2/v^{1/2}$, because then $dz = 2sds/v^{1/2} - s^2dv/2v^{3/2}$:[218]

$$-d\mu = \frac{2sds}{\sqrt{av}} - \frac{s^2 dv}{2\sqrt{av^3}} = \frac{dz}{\sqrt{a}}$$

The left-hand side and right-hand side of this equation can now be easily integrated:[219]

$$\int_{\mu=\mu_0}^{\mu} d\mu = -\int_{z=z_0}^{z} \frac{dz}{\sqrt{a}}$$

$$\mu = \mu_0 + \frac{z_0 - z}{\sqrt{a}} = \mu_0 + \frac{s_0^2}{\sqrt{av_0}} - \frac{s^2}{\sqrt{av}}$$

Realistic problems: The fundamental relation contains the maximum thermodynamic content, and so the ideal goal is to be able to express the internal energy as a function of the entropy, volume, and mole numbers (or the entropy in terms of the internal energy, volume, and mole numbers in the entropy representation). Measuring volume and deducing mole numbers are basic measurements, but expressing a relationship between internal energy and entropy is not as straightforward. It is a more practical expectation to determine a couple of the equations of state, from which the fundamental relation can be found up to an undetermined constant of integration.

Consider, for example, an ideal gas with a single chemical component. The ideal gas law, $PV = nRT$, can be combined with the internal energy, $U = c_V nT$, to yield two of the three equations of state. The equation of state for chemical potential is not as straightforward to deduce as the equations of state for temperature and pressure. The Gibbs-Duhem relation can be integrated to determine the third equation of state, but not without an undetermined constant of integration.

It is a more realistic expectation to be able to determine a couple of equations of state, from which the other equations of state and the fundamental relation can be pieced together – not completely, but with some undetermined constants (at best). The missing equation(s) of state can be determined, up to a point, by applying the Gibbs-Duhem relation, and the fundamental relation can be expressed in terms of the equations of state using Euler's equation. Alternatively, the fundamental relation can be expressed in terms of the known equations of state by integrating the differential form of the first law of thermodynamics, without first obtaining any missing equation(s) of state. Either way, at a minimum, an undetermined constant of integration can be expected.

---

[218] This integration 'trick' is not uncommon. If you are integrating the Gibbs-Duhem relation or the differential form of the first law of thermodynamics and have three differential variables, you should be looking for a way to combine two terms by such a substitution – since you can't integrate those three differentials term by term, and three variables won't be separable.

[219] As indefinite integrals, a single constant would result. Thus, the limits of these definite integrals are related.

Example. Two equations of state for a system are known: $T = au^{1/2}/v^{1/4}$, $P = u/2v$. Determine the fundamental relation.

These equations of state can be substituted into the differential form of the first law of thermodynamics (expressed for this single-component system in terms of molar extensive parameters):

$$du = Tds - Pdv = \frac{au^{\frac{1}{2}}}{v^{\frac{1}{4}}}ds - \frac{u}{2v}dv$$

In this case, it is easy to separate molar entropy from molar energy and molar volume:

$$ads = \frac{v^{\frac{1}{4}}}{u^{\frac{1}{2}}}du + \frac{u^{\frac{1}{2}}}{2v^{3/4}}dv$$

Similar to the previous example, the two terms on the right-hand side of the equation can be condensed to an implicit derivative of a product rule. Specifically, by defining a new variable, $z \equiv u^{1/2}v^{1/4}$, the equation can be expressed as:

$$ads = 2dz$$

This integral is trivial and yields the following fundamental relation:

$$a(s - s_0) = 2(z - z_0)$$
$$s = s_0 + \frac{2\left(u^{1/2}v^{1/4} - u_0^{1/2}v_0^{1/4}\right)}{a}$$

6.2 Equilibrium Thermodynamics

Closed system: The energy, volume, and mole numbers each remain constant for a closed system.

Adiabatic wall: There is no heat flux through an adiabatic wall.

Diathermal wall: Heat may flow through a diathermal wall.

Thermal equilibrium: Consider a closed system that consists of two subsystems, A and B, at different temperatures, separated by an immovable, diathermal wall, which the particles of the system cannot cross. The two subsystems can only exchange heat, as the volumes and mole numbers must remain constant. The total entropy must be additive over the two subsystems: $S_A(U_A, V_A, n_{A1}, n_{A2}, \cdots, n_{Ar}) + S_B(U_B, V_B, n_{B1}, n_{B2}, \cdots, n_{Br}) = S(U, V, n_1, n_2, \cdots, n_r)$. For an infinitesimal variation in the total entropy of the system,

$$\left(\frac{\partial S_A}{\partial U_A}\right)_{V,n_1,n_2,\cdots,n_r} dU_A + \left(\frac{\partial S_B}{\partial U_B}\right)_{V,n_1,n_2,\cdots,n_r} dU_B = dS$$

$$\frac{dU_A}{T_A} + \frac{dU_B}{T_B} = dS$$

Since the total internal energy of this closed system is conserved – i.e. $U_A + U_B = U$ is constant – the free parameters U_A and U_B must maximize the total entropy of the system at equilibrium.[220] Specifically, the entropy maximum postulate requires that $dS = 0$,[221] or

$$\frac{dU_A}{T_A} = -\frac{dU_B}{T_B}$$

Conservation of energy also requires that $dU_A + dU_B = 0$, from which it follows that $T_A = T_B$ for the final equilibrium state. That is, the temperatures of the two subsystems will be equal when thermal equilibrium is reached. The conceptual significance of the minus sign above is that, in order to conserve energy, an increase in the energy of one subsystem necessitates a decrease in the energy of the other.

Since no work (mechanical or chemical) was done, the first law requires that $dU = dU_A + dU_B = đQ = đQ_A + đQ_B$. Thus, conservation of energy, $dU_A = -dU_B$, implies that $đQ_A = -đQ_B$: The heat gained by one subsystem equals the heat lost by the other. The differential increase in entropy from the initial state is $dS = dU_A/T_A + dU_B/T_B$. With a quasistatic process as our guide, this becomes $dS = đQ_A/T_A + đQ_B/T_B = đQ_A(1/T_A - 1/T_B)$, where these initial temperatures are, unlike the final temperatures, not equal. Although $dS = 0$ at equilibrium, $dS > 0$ during the heat exchange and the total entropy of the system increases over the course of the process: $\Delta S > 0$. Therefore, if initially $T_A > T_B$, the heat flux $đQ_A$ must be negative ($\Delta Q_A < 0$) since $1/T_A - 1/T_B < 0$ (in order to make $\Delta S > 0$). That is, heat must flow from the higher-temperature system to the lower-temperature system.[222] Note that the entropy decreases ($\Delta S_A < 0$) for subsystem A, but increases overall.

Important Distinction. Although the equilibrium condition is $dS = 0$, the finite change in the total entropy from the initial equilibrium state to the final equilibrium state is, in general, $\Delta S \geq 0$. The condition $dS = 0$ applies specifically at equilibrium: At equilibrium, for a given internal energy, any unconstrained parameters will assume values that maximize the entropy. A more mathematically literal way of looking at this is: Once thermal equilibrium is reached, a small disturbance from equilibrium, δU, which results in a small transfer of energy from one system to the other, does not result in a change in the total entropy. Between the initial and final equilibrium states, however, dS can be nonzero, thereby resulting in a possible increase in the total entropy: $\Delta S \geq 0$.

---

[220] So the 'freedom' of these parameters is reduced (in this case, it is removed) by the entropy maximum postulate.
[221] Additionally, for the total entropy to indeed be a maximum, it is necessary to require that $d^2 S < 0$.
[222] Here, we have worked in the entropy representation and applied the entropy maximum postulate – i.e. for the given total internal energy, the free parameters are such that the entropy is maximized. We could instead work in the energy representation and apply the energy minimum postulate. In this case, we would 'pretend' that energy were free, and find the combination of free parameters that minimize the energy for a fixed total entropy. (In this case, this 'minimum' energy must be the total internal energy that the system has, which can't really vary.)

Thermodynamic temperature: As we have seen, from a minimum number of postulates, thermodynamics defines a mathematical quantity, T, as a partial derivative of the internal energy of a system with respect to its entropy for fixed volume and mole numbers. Based on the thermodynamic behavior of this mathematical quantity, we interpret this mathematical quantity conceptually as the absolute temperature of the system.

We have already seen from one postulate of thermodynamics that temperature is a positive quantity; and in agreement with the third law of thermodynamics (or the Nernst postulate), another postulate requires that the total entropy of the system approach zero as the absolute temperature approaches zero. We have now just seen that the thermodynamic definition of temperature corresponds to our intuitive notion for thermal equilibrium – i.e. two systems were placed in thermal contact and reached equilibrium when their temperatures were equal. Furthermore, we saw that heat tends to flow from systems with higher temperature to systems with lower temperature, and that this heat flow was associated with an increase in entropy (we will analyze this more carefully in Sec. 6.5). By employing a heat engine, thermodynamics even provides a means of determining the temperature of one system relative to that of another: Specifically, allowing a Carnot engine to operate for one cycle between the higher-temperature and lower-temperature systems, the temperature ratio can be related to the upper limit of experimental efficiency ($T_c/T_h = 1 - e$).

Temperature is also related to the quasistatic heat flux through the equation, $đQ = TdS$. It is the abstract notion of entropy that makes the intuitive connection between heat and temperature. Notice that this equation does not associate a temperature change with heat flow; rather, it associates a change in entropy with heat flow. That is, heat can be exchanged even if temperature is constant, and temperature can change even if there is no heat exchanged. Temperature change and heat change are related through the first law of thermodynamics, which accounts for internal energy changes, heat exchanges, mechanical work, and changing mole numbers.

Example. Three moles of a monatomic ideal gas at an initial temperature of 300 K and five moles of a diatomic ideal gas (for which $c_{2V} = 5R/2$) at an initial temperature of 400 K are separated by an immovable, diathermal wall, which is impermeable to the flow of matter, in a closed system. What is the equilibrium temperature?

We know an equation that relates the absolute temperature of an ideal gas to its internal energy: $U = nc_V T$. For the monatomic ideal gas, $c_{1V} = 3R/2$. Initially, $U_{10} = n_1 c_{1V} T_{10}$ and $U_{20} = n_2 c_{2V} T_{20}$; and finally, $U_1 = n_1 c_{1V} T_1$ and $U_{20} = n_2 c_{2V} T_2$. At equilibrium, $T_1 = T_2 \equiv T$. Conservation of energy requires: $U_{10} + U_{20} = U_1 + U_2$. Thus, $n_1 c_{1V} T_{10} + n_2 c_{2V} T_{20} = n_1 c_{1V} T + n_2 c_{2V} T$. The equilibrium temperature is therefore:

$$T = \frac{n_1 c_{1V} T_{10} + n_2 c_{2V} T_{20}}{n_1 c_{1V} + n_2 c_{2V}} = \frac{3(3R/2)300 \text{ K} + 5(5R/2)400 \text{ K}}{3(3R/2) + 5(5R/2)} = 374 \text{ K}$$

Mechanical equilibrium: Consider a closed system that consists of two subsystems, A and B, at different temperatures and pressures, separated by a movable, adiabatic wall, which is impermeable to the exchange of matter. No heat can be exchanged between the subsystems, but mechanical work can be done. Since neither subsystem exchanges heat nor does chemical work, the first law for each subsystem is:

$$dU_A = -đW_A = -P_A dV_A \ , \ \ dU_B = -đW_B = -P_B dV_B$$

The total internal energy, $U_A + U_B = U$, is conserved – i.e. $dU_A + dU_B = 0$ or $dU_A = -dU_B$. Combining this with the first law, $P_A dV_A = -P_B dV_B$. The total volume of the system, $V_A + V_B = V$, is fixed, such that $dV_A + dV_B = 0$ or $dV_A = -dV_B$. Together, these relations require that $P_A = P_B$. That is, the pressures of the two subsystems must be equal when mechanical equilibrium is reached.

For an infinitesimal variation in the total entropy of the system,

$$\left(\frac{\partial S_A}{\partial U_A}\right)_{V,n_1,n_2,\cdots,n_r} dU_A + \left(\frac{\partial S_B}{\partial U_B}\right)_{V,n_1,n_2,\cdots,n_r} dU_B + \left(\frac{\partial S_A}{\partial V_A}\right)_{U,n_1,n_2,\cdots,n_r} dV_A + \left(\frac{\partial S_B}{\partial V_B}\right)_{U,n_1,n_2,\cdots,n_r} dV_B = dS$$

$$\frac{dU_A}{T_A} + \frac{dU_B}{T_B} + \frac{P_A dV_A}{T_A} + \frac{P_B dV_B}{T_B} = dS$$

Normally, as we will see in the case where heat can be exchanged in addition to mechanical work done, we would impose conservation of energy and factor out a dU_A from the first two terms, and impose a fixed total volume and factor out a dV_A from the last two terms. Then the entropy maximum postulate, $dS = 0$, would impose conditions on the otherwise unconstrained parameters (namely, temperature and pressure). However, we will see that this purely mechanical process provides an exception to this technique. In particular, we will see that dS already equals zero as a result of $P_A = P_B$ – the condition for mechanical equilibrium which we have already derived without imposing the entropy maximum postulate. Since $dS = 0$ already, the temperature remains thermodynamically indeterminate in this purely mechanical process. (If, however, the wall is diathermal instead of adiabatic, then we will see that both the pressures and temperatures must be equal for the final state.)

Picking up from where we left off with the differential entropy expression (which conceptually represents an infinitesimal variation from the final state, which is an equilibrium state), and using the relations, $dU_A = -P_A dV_A$ and $dU_B = -P_B dV_B$, which we had obtained previously, we find that:

$$dS = \frac{dU_A}{T_A} + \frac{dU_B}{T_B} + \frac{P_A dV_A}{T_A} + \frac{P_B dV_B}{T_B} = \frac{dU_A}{T_A} + \frac{dU_B}{T_B} - \frac{dU_A}{T_A} - \frac{dU_B}{T_B} = 0$$

Thus, the entropy is already a maximum for equalized pressures, such that no additional constraint (namely, on temperature) may be placed. Our only conclusion regarding the equilibrium state reached is that the pressures of the two subsystems must be equal.[223,224]

---

[223] Notice that the relation $đQ = TdS$ is not helpful for this purely mechanical process. Not only is no heat exchanged between the system and surroundings, but there is not even any heat exchanged between subsystems. As always, the entropy is only well-defined for the equilibrium position, yet the finite change (ΔS) between equilibrium states can be calculated along a reversible path. Since no heat is exchanged ($\Delta Q = 0$), the entropy must be the same in the initial and final equilibrium states: $\Delta S = 0$. This is why the entropy is already maximized for purely mechanical equilibrium.

[224] Thermal equilibrium is a static equilibrium state, but pure mechanical equilibrium is dynamic. An oscillating pendulum passes through its equilibrium state twice per oscillation, but does not stop at equilibrium in the absence of damping forces.

If the two pressures are different in the initial equilibrium state, the wall will move toward the subsystem of lower pressure. If initially $P_A > P_B$, the wall moves toward subsystem B, thereby decreasing its volume ($\Delta V_B < 0$) while increasing the volume of subsystem A ($\Delta V_A > 0$). The reduced volume of subsystem B raises its pressure ($\Delta P_B > 0$); similarly, the pressure of subsystem A decreases ($\Delta P_A < 0$). Subsystem A expands, doing positive work, which decreases its internal energy, and just the opposite for subsystem B.

Thermal and mechanical equilibrium: Consider a closed system that consists of two subsystems, A and B, at different temperatures and pressures, separated by a movable, diathermal wall, which is impermeable to the exchange of matter. Now mechanical work can be done and heat can be exchanged. For an infinitesimal variation in the total entropy of the system,

$$\left(\frac{\partial S_A}{\partial U_A}\right)_{V,n_1,n_2,\cdots,n_r} dU_A + \left(\frac{\partial S_B}{\partial U_B}\right)_{V,n_1,n_2,\cdots,n_r} dU_B + \left(\frac{\partial S_A}{\partial V_A}\right)_{U,n_1,n_2,\cdots,n_r} dV_A + \left(\frac{\partial S_B}{\partial V_B}\right)_{U,n_1,n_2,\cdots,n_r} dV_B = dS$$

$$\frac{dU_A}{T_A} + \frac{dU_B}{T_B} + \frac{P_A dV_A}{T_A} + \frac{P_B dV_B}{T_B} = dS$$

In this case, conservation of energy, $dU_A = -dU_B$, and fixed total volume, $dV_A = -dV_B$, along with the entropy maximum postulate, $dS = 0$, yield two separate conditions:

$$dS = \left(\frac{1}{T_A} - \frac{1}{T_B}\right) dU_A + \left(\frac{P_A}{T_A} - \frac{P_B}{T_B}\right) dV_A = 0$$

$$1/T_A = 1/T_B \quad , \quad P_A/T_A = P_B T_B$$

That is, equilibrium is reached when the temperatures and pressures of the two subsystems equalize: $T_A = T_B$ and $P_A = P_B$.

Thermodynamic pressure: The mathematical definition of thermodynamic pressure as the partial derivative of the internal energy of the system with respect to the volume of the system for fixed entropy and mole numbers is naturally identified as the pressure when the differential form of the fundamental relation is compared to the first law of thermodynamics. In this way, the usual concept of pressure and its role in thermodynamic work are automatically preserved: $W = \int_{V=V_0}^{V} P dV$. We also see that thermodynamics conforms to the intuitive notion that pressures are equalized when mechanical equilibrium is attained. Also, just as heat has a tendency to be exchanged from a higher-temperature subsystem to a lower-temperature subsystem to attain thermal equilibrium, a movable wall separating two subsystems tends to expand the higher-pressure subsystem and contract the lower-pressure subsystem, which agrees with our intuitive notion of pressure and mechanical equilibrium.

Example. Three moles of a monatomic ideal gas at an initial temperature of 300 K and five moles of a diatomic ideal gas (for which $c_{2V} = 5R/2$) at an initial temperature of 400 K are separated by a movable, diathermal wall, which is impermeable to the flow of matter, in a closed system with a total volume of 12.0 m³. What are the equilibrium temperature and pressure?

Note that this is a variation of the previous example. What is new is that the wall is now movable and we are given the total volume; the other parameters are unchanged. Thus, the same equations apply to this problem: $U_{10} = n_1 c_{1V} T_{10}$, $U_{20} = n_2 c_{2V} T_{20}$, $U_1 = n_1 c_{1V} T_1$, and $U_{20} = n_2 c_{2V} T_2$, where $c_{1V} = 3R/2$. In this mechanical and thermal equilibrium problem, the final temperatures and pressures will be equal at equilibrium: $T_1 = T_2 \equiv T$ and $P_1 = P_2 \equiv P$. For this problem, we will also need to use the ideal gas law to relate the pressures and volumes: $P_{10} V_{10} = n_1 R T_{10}$, $P_{20} V_{20} = n_2 R T_{20}$, $P_1 V_1 = n_1 R T_1$, and $P_2 V_2 = n_2 R T_2$.

Conservation of energy requires: $U_{10} + U_{20} = U_1 + U_2$. Thus, the equilibrium temperature is the same as in the previous example: $T = 374$ K. The sum of the volumes of the subsystems must be constant: $V_{10} + V_{20} = V_1 + V_2 = 12.0$ m^3. Combining this with the equilibrium requirement $P_1 = P_2 \equiv P$,

$$V_1 + V_2 = 12.0 \text{ m}^3 = \frac{n_1 RT}{P} + \frac{n_2 RT}{P}$$

$$P = \frac{n_1 + n_2}{12.0 \text{ m}^3} RT = \frac{3 \text{ mol} + 5 \text{ mol}}{12.0 \text{ m}^3} (8.314 \text{ J/mol/K})(374 \text{ K}) = 2.07 \text{ kPa}$$

Diffusive equilibrium: Consider a closed system that consists of two subsystems, A and B, at different temperatures and chemical potentials, separated by an immovable, adiabatic wall, which is permeable to one chemical component (n_1), but impermeable to the other chemical components. No heat can be exchanged between the subsystems and no mechanical work can be done, but chemical work is done as the particles of one chemical component pass through the wall. Since neither subsystem exchanges heat nor does mechanical work, the first law for each subsystem is:

$$dU_A = đW_{cA} = \mu_{1A} dn_{1A} \quad , \quad dU_B = đW_{cB} = \mu_{1B} dn_{1B}$$

The total internal energy, $U_A + U_B = U$, is conserved – i.e. $dU_A + dU_B = 0$ or $dU_A = -dU_B$. Combining this with the first law, $\mu_{1A} dn_{1A} = -\mu_{1B} dn_{1B}$. The total mole number for chemical component 1, $n_{1A} + n_{1B} = n_1$, is fixed, such that $dn_{1A} + dn_{1B} = 0$ or $dn_{1A} = -dn_{1B}$. Together, these relations require that $\mu_{1A} = \mu_{1B}$. That is, the chemical potentials of the two subsystems must be equal when diffusive equilibrium (or matter flow equilibrium) is reached.

For an infinitesimal variation in the total entropy of the system,

$$\left(\frac{\partial S_A}{\partial U_A}\right)_{V,n_1,n_2,\cdots,n_r} dU_A + \left(\frac{\partial S_B}{\partial U_B}\right)_{V,n_1,n_2,\cdots,n_r} dU_B - \left(\frac{\partial S_A}{\partial n_{1A}}\right)_{U,V,n_{i\neq 1}} dn_{1A} - \left(\frac{\partial S_B}{\partial n_{1B}}\right)_{U,V,n_{i\neq 1}} dn_{1B} = dS$$

$$\frac{dU_A}{T_A} + \frac{dU_B}{T_B} - \frac{\mu_{1A} dn_{1A}}{T_A} - \frac{\mu_{1B} dn_{1B}}{T_B} = dS$$

Since $dU_A = \mu_{1A} dn_{1A}$ and $dU_B = \mu_{1B} dn_{1B}$ for pure diffusive equilibrium, dS is automatically zero without imposing any entropy maximum postulate. Pure diffusive equilibrium is similar to pure mechanical equilibrium in this regard.

Thermal and diffusive equilibrium: Consider a closed system that consists of two subsystems, A and B, at different temperatures and chemical potentials, separated by an immovable, diathermal wall, which is permeable to one chemical component (n_1), but impermeable to the other chemical components. Now but heat can be exchanged and chemical work can be done. For an infinitesimal variation in the total entropy of the system,

$$\left(\frac{\partial S_A}{\partial U_A}\right)_{V,n_1,n_2,\cdots,n_r} dU_A + \left(\frac{\partial S_B}{\partial U_B}\right)_{V,n_1,n_2,\cdots,n_r} dU_B - \left(\frac{\partial S_A}{\partial n_{1A}}\right)_{U,V,n_{i\neq 1}} dn_{1A} - \left(\frac{\partial S_B}{\partial n_{1B}}\right)_{U,V,n_{i\neq 1}} dn_{1B} = dS$$

$$\frac{dU_A}{T_A} + \frac{dU_B}{T_B} - \frac{\mu_{1A}dn_{1A}}{T_A} - \frac{\mu_{1B}dn_{1B}}{T_B} = dS$$

In this case, conservation of energy, $dU_A = -dU_B$, and fixed total mole number for chemical component 1, $dn_{1A} = -dn_{1B}$, along with the entropy maximum postulate, $dS = 0$, yield two separate conditions:

$$dS = \left(\frac{1}{T_A} - \frac{1}{T_B}\right) dU_A + \left(\frac{\mu_{1A}}{T_A} - \frac{\mu_{1B}}{T_B}\right) dn_{1A} = 0$$
$$1/T_A = 1/T_B \ , \ \mu_{1A}/T_A = \mu_{1B}/T_B$$

That is, equilibrium is reached when the temperatures and chemical potentials of the two subsystems equalize: $T_A = T_B$ and $\mu_{1A} = \mu_{1B}$.

Phase equilibrium: Consider a closed system that is not divided into subsystems, but where one chemical component coexists in two phases. In this case, n_1 can be divided into n_{1A} and n_{1B}, where A and B represent the two different phases. This problem is formally identical to the case of diffusive and thermal equilibrium, for which $T_A = T_B$ and $\mu_{1A} = \mu_{1B}$ when the equilibrium state is reached. Thus, $T_A = T_B$ and $\mu_{1A} = \mu_{1B}$ for phase equilibrium.

Chemical equilibrium: Consider a closed system that is not divided into subsystems, but where two chemical components (n_1 and n_2) undergo chemical reactions. For an infinitesimal variation in the total entropy of the system (where there is no longer any division between subsystems),

$$-\left(\frac{\partial S}{\partial n_1}\right)_{U,V,n_{i\neq 1}} dn_1 - \left(\frac{\partial S}{\partial n_2}\right)_{U,V,n_{i\neq 2}} dn_2 = dS$$

$$-\frac{\mu_1 dn_1}{T} - \frac{\mu_2 dn_2}{T} = dS$$

The entropy maximum postulate, $dS = 0$, requires that $\mu_1 dn_1 = -\mu_2 dn_2$. For $n_1 + n_2 = const.$, $dn_1 = -dn_2$ and $\mu_1 = \mu_2$ in phase equilibrium. However, in a typical chemical reaction, two chemical components combine to form additional chemical components. The chemical reaction must be balanced, though, so in this perspective, we can write, in general, $\sum_{i=1}^{r} \mu_i \nu_i = 0$ at equilibrium, where ν_i are the stoichiometric coefficients – i.e. the coefficients of the chemical reaction after moving all terms to the same side of the equation.

Example. For the production of carbon dioxide, $2C + O_2 \rightarrow CO_2$, what are the stoichiometric coefficients?

Bringing all terms to the right-hand side, the stoichiometric coefficients are: $\nu_C = -2$, $\nu_{O_2} = -1$, and $\nu_{CO_2} = 1$.

Thermodynamic chemical potential: We see that the partial derivative of internal energy with respect to the mole number for a given chemical component, holding the other independent extensive variables constant, which is analogous to the mathematical definitions of temperature and pressure, plays a role that is similar to other potentials – such as gravitational potential or electric potential. Masses tend to move from higher to lower gravitational potential, and positive charges tend to move from higher to lower electric potential. Similarly, particles tend to move from higher to lower chemical potential. Chemical potential tends to be negative, like gravitational and electrical potential for bound particles: Conceptually, binding potentials are negative in the sense that they need energy to become free, at which point they would have zero energy – or, put another way, the reference position is at infinity. So particles tend to move from less negative to more negative potential. As with other potentials, the negative gradient of the potential – in this case, $-\vec{\nabla}\mu$ – represents the attractive force by which the particle accelerates in the direction of more negative potential. When two different chemical potentials equalize in diffusive equilibrium, the net force associated with the matter flow becomes zero. In this regard, chemical potential is a sort of matter flow potential.

Radiative equilibrium: When a system exchanges heat with its surroundings – or when one subsystem exchanges heat with another subsystem – in the form of thermal radiation, radiative equilibrium is reached when the rate that the system (or subsystem) emits electromagnetic radiation equals the rate at which it absorbs electromagnetic radiation from its surroundings (or the other subsystem). According to Stefan's law, the rate at which thermal radiation is emitted depends upon the temperature of the object, while the rate at which it is absorbed depends upon the temperature of the surroundings (or the other subsystem). In the case of purely radiative process, the temperature of the system (or subsystem) decreases if its temperature exceeds that of its surroundings (or the other subsystem) and increases if its temperature is less than that of its surroundings (or the other subsystem). Thus, pure radiative equilibrium is attained when the emission and absorption rates, and therefore the temperatures of the system and surroundings (or the two subsystems), equalize.

Thermodynamic equilibrium: A system is in thermodynamic equilibrium if the system is in thermal equilibrium, mechanical equilibrium, diffusive equilibrium, and radiative equilibrium.[225]

Reversibility: A quasistatic process, carried out over a series of equilibrium states, may be reversible (i.e. if there are no frictional or dissipative forces present). In this case, entropy is well-defined. A process that evolves through non-equilibrium states is irreversible. The overall entropy increases between initial and final equilibrium states for an irreversible process.

---

[225] However, the term is sometimes used more loosely.

6.3 Thermodynamic Potentials

Alternative formulations: The formulation of thermodynamics in terms of the energy or entropy representations is not at all convenient for applications because temperature and pressure – two of the intensive parameters – are much easier to measure in the laboratory than entropy – one of the extensive parameters. In thermodynamics,[226] it would be much more convenient to develop a formulation where the fundamental relation is a function of temperature or pressure, rather than entropy. This turns out to be possible, and the mechanism is called a Legendre transformation. The result of transforming the fundamental relation in the energy representation[227] via a Legendre transformation is called a thermodynamic potential. Such thermodynamic potential formulations in which entropy is a dependent variable tend to be more convenient than the energy or entropy representations.

Single-variable Legendre transformation: Imagine a considerably simplified scenario in which the internal energy is a function of a single extensive parameter, the entropy. In this case, the fundamental relation would have the form $U = U(S)$ and the total derivative would be $dU = TdS$ where $T = dU/dS$.[228] This simple picture consists of two extensive parameters and one intensive parameter. In the energy representation, the internal energy is the dependent variable and the entropy is the only independent extensive variable.

Our goal is to perform a transformation to a thermodynamic potential, A, which is a function of the temperature rather than the entropy: $A = A(T)$. We would like to do this without any cost – i.e. we would like the fundamental relation in the thermodynamic potential representation, $A = A(T)$, to include all of the thermodynamic content contained in the fundamental relation in the energy representation, $U = U(S)$.

The connection between the two formulations is the relationship between temperature and entropy: $T = dU/dS$. That is, A is a function of dU/dS. While U is a function of S, A is a function of the slope of U. Yet the slope of U by itself is not equivalent to knowing U. Given a function, $f(x)$, its slope, $f'(x)$, is unambiguous – as a derivative has not ambiguity; but given the slope (derivative) of a function, $f'(x)$, the function is ambiguous – as an indefinite integral suffers a constant of integration.

So we desire to know more than just the slope of U. Knowing both the slope and intercept of the tangent of U is equivalent to knowing U because the additional specification of the intercept removes the ambiguity associated with the would-be constant of integration. This is the spirit with which the Legendre transformation is applied in thermodynamics.

---

[226] In statistical mechanics, however, there is a direct method available for computing entropy.

[227] If the main goal is to replace the entropy with one of the intensive parameters, the Legendre transformation needs to be applied in the energy representation such that one of the arguments – entropy – can be replaced. A Legendre transformation can be applied in the entropy representation, but doing so does not eliminate the entropy. This does have some uses, and the transforms of the entropy are called Massieu functions.

[228] In this simplified picture, there are no other extensive parameters (namely, volume and mole number), and so temperature is defined as a derivative rather than the usual partial derivative. This is obviously not the case in any real thermodynamic model. When we generalize this technique to multiple variables, temperature will again be expressed as its usual partial derivative.

The thermodynamic potential A is, in this single-variable case, the vertical intercept of the tangent line defined by U's slope. The fundamental relation in the thermodynamic potential representation is obtained by expressing the vertical intercept of the tangent line in terms of its slope. Consider a particular point, (S, U), in the U-S plane.[229] The slope of $U(S)$ at that point equals T evaluated at S. The tangent line passes through the point (S, U) and the vertical intercept, $(0, A)$. The slope of the tangent line equals the rise over the run of the tangent line, which is the temperature: $T = (U - A)/(S - 0) = (U - A)/S$. Solving for A gives $A = U - TS$. Thus, the tangent line passing through the point (S, U) has slope T and vertical intercept A.[230]

Given $U = U(S)$, following is the prescription for obtaining $A = A(T)$. First, take a derivative of U with respect to S; set the result equal to T. The thermodynamic potential is the vertical intercept, which means that $A = -TS + U$. Now you have three equations: the original equation, $U = U(S)$; its derivative set equal to T; and the equation for the thermodynamic potential, $A = -TS + U$. Eliminate U and S from these three equations. What remains is the new fundamental relation, $A = A(T)$.

Example. Given the function $U(S) = cS^3 + b$, where b and c are constants, determine $A = A(T)$.

Following the prescription for the Legendre transformation, we first set the derivative equal to the temperature: $T = dU/dS = 3cS^2$. We now have three equations:[231]

$$U = cS^3 + b \quad , \quad T = 3cS^2 \quad , \quad A = -TS + U$$

Algebraically, we wish to express the $-TS + U$ of the last equation in terms of T (and any constants) only. Thus, we should solve for S in the second equation: $S = (3c)^{-1/2} T^{1/2}$. Then we substitute this into the first equation: $U = 3^{-3/2} c^{-1/2} T^{3/2} + b$. Lastly, we substitute both U and S into the last equation:

$$A(T) = -\frac{T^{3/2}}{3^{1/2} c^{1/2}} + \frac{T^{3/2}}{3^{3/2} c^{1/2}} + b = \frac{T^{3/2}}{3^{1/2} c^{1/2}} \left(-1 + \frac{1}{3} \right) + b$$

$$A(T) = -\frac{2 T^{3/2}}{3^{3/2} c^{1/2}} + b$$

---

[229] In math, we conventionally put x on the horizontal axis and y on the vertical axis, and call this the xy plane. However, when we place P and V (or P and T) on the same diagram with P on the vertical axis, we call it a P-V diagram. So we call it the P-V plane, which goes against the convention of the xy plane.

[230] If this simplified picture with a single independent parameter is to satisfy all of the postulates of thermodynamics, then according to the Euler equation for this simplified picture, A would equal zero! However, we're just illustrating the main principles of the Legendre transformation here, which are easier to convey with a single independent variable in mind. Obviously, the examples that we provide for the single-variable Legendre transformation will not involve thermodynamically viable fundamental relations. Instead, they will illustrate the mathematical technique. We could have chosen to work with more general mathematical notation, such as $y(x)$, but working with $U(S)$ has the advantage of being much more tangible in terms of applying to thermodynamics.

[231] Notice that these are not plausible extensive and intensive variables. That is, U is not a first-order homogeneous function of S and T is not a zero-order homogeneous function of S. Nonetheless, this function serves to exemplify the mathematical method, for which the thermodynamic properties are irrelevant.

Inverse single-variable Legendre transformation: Let us now consider the inverse problem: Given the fundamental relation in the thermodynamic potential representation, $A = A(T)$, how do you determine the fundamental relation in the energy representation, $U = U(S)$?

Consider the equation $A = -TS + U$, which can also be expressed as $U = TS + A$. Observe that $T = dU/dS$ and $S = -dA/dT$. In the Legendre transformation, we use $U(S)$, $T = dU/dS$, and $A = -TS + U$ to transform from $U = U(S)$ to $A = A(T)$. In the inverse Legendre transformation, we use $A(T)$, $S = -dA/dT$, and $U = TS + A$ to transform from $A = A(T)$ to $U = U(S)$.

Given $A = A(T)$, following is the prescription for obtaining $U = U(S)$. First, take a derivative of A with respect to T; set the result equal to $-S$. The internal energy is equal to $U = TS + A$. Now you have three equations: the original equation, $A = A(T)$; its derivative set equal to $-S$; and the equation for the internal energy, $U = TS + A$. Eliminate A and T from these three equations. What remains is the new fundamental relation, $U = U(S)$.

Example. Given the function $A(T) = cT^2 + b$, where b and c are constants, determine $U = U(S)$.

Following the prescription for the inverse Legendre transformation, we first set the derivative equal to the negative of the entropy: $S = -dA/dT = -2cT$. We now have three equations:[232]

$$A = cT^2 + b \quad , \quad S = -2cT \quad , \quad U = TS + A$$

Algebraically, we wish to express the $TS + A$ of the last equation in terms of S (and any constants) only. Thus, we should solve for T in the second equation: $T = -S/2c$. Then we substitute this into the first equation: $A = S^2/(4c) + b$. Lastly, we substitute both A and T into the last equation:

$$U(S) = -\frac{S^2}{2c} + \frac{S^2}{4c} + b = -\frac{S^2}{4c} + b$$

Multi-variable Legendre transformation: In thermodynamics, the internal energy is not a function of a single extensive parameter, but of $r + 2 \geq 3$ independent extensive parameters: $U(S, V, n_1, n_2, \cdots, n_r)$. We may wish to replace k of these $r + 2$ extensive parameter(s), $S, V, n_1, n_2, \cdots, n_r$, by their conjugate variable(s) – $T, P, \mu_1, \mu_2, \cdots, \mu_r$, respectively – where $1 \leq k \leq r + 2$ – via a Legendre transformation to a thermodynamic potential. The following thermodynamic potentials prove to be rather useful in thermodynamics, statistical mechanics, and/or chemistry, and therefore are quite common:

- Helmholtz free energy, $F = F(T, V, n_1, n_2, \cdots, n_r)$, where S is replaced with T.
- Gibbs free energy, $G = G(T, P, n_1, n_2, \cdots, n_r)$, where S and V are replaced with T and P.
- Grand canonical potential, $\Psi = \Psi(T, V, \mu)$, where S and n are replaced with T and μ.[233]
- Enthalpy, $H = H(S, P, n_1, n_2, \cdots, n_r)$, where V is replaced with P.

---

[232] Notice that these are not plausible extensive and intensive variables. That is, U is not a first-order homogeneous function of S and T is not a zero-order homogeneous function of S. Nonetheless, this function serves to exemplify the mathematical method, for which the thermodynamic properties are irrelevant.

[233] The grand canonical potential, represented by the uppercase Greek letter psi (Ψ), is used when there is just a single chemical component.

The fundamental relation in the energy representation, $U(S,V,n_1,n_2,\cdots,n_r)$, represents an $(r+1)$-dimensional hypersurface. Each conjugate variable,

$$T \equiv \left(\frac{\partial U}{\partial S}\right)_{V,n_1,n_2,\cdots,n_r} \quad , \quad P \equiv -\left(\frac{\partial U}{\partial V}\right)_{S,n_1,n_2,\cdots,n_r} \quad , \quad \mu_i \equiv \left(\frac{\partial U}{\partial n_i}\right)_{S,V,n_{j\neq i}}$$

represents a partial slope of the tangent hyperplane. The k conjugate variables replacing the k extensive parameters in the Legendre transformation provide k partial slopes of the tangent hyperplane at any point $(S,V,n_1,n_2,\cdots,n_r)$. The thermodynamic potential is the vertical intercept of the tangent hyperplane. The fundamental relation in the thermodynamic potential representation expresses the vertical intercept of the tangent hyperplane as a function of its partial slopes with respect to the k extensive parameters replaced in the Legendre transformation.

Given $U(S,V,n_1,n_2,\cdots,n_r)$, following is the prescription for performing the Legendre transformation to replace k of the extensive parameters with their corresponding conjugate variables. First, take derivatives of U with respect to each of the k extensive parameters to be replaced; set the results equal to their conjugate variables (i.e. T, $-P$, or μ_i). The thermodynamic potential is the vertical intercept, which means that the thermodynamic potential equals U minus k terms of products of conjugate variables (e.g. $U-TS$ or $U-TS+PV$). Now you have $k+2$ equations: the original equation, $U = U(S,V,n_1,n_2,\cdots,n_r)$; k derivatives set equal to intensive parameters; and the equation for thermodynamic potential. Eliminate U and the k extensive parameters to be replaced from these $k+2$ equations. What remains is the new fundamental relation in the thermodynamic potential representation.

Inverse multi-variable Legendre transformation: Given the thermodynamic potential, following is the prescription for performing the Legendre transformation to replace all k of the intensive parameters with their corresponding conjugate variables. First, take derivatives of the thermodynamic potential with respect to each of the k intensive parameters; set the results equal to the negative of their conjugate variables (i.e. $-S$, $+V$, or $-N_i$). The thermodynamic potential is the vertical intercept, which means that the thermodynamic potential equals U minus k terms of products of conjugate variables (e.g. $U-TS$ or $U-TS+PV$): Solve for U in this equation. Now you have $k+2$ equations: the original equation for the thermodynamic potential; k derivatives set equal to intensive parameters; and the equation for thermodynamic potential. Eliminate the thermodynamic potential and the k intensive parameters from these $k+2$ equations. What remains is the fundamental relation in the energy representation.

Helmholtz free energy: The most obvious thermodynamic potential – obtained by replacing the entropy with the temperature – gives rise to the Helmholtz free energy, F. The fundamental relation in the Helmholtz representation is $F = F(T,V,n_1,n_2,\cdots,n_r)$. The Helmholtz potential is related to the internal energy by $F = U - TS$. In the Helmholtz representation, the entropy is given by $S = -\left(\frac{\partial F}{\partial T}\right)_{V,n_1,n_2,\cdots,n_r}$ and the total differential is $dF = -SdT - PdV + \sum_{i=1}^{r}\mu_i dn_i$. The Helmholtz representation is particularly useful for thermodynamic systems in contact with a thermal reservoir, and also makes many calculations much simpler in statistical mechanics.

Example. The fundamental relation for a system is $U = aS^3V^{-1}n^{-1}$, where a is a constant. Find the fundamental relation in the Helmholtz representation.

After finding the temperature, $T = \left(\frac{\partial U}{\partial S}\right)_{V,n} = 3aS^2V^{-1}n^{-1}$, there are three equations:

$$U = aS^3V^{-1}n^{-1} \quad , \quad T = 3aS^2V^{-1}n^{-1} \quad , \quad F = U - TS$$

We need to eliminate U and S to obtain the fundamental relation, $F = F(T,V,n)$. We can do this by solving for S in the middle equation: $S = 3^{-1/2}a^{-1/2}T^{1/2}V^{1/2}n^{1/2}$. We substitute this expression into the first equation: $U = 3^{-3/2}a^{-1/2}T^{3/2}V^{1/2}n^{1/2}$. Substituting these equations into the last expression,

$$F = \frac{T^{3/2}V^{1/2}n^{1/2}}{3^{3/2}a^{1/2}} - \frac{T^{3/2}V^{1/2}n^{1/2}}{3^{1/2}a^{1/2}} = -\frac{2T^{3/2}V^{1/2}n^{1/2}}{3^{3/2}a^{1/2}}$$

Example. The Helmholtz fundamental relation for a system is $F = -3 \cdot 4^{-4/3}a^{-1/3}T^{4/3}V^{2/3}n^{1/3}$, where a is a constant. Find the fundamental relation in the energy representation.

We first take a derivative to find the entropy, $S = -\left(\frac{\partial F}{\partial T}\right)_{V,n} = 4^{-1/3}a^{-1/3}T^{1/3}V^{2/3}n^{1/3}$. There are now three equations:

$$F = -3 \cdot 4^{-4/3}a^{-1/3}T^{4/3}V^{2/3}n^{1/3} \quad , \quad S = 4^{-1/3}a^{-1/3}T^{1/3}V^{2/3}n^{1/3} \quad , \quad U = F + TS$$

We need to eliminate F and T to obtain the fundamental relation, $U = U(S,V,n)$. We can do this by solving for T in the middle equation: $T = 4aS^3V^{-2}n^{-1}$. We substitute this expression into the first equation: $F = -3aS^4V^{-2}n^{-1}$. Substituting these equations into the last expression,

$$U = -\frac{3aS^4}{V^2n} + \frac{4aS^4}{V^2n} = \frac{aS^4}{V^2n}$$

Gibbs free energy: It is sometimes useful to replace entropy with the temperature and also replace volume with pressure. The corresponding thermodynamic potential is called the Gibbs free energy, G. The fundamental relation in the Gibbs representation is $G = G(T, P, n_1, n_2, \cdots, n_r)$. The Gibbs potential is related to the internal energy by $G = U - TS + PV$. In the Gibbs representation, the entropy is given by $S = -\left(\frac{\partial G}{\partial T}\right)_{P,n_1,n_2,\cdots,n_r}$, the volume is given by $V = \left(\frac{\partial G}{\partial P}\right)_{T,n_1,n_2,\cdots,n_r}$, and the total differential is $dG = -SdT + VdP + \sum_{i=1}^{r} \mu_i dn_i$. The Gibbs potential is well-suited for thermodynamic systems in contact with both pressure and thermal reservoirs, and is also useful for equilibrium conditions of chemical reactions.

Example. The fundamental relation for a system is $U = aS^3V^{-1}n^{-1}$, where a is a constant. Find the fundamental relation in the Gibbs representation.

We first find temperature, $T = \left(\frac{\partial U}{\partial S}\right)_{V,n} = 3aS^2V^{-1}n^{-1}$, and pressure, $P = -\left(\frac{\partial U}{\partial V}\right)_{S,n} = 3aS^2V^{-1}n^{-1}$. There are now four equations:

$$U = aS^3V^{-1}n^{-1} \quad , \quad T = 3aS^2V^{-1}n^{-1} \quad , \quad P = aS^3V^{-2}n^{-1} \quad , \quad G = U - TS + PV$$

We need to eliminate U, S, and V to obtain the fundamental relation, $G = G(T, P, n)$. We can do this by first solving for V in the third equation: $V = a^{1/2}S^{3/2}P^{-1/2}n^{-1/2}$. We next substitute this into the second equation and solve for S: $S = 3^{-2}a^{-1}T^2P^{-1}n$. Substituting this back into the equation for V, $V = 3^{-3}a^{-1}T^3P^{-2}n$. We substitute the equations for S and V into the first equation: $U = 3^{-3}a^{-3}T^3P^{-1}n$. Substituting these equations into the last expression,

$$G = \frac{T^3 n}{3^3 aP} - \frac{T^3 n}{3^2 aP} + \frac{T^3 n}{3^3 aP} = -\frac{T^3 n}{3^3 aP}$$

Grand canonical potential: A thermodynamic potential that is useful chiefly in statistical mechanics is the grand canonical potential, Ψ. The fundamental relation in the grand canonical representation is $\Psi = \Psi(T, V, \mu)$. Note that there is just one chemical component here. The grand canonical potential is related to the internal energy by $\Psi = U - TS - \mu n$. In the grand canonical representation, the entropy is given by $S = -\left(\frac{\partial \Psi}{\partial T}\right)_{V,\mu}$, the mole number is given by $n = \left(\frac{\partial \Psi}{\partial \mu}\right)_{T,V}$, and the total differential is $d\Psi = -SdT - PdV - Nd\mu$. The grand canonical potential is useful for a system in contact with energy and particle reservoirs.

Enthalpy: A thermodynamic potential that is particularly common in chemistry is the enthalpy, H. The fundamental relation in the enthalpy representation is $H = H(S, P, n_1, n_2, \cdots, n_r)$. The enthalpy is related to the internal energy by $H = U + PV$. In the enthalpy representation, the volume is given by $V = \left(\frac{\partial H}{\partial P}\right)_{S,n_1,n_2,\cdots,n_r}$ and the total differential is $dH = TdS + VdP + \sum_{i=1}^{r} \mu_i dn_i$.

The enthalpy of a system differs from the internal energy by the term PV, which can be conceptually interpreted as the energy stored in the surroundings. This is useful for chemical reactions that occur in open vessels, as in that case the system is pushing against atmospheric pressure. In such cases the atmosphere serves as a reservoir for pressure, which means that the process occurs at approximately constant pressure. For these reasons, enthalpy is very common in chemistry. For an isobaric process carried out with constant mole numbers, $dH = TdS = đQ$. Thus, in chemistry, the enthalpy is often interpreted as a potential for heat, where it is called the heat content and establishes the heat of reactions. Compare to the Gibbs free energy, which is useful for the equilibrium conditions of chemical reactions.

Complete Legendre transformation: It is interesting to note that the complete Legendre transformation – i.e. every independent extensive parameter is replaced by its conjugate variable – of the internal energy, which equals $U - TS + PV + \sum_{i=1}^{r} \mu_i n_i$, is zero according to the Euler equation.

6.4 Fundamental Relations for Simple Systems

Applied thermodynamics: When thermodynamics is applied to real systems, the starting point usually consists of one or more empirical relations between the intensive parameters. It is generally feasible to put together one or two equations of state, but there are generally one or more missing equations of state. If two or more equations of state are known, it is important to check that they are thermodynamically compatible – because not any set of equations of state that one may write down will satisfy the postulates of thermodynamics. As a partial check of compatibility, given two equations of state, we can check if internal energy and entropy are perfect differentials.

If just one equation of state is missing, the technique of integrating the first law or the Gibbs-Duhem relation with an implicit product rule antiderivative can be applied. If two equations of state are missing, another equation of state can still be found that is compatible with the known equations of state up to an undetermined function, and then the usual integration technique can be applied to find the fundamental relation. Thus, the fundamental relation for a real system is generally unknown up to an arbitrary integration constant at best, and up to an undetermined function in other cases.

Compatibility of equations of state: Consider the first law in molar form for a system with a single chemical component: $du = Tds - Pdv$. The mixed second-order partial derivatives of u with respect to s and v must be independent of the order in which they are applied in order for du to be a perfect differential:[234]

$$\frac{\partial^2 u}{\partial v \partial s} = \frac{\partial^2 u}{\partial s \partial v}$$

$$\left(\frac{\partial T}{\partial v}\right)_s = -\left(\frac{\partial P}{\partial s}\right)_v$$

Thus, the equations of state, $T(s,v)$ and $P(s,v)$ must satisfy this partial derivative relation in order to be thermodynamically compatible. However, this requires knowing the equations of state in terms of entropy. This is more plausible in statistical mechanics, where entropy can be calculated directly, but it is not convenient in thermodynamics. So let us reconsider compatibility in the entropy representation.

Isolating the entropy in the first law, $ds = du/T + Pdv/T$. In order for the entropy to be a perfect differential, the equations of state must satisfy:[235]

$$\frac{\partial^2 s}{\partial v \partial u} = \frac{\partial^2 s}{\partial u \partial v}$$

$$\left(\frac{\partial}{\partial v}\frac{1}{T}\right)_u = \left(\frac{\partial}{\partial u}\frac{P}{T}\right)_v$$

---

[234] In these mixed second-order partial derivatives, s is held constant when differentiating with respect to v, and vice-versa. In obtaining the second equation, we used the fact that $\left(\frac{\partial u}{\partial s}\right)_v = T$ and $\left(\frac{\partial u}{\partial v}\right)_s = -P$.

[235] Here we have used the fact that $\left(\frac{\partial s}{\partial u}\right)_v = 1/T$ and $\left(\frac{\partial s}{\partial v}\right)_u = P/T$.

This requires knowing the equations of state for temperature and pressure in terms of molar volume and molar internal energy, which is a much more reasonable expectation. We can use the relationship between the partial derivatives of $1/T$ and P/T to check for compatibility; and if a single equation of state is known, we can use this compatibility relation to determine a missing equation of state up to an undetermined function.

Example. Two equations for a system are known: $Pv^{1/2} = aT^2$ and $u = 2av^{1/2}T^2$. Are these two equations thermodynamically compatible?

First, solve for $1/T$ and P/T in terms of u and v; these equations will have the form of the equations of state in the entropy representation. From the second equation, $1/T = 2^{1/2}a^{1/2}v^{1/4}u^{-1/2}$. Substituting this in for one power of $1/T$ in the first equation, $P/T = 2^{-1/2}a^{1/2}v^{-3/4}u^{1/2}$. Now compatibility can be checked:

$$\left(\frac{\partial}{\partial v}\frac{1}{T}\right)_u = \left(\frac{\partial}{\partial v}\frac{2^{1/2}a^{1/2}v^{1/4}}{u^{1/2}}\right)_u = \frac{a^{1/2}}{2^{3/2}u^{1/2}v^{3/4}}$$

$$\left(\frac{\partial}{\partial u}\frac{P}{T}\right)_v = \left(\frac{\partial}{\partial u}\frac{a^{1/2}u^{1/2}}{2^{1/2}v^{3/4}}\right)_v = \frac{a^{1/2}}{2^{3/2}u^{1/2}v^{3/4}}$$

Since both of these partial derivatives are equal, the two equations of state are compatible.[236]

Missing two equations of state: Suppose that we have a system with a single chemical component for which a single equation of state is known: $T = T(s, v)$. We can compute $\left(\frac{\partial T}{\partial v}\right)_s$, which we know must be equal to $-\left(\frac{\partial P}{\partial s}\right)_v$ in order for the two equations of state to be compatible. The partial derivative $\left(\frac{\partial T}{\partial v}\right)_s$ will generally be a function of both s and v. We can solve for the second equation of state, $P = P(s, v)$, by integrating the following first-order differential equation, treating v as a constant:

$$\left(\frac{\partial T}{\partial v}\right)_s = -\frac{dP}{ds} \quad , \quad P = \lambda(v) - \int \left(\frac{\partial T}{\partial v}\right)_s ds$$

where $\lambda(v)$ is an undetermined[237] function of integration – since we treated v as a constant.

If instead we knew only $P = P(s, v)$, we could determine the second equation of state, $T = T(s, v)$ by integrating the following first-order differential equation, treating s as a constant:

$$\left(\frac{\partial P}{\partial s}\right)_v = -\frac{dT}{dv} \quad , \quad T = \lambda(s) - \int \left(\frac{\partial P}{\partial s}\right)_v dv$$

---

[236] For a complete check of compatibility, we should check that the postulates of thermodynamics are satisfied.
[237] The function $\lambda(v)$ is not completely arbitrary. The postulates of thermodynamics restrict the possibilities.

> **Example.** Find an equation of state that is compatible with $P = as^2 v^{-3/2}$, where a is a constant. Applying the appropriate integral, the second equation of state, $T = T(s, v)$, is:[238]
>
> $$T = \lambda(s) - \int \left(\frac{\partial P}{\partial s}\right)_v dv = \lambda(s) - \int \frac{2asdv}{v^{3/2}} = \lambda(s) - 2as \int \frac{dv}{v^{3/2}} = \frac{4as}{v^{1/2}} + \lambda(s)$$
>
> where s is treated as a constant in the integration.[239] The equation $T = 4as v^{-1/2}$ would be a compatible equation of state, as would the same equation plus an undetermined function of molar entropy, $\lambda(s)$.

In thermodynamics, if only one equation of state is known, it is more likely to be known as a function of internal energy than as a function of entropy. In this case, it is more practical to work in the entropy representation. Suppose that we have a system with a single chemical component for which a single equation of state is known: $1/T = 1/T(u, v)$. We can compute $\left(\frac{\partial}{\partial v}\frac{1}{T}\right)_u$, which we know must be equal to $\left(\frac{\partial}{\partial u}\frac{P}{T}\right)_v$ in order for the two equations of state to be compatible. The partial derivative $\left(\frac{\partial}{\partial v}\frac{1}{T}\right)_u$ will generally be a function of both u and v. We can solve for the second equation of state, $P/T = P/T(u, v)$, by integrating the following first-order differential equation, treating v as a constant:

$$\left(\frac{\partial}{\partial v}\frac{1}{T}\right)_u = \frac{d}{du}\frac{P}{T} \quad , \quad \frac{P}{T} = \lambda(v) + \int \left(\frac{\partial}{\partial v}\frac{1}{T}\right)_u du$$

If instead we knew only $P/T = P/T(u, v)$, we could determine the second equation of state, $1/T = 1/T(u, v)$ by integrating the following first-order differential equation, treating u as a constant:

$$\left(\frac{\partial}{\partial u}\frac{P}{T}\right)_v = \frac{d}{dv}\frac{1}{T} \quad , \quad \frac{1}{T} = \lambda(u) + \int \left(\frac{\partial}{\partial u}\frac{P}{T}\right)_v dv$$

> **Example.** Find an equation of state that is compatible with $T = au^{2/3} v^{-1/3}$, where a is a constant. Applying the appropriate integral, the second equation of state, $P/T = P/T(u, v)$, is:[240]
>
> $$\frac{P}{T} = \lambda(v) + \int \left(\frac{\partial}{\partial v}\frac{1}{T}\right)_u du = \lambda(v) + \int \left(\frac{\partial}{\partial v}\frac{v^{1/3}}{au^{2/3}}\right)_u du = \lambda(v) + \frac{1}{3av^{2/3}} \int \frac{du}{u^{2/3}} = \frac{u^{1/3}}{av^{2/3}} + f(v)$$
>
> where v is treated as a constant. Therefore, the equation $P/T = u^{1/3} v^{-2/3}/a$ would serve as a compatible equation of state, as would the same equation plus an undetermined function of molar volume, $\lambda(v)$.

---

[238] The function of integration, $\lambda(s)$, also includes any constants of integration.
[239] We're integrating a partial differential equation where s was held constant. Otherwise, we couldn't pull s out. Instead, we would have to express s as a function of v in order to perform the integration.
[240] This time we're integrating a partial differential equation where v was held constant.

Electromagnetic radiation: Consider an enclosed volume, V, in thermal equilibrium where the walls are maintained at constant temperature, T. Macroscopically, the system is a volume of electromagnetic radiation; microscopically, it is a photon gas. Thermal radiation is continuously absorbed and reemitted by the walls in the form of electromagnetic radiation. The photons exert pressure, P, on the walls during these collisions much like a gas of particles exerts a pressure on its container.

The energy density – internal energy per unit volume, ρ_T – provides a thermodynamic measure of the electromagnetic energy of the system. The energy density is related to the radiancy (or intensity) – power per unit area, R_T – of electromagnetic radiation that would escape through a small opening in the enclosure.[241,242] For simplicity, momentarily assume that the enclosure has a hemispherical shape and that a small circular hole is made in the center of the circular base. Since the enclosure was in thermal equilibrium, we expect that the electromagnetic radiation is distributed both homogeneously and isotropically. Consider photons heading toward the hole along the line connecting the top of the hemisphere to the hole, and compare these photons to photons heading toward the hole from some other direction: The hole is effectively larger in the former case and smaller in the latter case by a factor of $\cos\theta$. This is the polar angle of spherical coordinates if we place the origin at the hole and the positive z-axis passes through the top of the hemisphere. More photons escape from directions where θ is smaller. So while the energy density is homogeneous and isotropic, the number of photons contributing to the radiancy depends upon θ.

Thus, a greater fraction of the energy density contributes to the radiancy for smaller θ. The overall factor can be found by integrating the solid angle[243] over the hemisphere. Specifically, the following fraction compares the anisotropy (angular-dependence) of the energy heading toward the hole to the isotropy of the energy's distribution in the enclosure:

$$\frac{\int \cos\theta \, d\Omega}{\int d\Omega} = \frac{\int_{\theta=0}^{\frac{\pi}{2}} \int_{\varphi=0}^{2\pi} \cos\theta \sin\theta \, d\theta d\varphi}{\int_{\theta=0}^{\pi} \int_{\varphi=0}^{2\pi} \sin\theta \, d\theta d\varphi} = \frac{-2\pi \int_{\theta=0}^{\frac{\pi}{2}} \cos\theta \, d(\cos\theta)}{2\pi \int_{\theta=0}^{\pi} \sin\theta \, d\theta} = \frac{[-(\cos^2\theta)/2]_{\theta=0}^{\frac{\pi}{2}}}{[-\cos\theta]_{\theta=0}^{\pi}} = \frac{\pi}{4\pi} = \frac{1}{4}$$

---

[241] Since we are using P for pressure, let us use dU/dt for the instantaneous rate of energy flow, rather than also use P for power, in order to avoid confusion. Many thermodynamics and statistical mechanics texts use a lowercase u to represent the energy density, but we have been using this throughout for molar internal energy. So we are instead using the blackbody energy density notation of some modern physics texts, ρ_T, which matches the notation and language for radiancy, R_T. If you are familiar with the blackbody radiation problem from modern physics, you will be aware that there is a distinction between radiancy and spectral radiancy, for which the latter depends upon frequency; and similarly for energy density. However, this distinction will not impact our thermodynamic concerns.

[242] Ultimately, we will derive a fundamental relation for the electromagnetic radiation of the enclosure assuming that it is sealed and at equilibrium. Thus, this hole is hypothetical as far as the main problem is concerned. Experimentally, however, if the hole is made, measurements of the radiancy through the hole would agree with the energy density inside the enclosure.

[243] Recall that differential solid angle, $d\Omega$, is the angular part of the differential volume element in spherical coordinates: $d\Omega = \sin\theta \, d\theta d\varphi$. The total solid angle of a sphere is 4π steradians. This is a two-dimensional, or solid, angle. When you look at the sun (which you shouldn't do directly if you're fond of your retinas) and the moon in the sky from the earth, they appear about the same size to your eye – their solid angles (or 'apparent' sizes) are roughly the same.

The polar angle has half the range in the numerator as it does in the denominator because the energy represented in the denominator is isotropic throughout the volume, while the energy per unit volume escaping through the hole in the numerator involves only photons heading downward.

The relationship between the radiancy of electromagnetic energy escaping through the hole and the energy density of the enclosure involves this same factor of ¼: $R_T = c\rho_T/4$. There is also a factor of the speed of light, c, which is how fast the radiation propagates through the hole, as the radiancy is power per unit area: power is a derivative of energy with respect to time (there's the dt), and compare the per unit area of radiancy to the per unit volume of energy density (there's the dx). The factor of $c/4$ can also be understood conceptually as follows: There is a factor of ½ from selecting only photons with a downward component of velocity (associated with the smaller upper limit in the numerator's integral) and a factor of $c/2$ from averaging over the photon's exit velocities (associated with the inclusion of a factor of $\cos\theta$ in the numerator's integral).

This result is important because we know that $R_T = \sigma T^4$, according to Stefan's law.[244] Now we can apply Stefan's law to get an equation of state for the system (without that hypothetical hole): $\rho_T = 4\sigma AT^4/c$. In terms of our usual thermodynamic variables: $U = 4\sigma V T^4/c$. In the form of an equation of state, $1/T = (4\sigma V/cU)^{1/4}$.

It is curious that the internal energy is independent of N. The system consists of electromagnetic radiation, which does not consist of particles of matter that must be conserved, but consists of particles of radiation (photons) that are absorbed and reemitted during interactions with the walls of the enclosure. That is, if this were a gas of matter, we could, in principle, count the particles and label the total as N, which would not change in a closed system; but this does not apply to this system of electromagnetic radiation. This is a unique system for which the fundamental relation will depend upon only two extensive parameters: $S = S(U,V)$.

We can find a compatible equation of state via integration. This is a special system where we can do this without using molar variables, since its thermodynamics is independent of mole number:

$$\frac{P}{T} = \lambda(V) + \int \left(\frac{\partial}{\partial V}\frac{1}{T}\right)_U dU = \lambda(V) + \int \left(\frac{\partial}{\partial V}\frac{4^{1/4}\sigma^{1/4}V^{1/4}}{c^{1/4}U^{1/4}}\right)_U dU$$

$$\frac{P}{T} = \lambda(V) + \int \frac{\sigma^{1/4}}{4^{3/4}c^{1/4}U^{1/4}V^{3/4}} dU = \lambda(V) + \frac{4^{1/4}\sigma^{1/4}U^{3/4}}{3c^{1/4}V^{3/4}} = \lambda(V) + \frac{U}{3VT}$$

We can obtain the fundamental equation in this case very simply because the Euler relation for this unique system includes only three terms:

$$S = \frac{U}{T} + \frac{P}{T}V = \frac{U}{T} + \frac{U}{3T} + V\lambda(V) = \frac{4U}{3T} + V\lambda(V) = \frac{4^{5/4}\sigma^{1/4}U^{3/4}V^{1/4}}{3c^{1/4}} + V\lambda(V)$$

For this simple system, $\lambda(V)$ is not undetermined, but must be zero in order that a partial derivative of entropy with respect to volume at constant internal energy reproduce pressure divided by temperature.

---

[244] We derived $R_T = c\rho_T/4$ without any fractional losses to associate with emissivity, as the hypothetical hole provided a clear route for electromagnetic energy to escape cleanly. The hole represents a blackbody radiator – the cavity being a perfect emitter/absorber of thermal radiation.

Therefore, the fundamental relation in the entropy representation is:

$$S(U,V) = \frac{4^{5/4}\sigma^{1/4}U^{3/4}V^{1/4}}{3c^{1/4}}$$

The equations of state are:

$$\frac{1}{T}(U,V) = \frac{4^{1/4}\sigma^{1/4}V^{1/4}}{c^{1/4}U^{1/4}}$$
$$\frac{P}{T}(U,V) = \frac{4^{1/4}\sigma^{1/4}U^{3/4}}{3c^{1/4}V^{3/4}}$$

Observe that $P = U/(3V) = \rho_T/3$. That is, the pressure is equal to $1/3$ of the energy density.

We could have obtained this second equation of state from kinetic theory in analogy with the pressure on an ideal gas – i.e. by considering the pressure that the photons exert on the walls of the enclosure during their collisions. Recall that for an ideal gas, we found the pressure exerted on the walls to be $P_{i.g.} = \rho\overline{v_{i.g.}^2}/3$. In terms of the translational kinetic energy, $P_{i.g.} = m\overline{v_{i.g.}^2}/(3V) = 2K_{i.g.}^{trans}/(3V)$. Photons behave similar to the ideal gas as far as the pressure calculation is concerned, but have two possible polarizations,[245] such that the internal energy of the photon gas is twice the translational kinetic energy of the ideal gas.[246] Therefore, $P_{em} = U_{em}/(3V)$.

The fundamental relation in the energy representation is:

$$U(S,V) = \frac{3^{4/3}c^{1/3}S^{4/3}}{4^{5/3}\sigma^{1/3}V^{1/3}}$$

The equations of state in the energy representation are:

$$T(S,V) = \frac{3^{1/3}c^{1/3}S^{1/3}}{4^{2/3}\sigma^{1/3}V^{1/3}}$$
$$P(S,V) = \frac{3^{1/3}c^{1/3}S^{4/3}}{4^{5/3}\sigma^{1/3}V^{4/3}}$$

Observe that $P = ST/(4V)$.

Compare the equations $PV = ST/4$ and $U = 3PV = 3ST/4$ for a photon gas to the equations $PV = nRT$ and $U = nc_V T = PVc_V/R$ for an ideal gas. An isentropic photon gas thus behaves very much like an ideal gas, but when S is not constant, this makes a significant difference. Unlike an ideal gas, ΔU is generally nonzero for an isothermal process, as U depends upon S as well as T.

---

[245] Photons are spin-one bosons. Spin-one bosons have three possible polarizations, but light is a transverse wave, which removes one of the polarizations. Put another way, the electric (and magnetic) field of an electromagnetic wave can be linearly polarized in any direction in the (2D) plane perpendicular to the direction of propagation.

[246] Photons has zero rest-mass, but do have relativistic mass and carry energy and momentum. Photons have twice the number of degrees of freedom of a monatomic ideal gas. Therefore, according to the theorem of equipartition of energy, a photon gas has twice the internal energy of a monatomic ideal gas.

6 An Introduction to Thermodynamics

The Helmholtz free energy is proportional to the internal energy:

$$F = U - TS = U - \frac{4U}{3} = -\frac{U}{3}$$

$$F(T,V) = -\frac{4\sigma V T^4}{3c}$$

The Gibbs free energy is exactly zero, since it is the complete Legendre transformation of the internal energy for this model that has just two independent extensive variables. The enthalpy equals:

$$H = U + PV = U + \frac{U}{3} = \frac{4U}{3} = 4PV$$

$$H(S,P) = \frac{3^{1/4} c^{1/4} S P^{1/4}}{4^{1/4} \sigma^{1/4}}$$

Example. A photon gas expands isothermally until its volume doubles. Determine how much work is done by the gas and how much heat is exchanged with its surroundings.

Rewrite the pressure in terms of volume and temperature; this will permit integration over volume, since temperature is constant for an isothermal expansion:

$$P = \frac{U}{3V} = \frac{4\sigma T^4}{3c}$$

Since the pressure of a photon gas depends only upon temperature, an isothermal expansion is also isobaric. The work done by the gas is found by integration:

$$W = \int_{V=V_0}^{2V_0} P\, dV = \int_{V=V_0}^{2V_0} \frac{4\sigma T^4}{3c}\, dV = \frac{4\sigma T^4 V_0}{3c} = PV_0$$

Since pressure turns out to be constant, the change in internal energy is:

$$\Delta U = \Delta(3PV) = 3P(2V_0 - V_0) = 3PV_0$$

According to the first law, the heat exchanged with the surroundings is:

$$Q = \Delta U + W = 4PV_0$$

Since it is positive, heat must be added to the system for a photon gas to expand isothermally.

Ideal gas: The ideal gas law, $PV = nRT$, and the equation for internal energy that we derived in Chapter 3, $U = nc_V T$, provide two equations of state. The fundamental relation in the entropy representation can be found from the molar form of the first law:

$$ds = \frac{du}{T} + \frac{Pdv}{T} = \frac{c_V du}{u} + \frac{Rdv}{v}$$

At first this may not look like it could be a perfect differential, but it does suggest a logarithmic structure, and logarithms have some interesting properties. Indeed, the substitution $z \equiv \ln(u^{c_V} v^R)$ leads to $dz = c_V du/u + Rdv/v$.[247,248] We can therefore write $ds = dz$ and integrate both sides, with the following result for the definite integral:

$$s = s_0 + \ln\left[\left(\frac{u}{u_0}\right)^{c_V}\left(\frac{v}{v_0}\right)^R\right] = s_0 + c_V \ln\left(\frac{u}{u_0}\right) + R\ln\left(\frac{v}{v_0}\right)$$

The third equation of state can now be found from the Euler relation:

$$\frac{\mu}{T} = \frac{u}{T} - s + \frac{Pv}{T} = c_V - s_0 - c_V\ln\left(\frac{u}{u_0}\right) - R\ln\left(\frac{v}{v_0}\right) + R = \left(\frac{\mu}{T}\right)_0 - c_V \ln\left(\frac{u}{u_0}\right) - R\ln\left(\frac{v}{v_0}\right)$$

In the entropy representation, the fundamental relation for an ideal gas is:[249]

$$S(U,V,n) = \frac{nS_0}{n_0} + n\ln\left[\left(\frac{U}{U_0}\right)^{c_V}\left(\frac{V}{V_0}\right)^R\left(\frac{n_0}{n}\right)^{c_P}\right] = \frac{nS_0}{n_0} + nc_V\ln\left(\frac{U}{U_0}\right) + nR\ln\left(\frac{V}{V_0}\right) - nc_P\ln\left(\frac{n}{n_0}\right)$$

Recall that $c_P = c_V + R$. The equations of state are:

$$\frac{1}{T}(U,V,n) = \frac{nc_V}{U}$$
$$\frac{P}{T}(U,V,n) = \frac{nR}{V}$$

$$\frac{\mu}{T}(U,V,n) = \left(\frac{\mu}{T}\right)_0 - \ln\left[\left(\frac{U}{U_0}\right)^{c_V}\left(\frac{V}{V_0}\right)^R\left(\frac{n_0}{n}\right)^{c_P}\right] = \left(\frac{\mu}{T}\right)_0 - c_V\ln\left(\frac{U}{U_0}\right) - R\ln\left(\frac{V}{V_0}\right) + c_P\ln\left(\frac{n}{n_0}\right)$$

Observe that $\mu = (c_P - S/n)T$.

In the energy representation, the fundamental relation for an ideal gas is:

$$U(S,V,n) = U_0\left(\frac{V_0}{V}\right)^{\gamma-1}\left(\frac{n}{n_0}\right)^{\gamma} e^{(S-nS_0/n_0)/c_V}$$

Recall that $\gamma = c_P/c_V$. The equations of state are:

---

[247] One interesting property of logarithms is that d/dx of $\ln(ax)$ is independent of a: $\frac{d}{dx}\ln(ax) = \frac{1}{x}$. Compare this with $\frac{d}{dx}\sin(ax) = a\cos(ax)$ or $\frac{d}{dx}ax^2 = 2ax$. We also used $\ln(x^a) = a\ln x$ and $\ln(xy) = \ln x + \ln y$.
[248] We assume that the temperature range is limited such that c_V will be approximately constant.
[249] Note that $s = S/N$ and $s_0 = S_0/N_0$ such that $Ns_0 = NS_0/N_0$ – i.e. s_0 does not equal S_0/N, in general.

$$T(S,V,n) = T_0 \left(\frac{V_0}{V}\right)^{\gamma-1} \left(\frac{n}{n_0}\right)^{\gamma-1} e^{(S-nS_0/n_0)/c_V}$$

$$P(S,V,n) = P_0 \left(\frac{V_0}{V}\right)^{\gamma} \left(\frac{n}{n_0}\right)^{\gamma} e^{(S-nS_0/n_0)/c_V}$$

$$\mu(S,V,n) = \mu_0 \frac{c_P - \frac{S}{n}}{c_P - \frac{S_0}{n_0}} \left(\frac{V_0}{V}\right)^{\gamma-1} \left(\frac{n}{n_0}\right)^{\gamma-1} e^{(S-nS_0/n_0)/c_V}$$

The Helmholtz free energy of an ideal gas is:

$$F(T,V,n) = \frac{nTF_0}{n_0 T_0} - nT \ln\left[\left(\frac{T}{T_0}\right)^{c_V} \left(\frac{V}{V_0}\right)^R \left(\frac{n_0}{n}\right)^R\right]$$

The Gibbs free energy of an ideal gas equals:

$$G(T,P,n) = \frac{nTG_0}{n_0 T_0} - nT \ln\left[\left(\frac{T}{T_0}\right)^{c_P} \left(\frac{P_0}{P}\right)^R\right]$$

The enthalpy of an ideal gas can be expressed quite simply as $H = U + PV = nc_P T$. This form lends it a natural interpretation as heat for the case of constant pressure – as is the case for a system in contact with a pressure reservoir. However, in its proper form, the enthalpy is:

$$H(S,P,n) = H_0 \left(\frac{P}{P_0}\right)^{\frac{\gamma-1}{\gamma}} \left(\frac{n}{n_0}\right) e^{(S-nS_0/n_0)/c_P}$$

Looking at the equation of state for temperature in the energy representation, for example, it can be seen that this thermodynamic system does not satisfy the Nernst postulate. Therefore, the ideal gas model does not exhibit correct behavior at very low temperatures. In Chapter 3, we learned that real gases tend to behave as ideal gases for low densities. At very low temperatures, higher pressures and higher densities are expected (and there comes a point where a real gas makes a phase transition, in which case it can't be expected to be an 'ideal gas'). Thus, the ideal gas model is not, generally, an appropriate model for very low temperatures, and so the problem of not satisfying the Nernst postulate does not restrict the model from its usefulness in describing gases of moderate densities.

Example. One mole of a monatomic ideal gas expands quasistatically according to $P^2V^3 = const.$ until its volume quadruples. Determine what happens to its pressure, temperature, internal energy, entropy, and enthalpy. Also, derive an equation for the work done during the process.

The equation that characterizes the process can be expressed as $P^2V^3 = P_0^2 V_0^3$. Therefore, if the volume increases by a factor of four $V = 4V_0$, the pressure decreases by a factor of eight: $P = P_0(V_0/V)^{3/2} = P_0(1/4)^{3/2} = P_0/8$. From the ideal gas law, the temperature decreases by a factor of two: $T = (P/P_0)(V/V_0)T_0 = (1/8)(4)T_0 = T_0/2$.

Since $U = nc_V T$, the internal energy also halves: $U = U_0/2$. The increase in entropy is:

$$\Delta S = n \ln\left[\left(\frac{U}{U_0}\right)^{c_V} \left(\frac{V}{V_0}\right)^R \left(\frac{n_0}{n}\right)^{c_P}\right] = (1 \text{ mol}) \ln\left(\frac{4^R}{2^{3R/2}}\right) = (1 \text{ mol})R \ln\left(\frac{2^2}{2^{3/2}}\right) = \frac{R \ln 2}{2}(1 \text{ mol})$$

Since $H = nc_P T$, the enthalpy halves just like temperature and internal energy: $H = H_0/2$.

The work done by the gas can be found via integration, which can be performed by using the relation for the process to write pressure as a function of volume:

$$W = \int_{V=V_0}^{V} P\,dV = \int_{V=V_0}^{V} \frac{P_0 V_0^{3/2}}{V^{3/2}}\,dV = P_0 V_0^{3/2} \int_{V=V_0}^{V} V^{-3/2}\,dV = -2P_0 V_0^{3/2}\left(\frac{1}{V^{1/2}} - \frac{1}{V_0^{1/2}}\right)$$

$$W = \frac{2P_0 V_0^{3/2}}{V_0^{1/2}} - \frac{2PV^{3/2}}{V^{1/2}} = 2(P_0 V_0^{1/2} - PV^{1/2})$$

Mixture of Ideal Gases: For a mixture of r ideal gases, Gibbs's theorem states that the fundamental relation in the entropy representation can be found by adding together the entropies that the individual ideal gases would have if it were the only gas in the same container at the same temperature:

$$S(U, V, n) = S_1(U_1, V, n_1) + S_2(U_2, V, n_2) + \cdots + S_r(U_r, V, n_r)$$

$$S_i(U_i, V, n_i) = \frac{n_i S_{i0}}{n_{i0}} + n_i \ln\left[\left(\frac{U_i}{U_{i0}}\right)^{c_{Vi}} \left(\frac{V}{V_0}\right)^R \left(\frac{n_{i0}}{n_i}\right)^{c_{Pi}}\right]$$

$$\frac{1}{T}(U_i, V, n_i) = \frac{n_i c_{Vi}}{U_i}$$

$$\frac{P_i}{T}(U_i, V, n_i) = \frac{n_i R}{V}$$

$$V = V_1 = V_2 = \cdots = V_r$$

$$T = T_1 = T_2 = \cdots = T_r$$

$$n = n_1 + n_2 + \cdots + n_r$$

The Helmholtz representation is particularly well-suited to describe a mixture of ideal gases because the temperature is the same for each ideal gas and, like entropy, the Helmholtz potential is additive:

$$F(T, V, n) = F_1(T, V, n_1) + F_2(T, V, n_2) + \cdots + F_r(T, V, n_r)$$

$$F(T, V, n_i) = \frac{n_i T F_{i0}}{n_{i0} T_0} - n_i T \ln\left[\left(\frac{T}{T_0}\right)^{c_{Vi}} \left(\frac{V}{V_0}\right)^R \left(\frac{n_{i0}}{n_i}\right)^R\right]$$

Virial expansion: The ideal gas model agrees with all gases with moderately low densities, but deviates more with increasing density. For a given gas, a large enough volume provides a low enough density, which suggests that the ideal gas law might be improved by correcting the dependence on molar volume (which relates to the density of the gas). This is achieved through the virial expansion, which replaces $1/v$ with a power series:

$$\frac{P}{RT} = \frac{1}{v}\left[1 + \frac{B_2(T)}{v} + \frac{B_3(T)}{v^2} + \cdots\right]$$

The reason for expanding in terms of $1/v$ is that the corrections become more significant at lower densities, which is where the disagreement with the ideal gas law is larger. The functions $B_i(T)$ are the ith virial coefficients. The virial coefficients take into account the interactions between molecules that are neglected in the ideal gas model. These intermolecular forces become more important as the molecules are compressed to reside closer together on average. Given the virial coefficients, a second equation of state can be found that is compatible with the virial expansion equation of state.

Van der Waals fluid: A particularly useful virial expansion, referred to as the van der Waals fluid because it can accommodate the liquid phase in addition to the gaseous phase, is made using the following virial coefficients:

$$B_2(T) = b - \frac{a}{RT} \quad , \quad B_3(T) = b^2 \quad , \quad B_{i>3}(T) = 0$$

where a and b are constants. Substituting these virial coefficients into the virial expansion yields:

$$\frac{P}{T} = \frac{R}{v} + \frac{bR}{v^2} - \frac{a}{v^2 T} + \frac{b^2 R}{v^3} = \frac{v^2 + bv + b^2}{v^3}R - \frac{a}{v^2 T} = \frac{v^2 + bv + b^2}{v^3}\left(\frac{v-b}{v-b}\right)R - \frac{a}{v^2 T}$$

$$\frac{P}{T} = \frac{v^3 - b^3}{v^3(v-b)}R - \frac{a}{v^2 T} = \frac{R}{v-b} - \frac{a}{v^2 T} + \mathcal{O}(b^3/v^3)$$

The approximation applied in the last step arises from the fact that the terms of the virial expansion are small corrections to the ideal gas law – i.e. we assume that $B_2(T)/v$ and $B_3(T)/v^2$ are small compared to 1, which implies that $b \ll v$. Therefore, terms on the order of b^3/v^3 make a negligible contribution. The thermodynamics for the van der Waals model develops around the following equation:

$$\frac{P}{T} = \frac{R}{v-b} - \frac{a}{v^2 T}$$

Conceptually, we can interpret this equation as follows. The first term, $R/(v-b)$, in the van der Waals equation corrects for the fact that the molecules are not actually point-particles, but occupy a finite fraction of the volume. The coefficient b represents the molar volume of the molecules of the fluid. As anticipated, this term is more significant for more dense fluids. The second term, $a/(v^2 T)$, corrects for the fact that the molecules do interact with one another. In particular, the attraction of molecules to one another reduces the pressure that the outer-lying molecules exert on the container of the fluid. The attractive intermolecular forces depend upon distance – the closer the molecules, the greater the attraction. Therefore, the pressure is reduced more for more concentrated fluids. The quantity a/v^2 represents the amount by which the pressure is reduced.

Recalling that the molar internal energy of an ideal gas equals $c_V T$, we expect the molar internal energy of the van der Waals correction to the ideal gas to have the form $u = c_V T - a/v$, where the correction term, $-a/v$, represents the contribution to the internal energy from the intermolecular forces. Since the intermolecular forces diminish the pressure by an amount a/v^2, the equivalent effect on the molar internal energy is to reduce it by an amount a/v. This follows because pressure is force per unit area, potential energy is defined such that the negative change in potential energy equals the work done by the force, and molar work equals the integral of pressure over molar volume: $\Delta u = u - u_0 = -\int_{v=v_0}^{v}(-a/v^2)dv = -a/v$. The second equation of state is then:

$$\frac{1}{T} = \frac{c_V}{u + a/v}$$

This second equation of state permits the first equation of state to be expressed in the appropriate form for the entropy representation:

$$\frac{P}{T} = \frac{R}{v - b} - \frac{a c_V}{v^2 (u + a/v)}$$

It is important to check that these two equations of state are indeed compatible. This can be done by calculating the mixed second derivatives:[250]

$$\frac{\partial^2 s}{\partial v \partial u} = \left(\frac{\partial}{\partial v}\frac{1}{T}\right)_u = \left[\frac{\partial}{\partial v}\left(\frac{c_V}{u + a/v}\right)\right]_u = \frac{-c_V}{(u + a/v)^2}\left(-\frac{a}{v^2}\right) = \frac{a c_V}{(uv + a)^2}$$

$$\frac{\partial^2 s}{\partial u \partial v} = \left(\frac{\partial}{\partial u}\frac{P}{T}\right)_v = \left[\frac{\partial}{\partial u}\left(\frac{R}{v - b} - \frac{a c_V}{uv^2 + av}\right)\right]_v = \frac{a c_V v^2}{(uv^2 + av)^2} = \frac{a c_V}{(uv + a)^2}$$

The fundamental relation can be found from the molar differential form of the first law:

$$ds = \frac{du}{T} + \frac{Pdv}{T} = \frac{c_V du}{u + a/v} + \frac{Rdv}{v - b} - \frac{a c_V dv}{v^2(u + a/v)}$$

As usual, the right-hand side can be collapsed into a single (perfect) differential through a substitution motivated by looking for an inverse implicit produce rule: $z \equiv R \ln(v - b) + c_V \ln(u + a/v)$. It is easy to verify that dz does reproduce the right-hand side of the above equation, such that $ds = dz$. The molar form of the fundamental relation in the energy representation is then:

$$s = s_0 + R \ln\left(\frac{v - b}{v_0 - b}\right) + c_V \ln\left(\frac{u + a/v}{u_0 + a/v_0}\right)$$

The third equation of state can be obtained from the Euler relation:

---

[250] As with the ideal gas, we assume that c_V is approximately constant.

$$\frac{\mu}{T} = \frac{u}{T} - s + \frac{Pv}{T} = \left(\frac{\mu}{T}\right)_0 - \frac{uc_V}{\left(u + \frac{a}{v}\right)} - R \ln\left(\frac{v-b}{v_0 - b}\right) - c_V \ln\left(\frac{u + a/v}{u_0 + a/v_0}\right) + \frac{Rv}{v-b} - \frac{ac_V}{v\left(u + \frac{a}{v}\right)}$$

The fundamental relation in the entropy representation is:

$$S(U, V, n) = \frac{nS_0}{n_0} + nR \ln\left(\frac{V/n - b}{V_0/n_0 - b}\right) + nc_V \ln\left(\frac{U/n + an/V}{U_0/n_0 + an_0/V_0}\right)$$

The corresponding equations of state are:

$$\frac{1}{T}(U, V, n) = \frac{c_V}{U/n + an/V}$$

$$\frac{P}{T}(U, V, n) = \frac{R}{\frac{V}{n} - b} - \frac{ac_V n^2}{V^2\left(\frac{U}{n} + \frac{an}{V}\right)}$$

$$\frac{\mu}{T}(U, V, n) = \left(\frac{\mu}{T}\right)_0 + \frac{c_V(U/n - an/V)}{\frac{U}{n} + \frac{an}{V}} - R \ln\left(\frac{V/n - b}{V_0/n_0 - b}\right) - c_V \ln\left(\frac{U/n + an/V}{U_0/n_0 + an_0/V_0}\right) + \frac{RV}{V - bn}$$

The fundamental relation in the energy representation can be obtained by solving for U in the fundamental relation in the entropy representation, and the corresponding equations of state can be found by substituting the expression for U in the equations of state in the entropy representation (and, in the case of two of the equations, eliminating temperature). The equation of state for temperature in the energy representation can be used to trade S for T in order to transform to the Helmholtz representation. The Gibbs and enthalpy representations can similarly be found, but the swap from pressure to volume leads to some cumbersome expressions.

Like the ideal gas, the Nernst postulate is not satisfied; and like the ideal gas, there is a problem at very high densities. The van der Waals model does not apply to very high densities, such as solids, but unlike the ideal gas law, it extends to include liquids. As a result, the van der Waals fluid exhibits a richer thermodynamics, including a gaseous-liquid phase transition.

Example. Derive an equation for the work done by a van der Waals fluid with constant mole number along an isotherm and, separately, along an isentrope.

Temperature is constant for an isothermal process, which means that the work integral can be performed by substituting an expression for P in terms of T, V, and n:

$$\frac{P}{T} = \frac{R}{v - b} - \frac{a}{v^2 T} = \frac{R}{V/n - b} - \frac{an^2}{V^2 T}$$

$$P = \frac{nRT}{V - bn} - \frac{an^2}{V^2}$$

$$W = \int_{V=V_0}^{V} P \, dV = \int_{V=V_0}^{V} \left(\frac{nRT}{V - bn} - \frac{an^2}{V^2}\right) dV = nRT \ln\left(\frac{V - bn}{V_0 - bn}\right) + an^2\left(\frac{1}{V} - \frac{1}{V_0}\right)$$

Observe that the equations for the isotherm and for the work done simplify to those of the ideal gas in the limit that a and b both approach zero.

Entropy is constant along an isentrope, such that the fundamental relation in the entropy representation yields:

$$nR \ln\left(\frac{V-bn}{V_0-bn}\right) = -nc_V \ln\left(\frac{U+an^2/V}{U_0+an^2/V_0}\right)$$

$$\frac{T}{T_0} = \frac{U+an^2/V}{U_0+an^2/V_0} = \left(\frac{V-bn}{V_0-bn}\right)^{-R/c_V}$$

This expression for temperature can be substituted into the equation for pressure in order to perform the work integral:

$$P = \frac{nRT}{V-bn} - \frac{an^2}{V^2} = \frac{nRT_0}{V-bn}\left(\frac{V-bn}{V_0-bn}\right)^{-\frac{R}{c_V}} - \frac{an^2}{V^2} = RT_0\frac{(V-bn)^{-\gamma}}{(V_0-bn)^{-R/c_V}} - \frac{an^2}{V^2}$$

$$W = \int_{V=V_0}^{V} P\,dV = \int_{V=V_0}^{V}\left[RT_0\frac{(V-bn)^{-\gamma}}{(V_0-bn)^{-R/c_V}} - \frac{an^2}{V^2}\right]dV$$

$$W = \frac{RT_0}{(1-\gamma)}\left[\frac{(V-bn)^{-\gamma}}{(V_0-bn)^{-\gamma}} - 1\right] + an^2\left(\frac{1}{V} - \frac{1}{V_0}\right)$$

Again, these reduce to the ideal gas form for the low-density limits – e.g. PV^γ becomes constant for the isentrope as a and b both approach zero, which agrees with a quasistatic adiabat for an ideal gas.

6.5 Extremum Principles

Reversibility: For a thermodynamic process that is carried out quasistatically, the system slowly evolves through a succession of small departures from equilibrium, where equilibrium is allowed to settle in prior to the subsequent departure. Thermodynamic quantities cannot be represented for non-equilibrium states – e.g. temperature and pressure do not have well-defined values representative of the system as a whole during non-equilibrium states. Thus, the expressions TdS for the heat exchange and PdV for the mechanical work are only valid for a thermodynamic process that is carried out via a large number of closely-spaced equilibrium states – i.e. quasistatically. Furthermore, the process is only reversible in the limit that the net entropy change of the quasistatic process approaches zero.

Thus, a process for which the entropy increases is irreversible. Spontaneous processes – such as free expansion (e.g. a wall is suddenly removed, allowing a gas to expand) – are irreversible, as any suddenly unconstrained parameters adjust themselves so as to maximize the entropy for a fixed internal energy. For example, the spontaneous flow of heat from a higher-temperature subsystem to a lower-temperature subsystem is irreversible.

However, this does not mean that heat cannot be made to flow from a lower-temperature subsystem to a higher-temperature subsystem – just that there is a hidden cost of doing so. The overall entropy (system plus surroundings) must increase during an irreversible process, according to the second law of thermodynamics, so the hidden cost of decreasing the entropy of a system is to increase the entropy of its surroundings more than the entropy of the system is diminished. In this spirit, the initial and final equilibrium states of an irreversible process can be made to occur in reverse order by paying a cost. The reversal of initial and final equilibrium states for an irreversible process is a local reversal – since the overall increase in entropy in the system plus surroundings cannot be reversed. Thus, the entropy of the universe must increase in order to accommodate a local reversal of initial and final equilibrium states for an irreversible process. Compare to a reversible process, for which the overall entropy is constant whether the process is carried out forwards or backwards.

Relaxation times: The time it takes for a system to return to equilibrium after a small perturbation is called the relaxation time, τ. If a thermodynamic process occurs over a time period that is very long compared to τ, the process may be reversible; but a thermodynamic process that occurs over a short time period compared to τ is irreversible, since it cannot be quasistatic.

Useful work output: Consider a system that is in initial equilibrium state i. One or more internal constraints can be removed, in which case the unconstrained parameters adjust themselves so as to maximize the entropy for a fixed internal energy. The tendency of the system to maximize its entropy for a fixed internal energy can be harnessed in order to produce useful work. The maximum amount of work that can be harnessed in this way is governed by the maximum work theorem.

Maximum work theorem: Consider a system that is in a particular initial equilibrium state, i, which is to be brought to a particular final equilibrium state, f, and where work is to be delivered during the thermodynamic process that connects these two equilibrium states. According to the maximum work theorem, the amount of work that can be delivered is maximum if the process is reversible – and it is independent of the nature of the reversible process (since $\Delta S_{overall} = 0$ for any reversible process, the details are irrelevant). Furthermore, the heat delivered during the process is minimum when the work delivered is maximum. In the case where the maximum work (and hence minimum heat) is delivered, the work is said to be delivered to a reversible work source and the heat is said to be delivered to a reversible heat source.

Reversible work source: In order to be reversible, the work source must be enclosed by walls that are both adiabatic (so as not to exchange heat with its surroundings) and impermeable to the flow of matter (otherwise, chemical work would be done during the particle exchange) and the work source must also be non-frictional. Additionally, the processes must be quasistatic. Essentially, mechanical work is done as a movable wall slowly (compared to the characteristic relaxation time) increases the volume.

Reversible heat source: In order to be reversible, the heat source must be enclosed by walls that are both immovable (such that no mechanical work is done) and impermeable to the flow of matter (so no chemical work is done, either). Also, the heat must be exchanged quasistatically.

Thermal reservoir: If the reversible heat source is very large in size compared to the system, then the temperature of the reversible heat source will not change noticeably regardless of how much heat may be exchanged between the system and the reversible heat source. In this case, the reversible heat source is called a thermal reservoir. For example, the atmosphere is a thermal reservoir relative to your house. If you open the door, your house will exchange heat with the atmosphere. Your house may change temperature considerably as a result of this heat exchange, but opening your door is not going to change the temperature of the atmosphere.

Applying the maximum work theorem: Suppose that a system in equilibrium state i is to decrease its internal energy to equilibrium state f, and that this energy difference, $\Delta U_{system} = U_f - U_i < 0$, is to be used to deliver work, W_{RWS}, to a reversible work source and heat, Q_{RHS}, to a reversible heat source. Conservation of energy requires that $W_{RWS} + Q_{RHS} = -\Delta U_{system}$.[251] The overall entropy change (system plus surroundings) equals zero for a reversible process: $\Delta S_{system} + \int_i^f \frac{dQ_{RHS}}{T_{RHS}} = 0$. In this case, the work, W_{RWS}, delivered to the reversible work source is maximum.

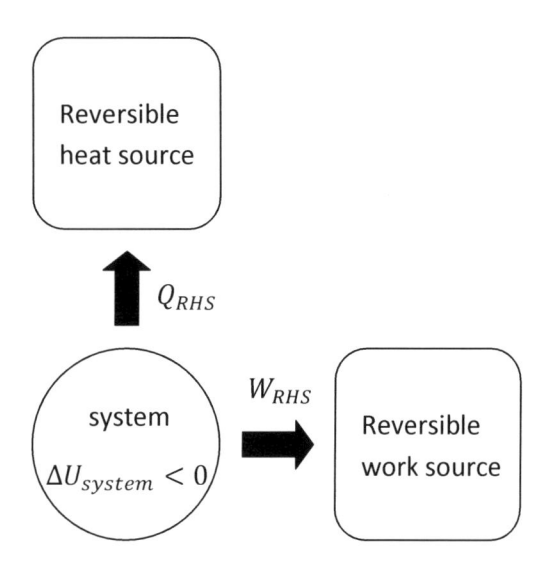

If the reversible heat source is a thermal reservoir, T_{RHS} is constant. For a thermal reservoir, the equations reduce to $Q_{RHS} = -T_{RHS}\Delta S_{system}$ and $W_{RWS} - T_{RHS}\Delta S_{system} = -\Delta U_{system}$.

Rather than extracting energy from the system to do work, work may be done to supply energy to a system. In this case, and other variations, the same equations apply with appropriate sign changes – e.g. if energy is supplied to the system, $\Delta U_{system} > 0$. For work that delivers energy to the system, the work done is least instead of greatest (yet still maximum work from the point of view of being less negative).

---

[251] The first law of thermodynamics states that the internal energy change for a system equals the heat absorbed by the system minus the work done by the system: $\Delta U = Q - W$. Compare the signs of the first law of thermodynamics with the signs of $W_{RWS} + Q_{RHS} = -\Delta U_{system}$. The reason for the difference is that in this case, ΔU_{system}, relates to a system, while W_{RWS} and Q_{RHS} relate to sources that are external to the system.

Whether energy is drawn from the system ($\Delta U_{system} < 0$), supplied to the system ($\Delta U_{system} > 0$), or the energy of the system is unchanged ($\Delta U_{system} = 0$), the first law of thermodynamics still applies to the system: $\Delta U_{system} = Q_{system} - W_{system}$ on the finite scale, and $dU_{system} = đQ_{system} - đW_{system}$ on the differential scale. Thus, the system may be doing work of its own, W_{system}, which is different from the work being delivered to (or supplied by) the reversible work source, W_{RWS}. It similarly important not to confuse heat absorbed by the system, Q_{system}, with heat delivered to (or from) the reversible heat source, Q_{RHS}. Conservation of energy requires that $\Delta U_{system} = Q_{system} - W_{system}$ for the system, and also that $W_{RWS} + Q_{RHS} = -\Delta U_{system}$ for the reversible heat and work sources and the system combined.

Example. A monatomic ideal gas with constant n has initial pressure P_0, volume V_0, and temperature T_0. What is the maximum work that can be delivered to a reversible work source by increasing the volume to $4V_0$ – at the original temperature, T_0 – if a thermal reservoir at temperature T_{res} is available?

Observe that $\Delta U_{system} = 0$, since $U = nc_V T$ for an ideal gas and $T = T_0$. In this case, the work delivered to the reversible heat source is supplied directly from the thermal reservoir: $W_{RWS} = -Q_{RHS}$. The work delivered to the reversible work source is maximum if the process is reversible, meaning that $Q_{RHS} = -T_{res} \Delta S_{system}$.

Since ΔS_{system}, like ΔU_{system}, is path-independent, we may compute ΔS_{system} for any reversible path with the same initial and final equilibrium state. Since the initial and final temperature of the system are the same ($T = T_0$), it is convenient to choose a reversible isotherm. For this choice, $\Delta S_{system} = \int_i^f \frac{đQ_{system}}{T_0} = Q_{system}/T_0$ and $Q_{system} = W_{system}$ (according to the first law, since we already know that $\Delta U_{system} = 0$). The work done by the system along the isotherm is:

$$W_{system} = \int_{V=V_0}^{4V_0} P dV = nRT_0 \int_{V=V_0}^{4V_0} \frac{dV}{V} = nRT_0 \ln 4$$

Putting this all together, $W_{RWS}^{max} = -Q_{RHS} = T_{res} W_{system}/T_0 = nRT_{res} \ln 4$.

Heat engine: If the system is an energy source (e.g. a furnace), the reversible heat source is at a much lower-temperature (e.g. the atmosphere), and the reversible work source is machinery to be operated as heat flows from the higher-temperature energy source to the lower-temperature reversible heat source, these three components (energy source, reversible heat source, and reversible work source) form a heat engine. The fact that some heat must be delivered – i.e. that heat input cannot be converted purely into work output, but must result in some exhaust – at lower temperature is a consequence of the second law of thermodynamics. That is, the heat engine cannot be perfectly efficient because the overall entropy of the system plus surroundings cannot decrease.

The heat engine delivers the maximum work when the heat output is minimum. In this case, the efficiency of the heat engine is maximum. We can see that there is a connection between the maximum work theorem, the second law of thermodynamics, the entropy maximum principle, and the efficiency of a heat engine.

The maximum work theorem and the second law: Consider a system in equilibrium state i which is to decrease its internal energy to equilibrium state f, such that this energy difference, $\Delta U_{system} = U_f - U_i < 0$, may be used to deliver work, W_{RWS}, to a reversible work source and heat, Q_{RHS}, to a reversible heat source. Both the internal energy change of the system, ΔU_{system}, and the entropy change for the system, ΔS_{system}, are constant because the initial and final states, i and f, have been specified and these changes are path-independent. Conservation of energy between the system and the reversible heat and work sources requires that $W_{RWS} + Q_{RHS} = -\Delta U_{system}$. Since ΔU_{system} is fixed, the freedom is what fraction of the energy goes to the reversible work source instead of the reversible heat source. The maximum delivery of work will correspond to the least delivery of heat.

The entropy increase associated with the heat deposited into the reversible heat source is $\Delta S_{RHS} = \int_i^f \frac{đQ_{RHS}}{T_{RHS}}$. The overall entropy (system plus reversible heat source) must not decrease according to the second law of thermodynamics: $\Delta S_{system} + \Delta S_{RHS} \geq 0$. Since ΔS_{system} is fixed, this inequality places a requirement on the minimum entropy increase of the reversible heat source, $\Delta S_{RHS}^{min} = -\Delta S_{system}$. This minimum, ΔS_{RHS}^{min}, corresponds to a reversible process, for which the overall entropy increase is minimum (zero!). The value of ΔS_{RHS}^{min} also corresponds to the minimum amount of heat deposited into the reversible heat source, Q_{RHS}^{min}, which in turn corresponds to the maximum amount of work deposited into the reversible work source, W_{RWS}^{max}. Thus, the maximum work is indeed delivered to the reversible work source when the process that decreases the internal energy of the system is reversible; and this does indeed correspond to the minimum heat deposited into the reversible heat source. Here, we see the relationship between the maximum work theorem and the second law of thermodynamics, and you should already have observed the connection between the second law of thermodynamics and the entropy maximum principle.

The second law of thermodynamics is also related to the efficiency of a heat engine. We can also see this from the maximum work theorem. On the differential scale, conservation of energy becomes $đW_{RWS} + đQ_{RHS} = -dU_{system}$ and the overall entropy inequality becomes $dS_{system} + dS_{RHS} = dS_{system} + đQ_{RHS}/T_{RHS} \geq 0$. Combining these equations, $đQ_{RHS} = -dU_{system} - đW_{RWS} \geq -T_{RHS} dS_{system}$ or $đW_{RWS} \leq T_{RHS} dS_{system} - dU_{system}$.[252] The maximum work delivered to the reversible work source is thus $đW_{RWS}^{max} = T_{RHS} dS_{system} - dU_{system}$. For the system, $dU_{system} = đQ_{system} - đW_{system}$ and $dS_{system} = T_{system} Q_{system}$. Therefore, the maximum delivery of work to the reversible work source can be expressed as:

$$đW_{RWS}^{max} = -đW_{system} - đQ_{system}\left(1 - \frac{T_{RHS}}{T_{system}}\right)$$

which shows that the maximum delivery of work to the reversible work source corresponds to the maximum engine efficiency, $e_{max} = 1 - T_{RHS}/T_{system}$.

Entropy maximum principle: For a fixed value of the internal energy, equilibrium is attained when the free internal parameters adjust themselves so as to maximize the total entropy. The entropy maximum principle is related to the maximum work theorem and the second law of thermodynamics.

---

[252] Recall that multiplying both sides of an inequality reverses the inequality: e.g. $-x < -5 \implies x > 5$.

Heat flow and the second law: The entropy maximum principle of thermodynamics, the second law of thermodynamics, the maximum work theorem, and the efficiency of a heat engine are all related, and they also relate to the direction of heat flow. In order to see this, reconsider the problem of pure thermal equilibrium: A closed system is divided into two subsystems, A and B, at different initial temperatures, T_A and T_B, by an immovable, diathermal wall, which the particles of the system cannot cross. At equilibrium, the two temperatures are equal: $T_e \equiv T_{Ae} = T_{Be}$. Let us explore the direction of the heat flow assuming that the process by which thermal equilibrium is attained is quasistatic.

Since the two subsystems can only exchange heat, conservation of energy requires that the heat absorbed by one subsystem equals the heat released by the other: $Q_A + Q_B = 0$. This heat exchange can be expressed in terms of the heat capacities, C_A and C_B:[253]

$$\int_{T_A=T_A}^{T_e} C_A dT_A = - \int_{T_B=T_B}^{T_e} C_B dT_B$$

The total entropy change equals:

$$\Delta S_{overall} = \Delta S_A + \Delta S_B = \int_A^e \frac{đQ_A}{T_A} + \int_B^e \frac{đQ_B}{T_B} = \int_{T_A=T_A}^{T_e} \frac{C_A}{T_A} dT_A + \int_{T_B=T_B}^{T_e} \frac{C_B}{T_B} dT_B$$

Consider the simple case where the heat capacities are independent of temperature. In this case, the integrals are trivial. The conservation of energy integrals lead to

$$C_A(T_e - T_A) = -C_B(T_e - T_B)$$
$$T_e = \frac{C_A T_A + C_A T_B}{C_A + C_B}$$

and the total entropy integrals become

$$\Delta S_{overall} = C_A \ln\left(\frac{T_e}{T_A}\right) + C_B \ln\left(\frac{T_c}{T_B}\right)$$

If subsystem B is initially at higher temperature ($T_B > T_A$), then clearly (from the equation for T_e above) the equilibrium temperature lies in the range $T_A < T_e < T_B$, showing that $Q_A > 0$ and $Q_B < 0$ – i.e. heat flows from the higher-temperature subsystem to the lower-temperature subsystem. The total entropy equation shows that $\Delta S_A > 0$ and $\Delta S_B < 0$ – i.e. the entropy of the lower-temperature subsystem increases and that of the higher-temperature subsystem decreases. We can see that the overall entropy increases by setting the derivative of $\Delta S_{overall}$ with respect to T_A (for example) equal to zero:

---

[253] Note that this is the heat capacity, $đQ = CdT$, and not the specific or molar specific heat capacity. The heat capacity is useful for general arguments, such as this discussion of heat flow, for which we do not want to complicate matters by introducing mass or mole number. When working out computations for specific problems, however, the specific or molar specific heat capacity is generally more practical.

$$\frac{d}{dT_A}\left[C_A \ln\left(\frac{T_e}{T_A}\right) + C_B \ln\left(\frac{T_e}{T_B}\right)\right] = \frac{d}{dT_A}[C_A \ln(T_e) - C_A \ln(T_A) + C_B \ln(T_e) - C_B \ln(T_B)] = 0$$

$$\frac{C_A}{T_e}\frac{dT_e}{dT_A} - \frac{C_A}{T_A} + \frac{C_B}{T_e}\frac{dT_e}{dT_A} = 0$$

$$\frac{C_A}{T_e}\frac{C_A}{C_A + C_B} - \frac{C_A}{T_A} + \frac{C_B}{T_e}\frac{C_A}{C_A + C_B} = 0$$

$$\frac{C_A}{T_e} - \frac{C_A}{T_A} = 0$$

$$T_e = T_A$$

where we treat the independent variable, T_B, as a constant when taking a derivative with respect to T_A. This shows that $T_e = T_A$ is an extremum of $\Delta S_{overall}$. A second derivative reveals that it is a relative minimum. Comparison with the endpoints – i.e. the limits of $\Delta S_{overall}$ as T_A approaches zero or infinity – shows that the absolute minimum of $\Delta S_{overall}$ equals zero, and only occurs when $T_e = T_A$. The condition $T_e = T_A$ also implies that $T_B = T_A$, however, which contradicts our original assumption. We therefore conclude that $\Delta S_{overall} > 0$, as expected from the second law of thermodynamics.

Energy minimum principle: For a fixed value of the entropy, equilibrium is attained when the free internal parameters adjust themselves so as to minimize the internal energy. The energy minimum principle is the energy representation's equivalent of the entropy maximum principle of the entropy representation. Both principles apply to the equilibrium state that develops, but in practice one thermodynamic representation and corresponding extremum principle must be applied. That is, either the internal energy or the entropy is held constant while an extremum is applied to the other – so, for example, trying to vary both the internal energy and the entropy simultaneously in the same solution would be different from both extremum principles.

Entropy maximum and energy minimum principles: The energy minimum principle follows from the entropy maximum principle (and vice-versa). Consider, for illustration, the simple case where volume is a free parameter and the mole numbers are fixed. According to the entropy maximum principle, the volume will adjust itself so as to maximize the entropy for a given internal energy. The entropy is an extremum for a fixed internal energy if $\left(\frac{\partial S}{\partial V}\right)_{U,n_i} = 0$, and maximum if $\left(\frac{\partial^2 S}{\partial V^2}\right)_{U,n_i} < 0$. We wish to show that these conditions imply the energy minimum principle, so we need to rewrite these partial derivatives in terms of $\left(\frac{\partial U}{\partial V}\right)_{S,n_i}$ and $\left(\frac{\partial^2 U}{\partial V^2}\right)_{S,n_i}$ in order to see if the internal energy is indeed a minimum for a fixed value of the entropy.

Recall that we can apply a mathematical identity in order to write $\left(\frac{\partial S}{\partial V}\right)_{U,n_i}$ in terms of $\left(\frac{\partial U}{\partial V}\right)_{S,n_i}$: Namely, $\left(\frac{\partial S}{\partial V}\right)_{U,n_i}\left(\frac{\partial V}{\partial U}\right)_{S,n_i}\left(\frac{\partial U}{\partial S}\right)_{V,n_i} = -1$. Therefore, the condition $\left(\frac{\partial S}{\partial V}\right)_{U,n_i} = 0$ implies that $\left(\frac{\partial U}{\partial V}\right)_{S,n_i}\left(\frac{\partial S}{\partial U}\right)_{V,n_i} = 0$. Since $\left(\frac{\partial U}{\partial S}\right)_{V,n_i} = T$ and $T > 0$, according to the third law of thermodynamics, we conclude that $\left(\frac{\partial U}{\partial V}\right)_{S,n_i} = 0$. That is, the entropy maximum principle implies that the internal energy is an extremum for fixed entropy when entropy is a maximum for fixed internal energy.

Now we check the second derivative, $\left(\frac{\partial^2 U}{\partial V^2}\right)_{S,n_i}$, to see if the internal energy is indeed a minimum for fixed entropy when the entropy is a maximum for fixed internal energy. We begin with the condition on the second derivative of the entropy: $\left(\frac{\partial^2 S}{\partial V^2}\right)_{U,n_i} < 0$. We can apply the same mathematical identity to remove the internal energy from the subscript:

$$\left(\frac{\partial^2 S}{\partial V^2}\right)_{U,n_i} = \frac{\partial}{\partial V}\left(\frac{\partial S}{\partial V}\right)_{U,n_i} = -\frac{\partial}{\partial V}\left[\left(\frac{\partial U}{\partial V}\right)_{S,n_i}\left(\frac{\partial S}{\partial U}\right)_{V,n_i}\right] < 0$$

$$-\left(\frac{\partial S}{\partial U}\right)_{V,n_i}\left(\frac{\partial^2 U}{\partial V^2}\right)_{S,n_i} - \left(\frac{\partial U}{\partial V}\right)_{S,n_i}\frac{\partial}{\partial V}\left(\frac{\partial S}{\partial U}\right)_{V,n_i} < 0$$

The second term is zero since we previously found that $\left(\frac{\partial U}{\partial V}\right)_{S,n_i} = 0$. Since $\left(\frac{\partial S}{\partial U}\right)_{V,n_i} = 1/T$, this inequality becomes:

$$-\frac{1}{T}\left(\frac{\partial^2 U}{\partial V^2}\right)_{S,n_i} < 0$$

Because $T > 0$, this implies that

$$\left(\frac{\partial^2 U}{\partial V^2}\right)_{S,n_i} > 0$$

since multiplying both sides of an inequality by a negative number changes the direction of the inequality. Similar results are obtained for other combinations of free internal parameters, such that, in general, the entropy maximum principle implies the energy minimum principle.

Extremum principles: Because the entropy maximum principle implies the energy minimum principle, and vice-versa, the two representations remain on an equal footing – i.e. equivalent extremum principles apply to both. There are similar extremum principles in the thermodynamic potential formulations, also.

Helmholtz free energy minimum principle: The fundamental relation in the Helmholtz representation is obtained from the fundamental relation in the energy representation via a Legendre transform that swaps the roles of the entropy and temperature. For a system that is in contact with a thermal reservoir, the internal energy of the system plus reservoir is minimum, which translates to a minimum of the Helmholtz potential for a fixed temperature of the system equal to the temperature of the reservoir. The Helmholtz representation is thus well-suited to problems where a system is in contact with a thermal reservoir.

Enthalpy minimum principle: The fundamental relation in the enthalpy representation is obtained from the fundamental relation in the energy representation via a Legendre transform that swaps the roles of the pressure and volume. For a system that is in contact with a pressure reservoir, the internal energy of the system plus reservoir is minimum, which translates to a minimum of the enthalpy for a fixed pressure of the system equal to the pressure of the reservoir. The enthalpy representation is thus well-suited to problems where a system is in contact with a pressure reservoir.

Gibbs free energy minimum principle: The fundamental relation in the Gibbs representation is obtained from the fundamental relation in the energy representation via a Legendre transform that swaps the roles of the entropy and temperature as well as the pressure and volume. For a system that is in contact with both thermal and pressure reservoirs, the internal energy of the system plus reservoir is minimum, which translates to a minimum of the Gibbs free energy for fixed temperature and pressure of the system equal to those of the respective reservoirs. The Gibbs representation is thus well-suited to problems where a system is in contact with both thermal and pressure reservoirs, such as chemical reactions occurring in open vessels (for which the ambient air serves to maintain both constant temperature and air pressure in the system).

The Gibbs free energy is related to the internal energy by $G = U - TS + PV$. According to the Euler relation, $U - TS + PV - \sum_{i=1}^{r} n_i \mu_i = 0$. Therefore, the Gibbs free energy may be expressed as $G = \sum_{i=1}^{r} n_i \mu_i$. For a single-component system this is simply, $G = n\mu$. That is, the molar Gibbs free energy equals the chemical potential in the case of a single-component system: $g = \mu$. The Gibbs free energy is well-suited for diffusive equilibrium problems – since the chemical potentials are balanced in diffusive equilibrium. The Gibbs representation is similarly useful for chemical equilibrium problems.

6.6 The Maxwell Relations

Second derivatives: The fundamental relation expresses a thermodynamic relationship among extensive parameters, which are zero-order homogeneous functions. The total differential of the fundamental relation is identified with the first law of thermodynamics, provided that the intensive parameters are defined as first derivatives of extensive parameters. These intensive parameters are first-order homogeneous functions. We shall now consider the significance of the second derivatives.

Coefficient of thermal expansion: In thermal physics, the coefficient of linear expansion for a solid is defined as $dL = \alpha L_0 dT$. Length is not one of the fundamental thermodynamic variables, but thermal expansion can also be expressed in terms of volume for isotropic substances: $dV = \beta V_0 dT$. Recall that $\beta = 3\alpha$. Technically, the behavior of β depends upon which quantities, if any, are held constant as temperature is varied. Therefore, in our thermodynamic considerations, we now improve upon the definition of thermal expansion to specify that pressure and mole numbers are held constant:[254,255]

---

[254] In thermal physics, the relation $dV = \beta V_0 dT$ was assumed to hold approximately for small temperature changes, in which the distinction between V_0 and V would not be significant.

$$\beta = \frac{1}{V}\left(\frac{\partial V}{\partial T}\right)_{P,n_i}$$

Since temperature is itself a derivative, $T = (\partial U/\partial S)_{V,n_i}$, the coefficient of thermal expansion is a second derivative of the extensive parameters:[256]

$$\beta = \frac{1}{V}\frac{1}{\left[\frac{\partial}{\partial V}\left(\frac{\partial U}{\partial S}\right)_{V,n_i}\right]_{S,n_i}} = \frac{1}{V}\frac{1}{\left(\frac{\partial^2 U}{\partial V \partial S}\right)}$$

Example. Derive an expression for β for a van der Waals fluid.

The coefficient of thermal expansion, β, involves a partial derivative of V with respect to T at constant P. It is therefore useful to combine equations of state together to express P in terms of V and T: $P = \frac{nRT}{V-bn} - \frac{an^2}{V^2}$. It is also convenient to obtain $\left(\frac{\partial V}{\partial T}\right)_{P,n_i}$ as the reciprocal of $\left(\frac{\partial T}{\partial V}\right)_{P,n_i}$:

$$T = \left(P + \frac{an^2}{V^2}\right)\frac{V-bn}{nR} = \frac{PV}{nR} - \frac{Pb}{R} + \frac{an}{RV} - \frac{abn^2}{RV^2}$$

$$\left(\frac{\partial T}{\partial V}\right)_{P,n_i} = \left[\frac{\partial}{\partial V}\left(\frac{PV}{nR} - \frac{Pb}{R} + \frac{an}{RV} - \frac{abn^2}{RV^2}\right)\right]_{P,n_i} = \frac{P}{nR} - \frac{an}{RV^2} + \frac{2abn^2}{RV^3}$$

$$\beta = \frac{1}{V}\left(\frac{\partial V}{\partial T}\right)_{P,n_i} = \left(\frac{Pv}{R} - \frac{a}{Rv} + \frac{2ab}{Rv^2}\right)^{-1}$$

Molar specific heat capacity: In thermal physics, the molar specific heat capacities[257] at constant volume and pressure are defined by $đQ = nc_{P,V}dT$. For a quasistatic process – which we are primarily concerned with in thermodynamics, as it most naturally describes equilibrium states and reversible processes – the differential heat exchange is related to the entropy via $đQ = TdS$. In this case, the molar heat capacities can also be expressed as partial derivatives:

$$c_{P,V} = \frac{1}{n}\left(\frac{đQ}{dT}\right)_{P,V;n_i} = \frac{T}{n}\left(\frac{\partial S}{\partial T}\right)_{P,V;n_i}$$

As with the coefficient of thermal expansion, molar specific heat capacity is a second derivative of the extensive variables.

---

[255] In thermal physics, it is common to distinguish between linear expansion, α, and volume expansion, $\beta = 3\alpha$. However, in thermodynamics, it is more common to use α for volume expansion. Since this text includes both thermal physics and thermodynamics, we will preserve this distinction and continue using β for volume expansion.

[256] Since internal energy and entropy are perfect differentials, the thermodynamics variables satisfy the mathematical relationship, $\left(\frac{\partial z}{\partial x}\right)_y = 1/\left(\frac{\partial x}{\partial z}\right)_y$.

[257] Observe that the molar specific heat capacity is a more natural thermodynamic quantity than specific heat capacity as mole number is a fundamental extensive parameter, whereas mass is not.

Compressibility: Recall our definition of the bulk modulus from Sec. 3.7: $B = -\frac{dP}{dV/V} = -V\frac{dP}{dV}$. An increase in the pressure applied to an isotropic substance deforms the substance, decreasing its volume. In thermodynamics, we will work with the inverse bulk modulus, $\kappa = 1/B$, which is called the compressibility, and we will distinguish between partial derivatives at constant temperature – the isothermal compressibility, κ_T – and constant entropy – the adiabatic compressibility, κ_S:

$$\kappa_{T,S} = -\frac{1}{V}\left(\frac{\partial V}{\partial P}\right)_{T,S;n_i}$$

Like the coefficient of thermal expansion and molar specific heat capacities, the compressibility is a second derivative of the extensive variables.

Mixed second derivatives: The mixed second derivatives in thermodynamics are reversible, as a consequence of the internal energy and entropy being perfect differentials. For example,

$$\left(\frac{\partial^2 U}{\partial V \partial S}\right) = \left(\frac{\partial^2 U}{\partial S \partial V}\right) \Rightarrow \left[\frac{\partial}{\partial V}\left(\frac{\partial U}{\partial S}\right)_{V,n_i}\right]_{S,n_i} = \left[\frac{\partial}{\partial S}\left(\frac{\partial U}{\partial V}\right)_{S,n_i}\right]_{V,n_i} \Rightarrow \left(\frac{\partial T}{\partial V}\right)_{S,n_i} = -\left(\frac{\partial P}{\partial S}\right)_{V,n_i}$$

Partial derivatives of thermodynamic potentials: The partial derivatives of thermodynamics potentials with respect to their natural variables (e.g. F is a function of T, V, and n_i) can be evaluated by expressing the first law in terms of the thermodynamic potential. For example, the Helmholtz potential is related to the internal energy by $F = U - TS$.[258] The total differential of the Helmholtz potential is $dF = dU - TdS + SdT$. Combining this with the first law, $dF = TdS - PdV + \sum_{i=1}^{r}\mu_i n_i - TdS - SdT$, which reduces to $dF = -SdT - PdV + \sum_{i=1}^{r}\mu_i n_i$. From this, the partial derivatives of the Helmholtz potential are:

$$S = -\left(\frac{\partial F}{\partial T}\right)_{V,n_i} \quad, \quad -P = \left(\frac{\partial F}{\partial V}\right)_{T,n_i} \quad, \quad \mu_i = \left(\frac{\partial F}{\partial n_i}\right)_{T,V}$$

As usual, the mixed second derivatives are equal. For example,

$$\left(\frac{\partial^2 F}{\partial V \partial T}\right) = \left(\frac{\partial^2 F}{\partial T \partial V}\right) \Rightarrow \left[\frac{\partial}{\partial V}\left(\frac{\partial F}{\partial T}\right)_{V,n_i}\right]_{T,n_i} = \left[\frac{\partial}{\partial T}\left(\frac{\partial F}{\partial V}\right)_{T,n_i}\right]_{V,n_i}$$

$$\Rightarrow \left(\frac{\partial S}{\partial V}\right)_{T,n_i} = \left(\frac{\partial P}{\partial T}\right)_{V,n_i}$$

---

[258] It is easy to get these signs straight if you know the Euler relation, $U = TS - PV + \sum_{i=1}^{r}\mu_i n_i$, and remember that the complete Legendre transformation is zero because of the Euler relation. That is, the complete Legendre transformation is $U - TS + PV - \sum_{i=1}^{r}\mu_i n_i = 0$. A partial Legendre transformation has the same signs as the relevant terms of the complete Legendre transformation.

Maxwell relations: The Maxwell relations express equality among second derivatives. For example, the relation $\left(\frac{\partial S}{\partial P}\right)_{T,n_i} = -\left(\frac{\partial V}{\partial T}\right)_{P,n_i}$, which expresses equality of mixed second derivatives in the Gibbs representation, $\left(\frac{\partial^2 G}{\partial P \partial T}\right) = \left(\frac{\partial^2 G}{\partial T \partial P}\right)$, is a Maxwell relation. As thermodynamics is designed around perfect differentials, the inverse mixed second derivatives must also be equal – e.g. $\left(\frac{\partial P}{\partial S}\right)_{T,n_i} = -\left(\frac{\partial T}{\partial V}\right)_{P,n_i}$.

Thermodynamic square: Consider the set of thermodynamic potentials that can be obtained from the internal energy via Legendre transforms involving exchanges between two pairs of conjugate variables. This forms a set of Maxwell relations that can be represented on what is referred to as the thermodynamic square. For example, the set of thermodynamic potentials related through Legendre transforms involving the two pairs of conjugate variables S, T and V, P include U, F, G, and H. For this set of thermodynamic potentials, let us consider the mixed second derivatives of U, F, G, and H with respect to S or T and V or P for a single-component system.

Internal energy is a function of entropy, volume, and mole number: $U(S, V, n)$. In the energy representation, the Maxwell relation involving entropy and volume is $\left(\frac{\partial T}{\partial V}\right)_{S,n} = -\left(\frac{\partial P}{\partial S}\right)_{V,n}$. The roles of entropy and temperature are swapped in the Legendre transform to the Helmholtz representation: $F(T, V, n)$. In the Helmholtz representation, the analogous Maxwell relation is $\left(\frac{\partial S}{\partial V}\right)_{T,n} = \left(\frac{\partial P}{\partial T}\right)_{V,n}$, where the conjugate variables S and T simply swap as the cost of a minus sign. The Gibbs representation can be obtained directly from the Helmholtz representation through a Legendre transformation that swaps the roles of volume and pressure: $G(T, P, n)$. The similar Maxwell relation in the Gibbs representation is $\left(\frac{\partial S}{\partial P}\right)_{T,n} = -\left(\frac{\partial V}{\partial T}\right)_{P,n}$, where this time the conjugate variables V and P swap at the cost of a minus sign. The enthalpy can be obtained from the Gibbs via a Legendre transformation that swaps the roles of temperature and entropy: $H(S, P, n)$. Predictably, the corresponding Maxwell relation is $\left(\frac{\partial T}{\partial P}\right)_{S,n} = \left(\frac{\partial V}{\partial S}\right)_{P,n}$. The enthalpy can be returned to the energy representation, retrieving the original Maxwell relation, completing a cycle through the two pairs of conjugate variables – S and T, and P and V.

The prior relationships can be incorporated into a diagram, which is called the thermodynamic square. The four representations – U, F, G, and H – are placed on the sides of the square, and the two pairs of conjugate variables – S and T, and V and P – are placed on the corners, such that each potential is flanked by its natural variables (e.g. U is a flanked by S and V). The corner variables must be shared by both adjacent potentials, which means that U and G (and, similarly, F and H) must reside on opposite sides of the thermodynamic square, and that the conjugate variables must reside at opposite corners. Two arrows are added across the long diagonals to indicate whether the coefficients of the total differentials are positive or negative – e.g. in $dU = TdS - PdV + \mu dn$, T is positive and P is negative: If an arrow points away from the variable for which the partial derivative is taken with respect to, the coefficient is positive; if instead the arrow points toward it, the coefficient is negative. The variable held constant naturally appears adjacent to the variable for which the partial derivative is taken with respect to.

For example, the thermodynamic square states that a partial derivative of V with respect to S is taken with P held constant, and comes with a positive coefficient as an arrow points away from the S: $\left(\frac{\partial V}{\partial S}\right)_{P,n}$. The corresponding partial derivative on the opposite side of the thermodynamic square is $\left(\frac{\partial T}{\partial P}\right)_{S,n}$. This yields the Maxwell relation, $\left(\frac{\partial V}{\partial S}\right)_{P,n} = \left(\frac{\partial T}{\partial P}\right)_{S,n}$.

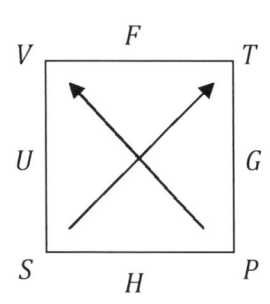

```
        F
  V ┌───────┐ T
    │  ╲ ╱  │
  U │   ╳   │ G
    │  ╱ ╲  │
  S └───────┘ P
        H
```

The thermodynamic square provides an easy means of rewriting the mixed second derivatives in different representations. The thermodynamic square can be applied to any set of four potentials related by Legendre transformations of two pairs of conjugate variables.

Example. Read off the Maxwell relation involving partial derivatives with respect to volume and entropy for the thermodynamic square illustrated above.

One is a partial derivative of T with respect to V at constant S; negative because the arrow points toward the V: $-\left(\frac{\partial T}{\partial V}\right)_{S,n}$. Its partner is a partial derivative of P with respect to S at constant V; positive because the arrow points away from the S: $\left(\frac{\partial T}{\partial S}\right)_{V,n}$. Thus, $\left(\frac{\partial T}{\partial V}\right)_{S,n} = -\left(\frac{\partial P}{\partial S}\right)_{V,n}$.

Thermodynamic octahedron: The thermodynamic square can easily be generalized to the set of thermodynamic potentials related by Legendre transformations of 3 pairs of conjugate variables. In this case, there will be 6 corners – one for each conjugate variable – and 8 sides – one for each thermodynamic potential. For example, for a single-component system, the 3 pairs of conjugate variables include S and T, V and P, and N and μ, and the 8 thermodynamic potentials include $U(S,V,n)$, $F(T,V,n)$, $G(T,P,n)$, $H(S,P,n)$, $A_\mu(S,V,\mu)$, $\Psi(T,V,\mu)$, $A_{P,\mu}(S,P,\mu)$, and $A_{T,P,\mu}(T,P,\mu)$. The symbol A represents the less common thermodynamic potentials, and its subscript designates which arguments have undergone a Legendre transformation compared to $U(S,V,n)$. The complete Legendre transformation equals zero: $A_{T,P,\mu}(T,P,\mu) = 0$.

The geometric shaped form by the 3 pairs of conjugate variables will have 6 corners, with the pairs S and T, V and P, and N and μ at opposite corners. There will be 8 triangular sides, as each thermodynamic potential is a function of one-half of the conjugate variables. This leaves 12 edges, as each corner connects to all other corners except its conjugate variable. The resulting shape is called an octahedron. In this case, 3 arrows drawn across the long diagonals indicate which variables have positive or negative coefficients.

The thermodynamic octahedron contains 3 thermodynamic sub-squares: $S, V, T,$ and P; $S, n, T,$ and μ; and $V, n, P,$ and μ. The Maxwell relations are obtained in the same way from these thermodynamic sub-squares as they are for the usual thermodynamic square.

For example, within the $V, n, P,$ and μ thermodynamic sub-square, the thermodynamic octahedron states that a partial derivative of V with respect to n is taken with P held constant, and comes with a positive coefficient as an arrow points away from the n: $\left(\frac{\partial V}{\partial n}\right)_P$. The corresponding partial derivative on the opposite side of this thermodynamic sub-square is $\left(\frac{\partial \mu}{\partial P}\right)_n$. This yields the Maxwell relation, $\left(\frac{\partial V}{\partial n}\right)_P = \left(\frac{\partial \mu}{\partial P}\right)_n$, where, additionally, either S or T (the conjugate variables outside of the thermodynamic sub-square), as appropriate, is held constant. That is, it is true that both $\left(\frac{\partial V}{\partial n}\right)_{S,P} = \left(\frac{\partial \mu}{\partial P}\right)_{S,n}$ and $\left(\frac{\partial V}{\partial n}\right)_{T,P} = \left(\frac{\partial \mu}{\partial P}\right)_{T,n}$.[259]

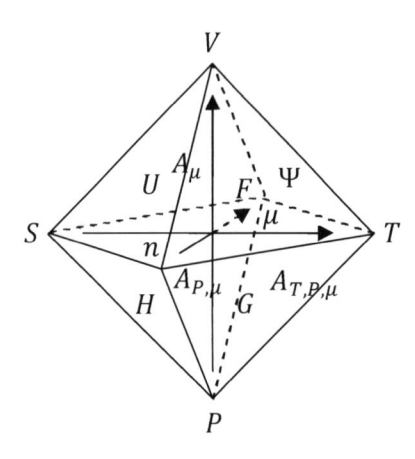

Thermodynamic cross-polytope: For N pairs of conjugate variables, there will be $2N$ corners and $2N(N-1)$ edges, forming an N-dimensional cross-polytope. This is the higher-dimensional generalization of the thermodynamic square.

Independent second derivatives: Not all of the second derivatives in thermodynamics are independent: In fact, all of the second derivatives can be expressed in terms of a relatively small number of second derivatives. For single-component systems, there are just three independent second derivatives. The conventional choice is to express all of the second derivatives in terms of the coefficient of thermal expansion, $\beta = \frac{1}{V}\left(\frac{\partial V}{\partial T}\right)_{P,n_i}$, the molar specific heat capacity at constant pressure, $c_P = \frac{T}{n_i}\left(\frac{\partial S}{\partial T}\right)_{P,n_j}$, and the isothermal compressibility, $\kappa_T = -\frac{1}{V}\left(\frac{\partial V}{\partial P}\right)_{T,n_i}$.

---

[259] In the figure, $U(S,V,n)$, $F(T,V,n)$, $G(T,P,n)$, and $H(S,P,n)$, lie at the centers of triangular faces on the front half of the octahedron, while $A_\mu(S,V,\mu)$, $\Psi(T,V,\mu)$, $A_{P,\mu}(S,P,\mu)$, and $A_{T,P,\mu}(T,P,\mu)$ lie at the centers of triangular faces on the back half of the octahedron.

It is interesting to note that these three quantities, which all involve partial derivatives with respect to pressure or temperature, can all be expressed in terms of second derivatives of the Gibbs potential:

$$\frac{\partial^2 G}{\partial T^2} = \left[\frac{\partial}{\partial T}\left(\frac{\partial G}{\partial T}\right)_{P,n}\right]_{P,n} = -\left(\frac{\partial S}{\partial T}\right)_{P,n} = -\frac{nc_P}{T}$$

$$\frac{\partial^2 G}{\partial T \partial P} = \left[\frac{\partial}{\partial T}\left(\frac{\partial G}{\partial P}\right)_{T,n}\right]_{P,n} = \left(\frac{\partial V}{\partial T}\right)_{P,n} = \beta V$$

$$\frac{\partial^2 G}{\partial P^2} = \left[\frac{\partial}{\partial P}\left(\frac{\partial G}{\partial P}\right)_{T,n}\right]_{T,n} = \left(\frac{\partial V}{\partial P}\right)_{T,n} = -\kappa_T V$$

Reduction of derivatives: The process of expressing a derivative in thermodynamics in terms of a few basic quantities is referred to as the reduction of derivatives. Given a partial derivative, our goal is to rewrite it in terms of β, c_P, κ_T, and/or intensive or extensive parameters like P and V. This means expressing the derivative in terms of $\left(\frac{\partial V}{\partial T}\right)_{P,n}$, $\left(\frac{\partial S}{\partial T}\right)_{P,n}$, and $\left(\frac{\partial V}{\partial P}\right)_{T,n}$, which are directly related to β, c_P, and κ_T. Following are some techniques for reducing the derivatives for a single-component system:[260]

- From the definitions of β, c_P, c_V, κ_T, and κ_S, replace $\left(\frac{\partial V}{\partial T}\right)_{P,n}$ with βV, $\left(\frac{\partial S}{\partial T}\right)_{P,n}$ with nc_P/T, $\left(\frac{\partial S}{\partial T}\right)_{V,n}$ with nc_V/T, $\left(\frac{\partial V}{\partial P}\right)_{T,n}$ with $-\kappa_T V$, and $\left(\frac{\partial V}{\partial P}\right)_{S,n}$ with $-\kappa_S V$.

- Since internal energy and entropy are perfect differentials, the order of operations of a partial derivative can be reversed through reciprocation. For example, $\left(\frac{\partial S}{\partial T}\right)_{P,n} = 1/\left(\frac{\partial T}{\partial S}\right)_{P,n}$. Therefore, $\left(\frac{\partial T}{\partial S}\right)_{P,n} = T/(nc_V)$.

- The Maxwell relations are particularly useful in the process of derivative reduction. For example, since $\left(\frac{\partial V}{\partial T}\right)_{P,n} = -\left(\frac{\partial S}{\partial P}\right)_{T,n}$, it follows that $\left(\frac{\partial S}{\partial P}\right)_{T,n} = -\beta V$.

- The chain rule can also be useful. Consider, for example, $\left(\frac{\partial V}{\partial S}\right)_{P,n}$. This can be expressed in terms of $\left(\frac{\partial V}{\partial T}\right)_{P,n}$ via application of the chain rule: $\left(\frac{\partial V}{\partial S}\right)_{P,n} = \left(\frac{\partial V}{\partial T}\right)_{P,n}\left(\frac{\partial T}{\partial S}\right)_{P,n}$. This shows that $\left(\frac{\partial V}{\partial S}\right)_{P,n} = \beta V T/(nc_P)$. The chain rule is especially useful when there is an S that would otherwise be difficult to deal with: Using the chain rule, a partial of S with respect to a variable other than temperature can be written in terms of c_P or c_V by introducing ∂T. This is precisely what we did with $\left(\frac{\partial V}{\partial S}\right)_{P,n}$.

---

[260] The technique for reducing derivatives can be extended to multi-component systems. One difference is that a larger set of basic quantities is needed. Another difference is that the Gibbs-Duhem relation will not be able to eliminate the chemical potentials.

- A variable can be removed from the subscript using a Maxwell relation or using the mathematical identity, $\left(\frac{\partial x}{\partial y}\right)_{z,n} \left(\frac{\partial y}{\partial z}\right)_{x,n} \left(\frac{\partial z}{\partial x}\right)_{y,n} = -1$. As an example, consider $\left(\frac{\partial T}{\partial P}\right)_{V,n}$. None of the standard second derivatives – i.e. β, c_P, c_V, κ_T, or κ_S – have V held constant. In this case, the Maxwell relation, $\left(\frac{\partial T}{\partial P}\right)_{V,n} = \left(\frac{\partial V}{\partial S}\right)_{T,n}$, seems to offer no improvement. However, V can be removed from the subscript using the identity above: $\left(\frac{\partial T}{\partial P}\right)_{V,n} = -\left(\frac{\partial T}{\partial V}\right)_{P,n} \left(\frac{\partial V}{\partial P}\right)_{T,n}$. It is now evident that $\left(\frac{\partial T}{\partial P}\right)_{V,n} = \kappa_T/\beta$. This identity is particularly useful when there is a thermodynamic or chemical potential in the subscript.

- The Gibbs-Duhem relation, $d\mu = -sdT + vdP$, can be used to reduce a partial derivative involving chemical potential. The two most obvious cases are: $\left(\frac{\partial \mu}{\partial T}\right)_{P,n} = -S/n$ and $\left(\frac{\partial \mu}{\partial P}\right)_{T,n} = V/n$. Other partial derivatives can also be obtained from the Gibbs-Duhem relation, such as $\left(\frac{\partial \mu}{\partial S}\right)_{V,n} = -\frac{S}{n}\left(\frac{\partial T}{\partial S}\right)_{V,n} + \frac{V}{n}\left(\frac{\partial P}{\partial S}\right)_{V,n}$ and $\left(\frac{\partial \mu}{\partial V}\right)_{S,n} = -\frac{S}{n}\left(\frac{\partial T}{\partial V}\right)_{S,n} + \frac{V}{n}\left(\frac{\partial P}{\partial V}\right)_{S,n}$.

- Partial derivatives involving thermodynamic potentials can similarly be reduced using the first law. For example, from $dF = -SdT - PdV$ (for constant n), it follows that $\left(\frac{\partial F}{\partial S}\right)_{V,n} = -S\left(\frac{\partial T}{\partial S}\right)_{V,n} - P\left(\frac{\partial V}{\partial S}\right)_{V,n}$, which can be further reduced using previous techniques. As another example, $\left(\frac{\partial F}{\partial V}\right)_{S,n} = -S\left(\frac{\partial T}{\partial V}\right)_{S,n} - P$.

- A partial derivative that involves U and a thermodynamic potential – e.g. $\left(\frac{\partial U}{\partial T}\right)_{F,n}$ – can also be reduced using the first law. For example, since $dU = TdS - PdV$ for constant n, $\left(\frac{\partial U}{\partial T}\right)_{F,n} = T\left(\frac{\partial S}{\partial T}\right)_{F,n} - P\left(\frac{\partial V}{\partial T}\right)_{F,n}$.

- These techniques can be applied to reduce a partial derivative to a combination of β, c_P, c_V, κ_T, κ_S, and/or intensive and extensive parameters like T and V. So lastly we need to remove c_V and κ_S, which are not independent. The following relationships provide the means to exchange c_V and κ_S for β, c_P, and κ_T:

$$c_P = c_V + \frac{\beta^2 TV}{n\kappa_T} \quad , \quad \kappa_T = \kappa_S + \frac{\beta^2 TV}{nc_P}$$

Example. Reduce the derivative $\left(\frac{\partial T}{\partial U}\right)_{V,n}$ for a single-component system with constant n.

As none of the desired final derivatives involve internal energy, thermodynamic potentials, or chemical potential, it is necessary to remove these quantities from the expression. In this case, this can be done very efficiently via the chain rule: $\left(\frac{\partial T}{\partial U}\right)_{V,n} = \left(\frac{\partial S}{\partial U}\right)_{V,n}\left(\frac{\partial T}{\partial S}\right)_{V,n} = 1/(nc_V)$. However, the final expression should involve only β, c_P, κ_T, and/or intensive and extensive parameters. Thus, we need to make a substitution: $\left(\frac{\partial T}{\partial U}\right)_{V,n} = \left(nc_P - \frac{\beta^2 TV}{\kappa_T}\right)^{-1}$.

Example. Reduce the derivative $\left(\frac{\partial P}{\partial T}\right)_{S,n}$ for a single-component system with constant n.

Except for the derivative $\left(\frac{\partial V}{\partial P}\right)_{S,n}$, which equals $-\kappa_S V$, it is necessary to remove S from the subscript (after removing any potentials – thermodynamic or chemical – from the expression). In this case, the Maxwell relation serves this purpose: $\left(\frac{\partial P}{\partial T}\right)_{S,n} = \left(\frac{\partial S}{\partial V}\right)_{P,n}$. This derivative can now be related to both β and c_P via the chain rule: $\left(\frac{\partial S}{\partial V}\right)_{P,n} = \left(\frac{\partial S}{\partial T}\right)_{P,n}\left(\frac{\partial T}{\partial V}\right)_{P,n}$. Therefore, $\left(\frac{\partial P}{\partial T}\right)_{S,n} = \frac{nc_P}{\beta TV}$.

Example. Reduce the derivative $\left(\frac{\partial S}{\partial P}\right)_{\mu,n}$ for a single-component system with constant n.

The chemical potential needs to be removed from the expression. In order to do this, it must first be removed from the subscript. A mathematical identity can be used to remove μ from the subscript: $\left(\frac{\partial S}{\partial P}\right)_{\mu,n} = -\left(\frac{\partial S}{\partial \mu}\right)_{P,n}\left(\frac{\partial \mu}{\partial P}\right)_{S,n}$. In this case, the Maxwell relations are not helpful for eliminating μ because they exchange μ for S. The Gibbs-Duhem relation, however, is generally the best means of removing μ from the derivative. Observe that $\left(\frac{\partial \mu}{\partial S}\right)_{P,n} = -\frac{S}{n}\left(\frac{\partial T}{\partial S}\right)_{P,n}$ and $\left(\frac{\partial \mu}{\partial P}\right)_{S,n} = -\frac{S}{n}\left(\frac{\partial T}{\partial P}\right)_{S,n} + \frac{V}{n}$.[261] The former reduces to $\left(\frac{\partial \mu}{\partial S}\right)_{P,n} = -TS/(n^2 c_P)$. The entropy must be removed from the subscript of the remaining derivative: $\left(\frac{\partial T}{\partial P}\right)_{S,n} = -\left(\frac{\partial T}{\partial S}\right)_{P,n}\left(\frac{\partial S}{\partial P}\right)_{T,n} = -\frac{T}{nc_P}\left(\frac{\partial S}{\partial P}\right)_{T,n} = \frac{T}{nc_P}\left(\frac{\partial V}{\partial T}\right)_{P,n} = \frac{\beta TV}{nc_P}$ such that $\left(\frac{\partial \mu}{\partial P}\right)_{S,n} = -\frac{\beta TSV}{n^2 c_P} + \frac{V}{n}$. Finally, $\left(\frac{\partial S}{\partial P}\right)_{\mu,n} = \frac{n^2 c_P}{TS}\left(-\frac{\beta TSV}{n^2 c_P} + \frac{V}{n}\right)^{-1}$.

Example. Reduce the derivative $\left(\frac{\partial V}{\partial P}\right)_{F,n}$ for a single-component system with constant n.

First, remove the Helmholtz potential, F, from the subscript:[262] $\left(\frac{\partial V}{\partial P}\right)_{F,n} = -\left(\frac{\partial V}{\partial F}\right)_{P,n}\left(\frac{\partial F}{\partial P}\right)_{V,n}$. Now look at the first law in the Helmholtz representation, which is $dF = -SdT - PdV$ for constant n. The two derivatives can be found from this: $\left(\frac{\partial F}{\partial V}\right)_{P,n} = -S\left(\frac{\partial T}{\partial V}\right)_{P,n} - P = -S/(\beta V) - P$ and $\left(\frac{\partial F}{\partial P}\right)_{V,n} = -S\left(\frac{\partial T}{\partial P}\right)_{V,n} = S\left(\frac{\partial T}{\partial V}\right)_{P,n}\left(\frac{\partial V}{\partial P}\right)_{T,n} = -S\kappa_T/\beta$. Therefore, $\left(\frac{\partial V}{\partial P}\right)_{F,n} = \frac{S\kappa_T}{\beta}\left(-\frac{S}{\beta V} - P\right)^{-1}$.

---

[261] Study this carefully. If you are new at this technique of obtaining a partial derivative from a differential, you probably will not feel convinced. Until you get used to this, you might feel more confident using the chain rule first in order to express these partial derivatives in terms of more obvious partial derivatives. Specifically, compare with these expressions: $\left(\frac{\partial \mu}{\partial S}\right)_{P,n} = \left(\frac{\partial \mu}{\partial T}\right)_{P,n}\left(\frac{\partial T}{\partial S}\right)_{P,n} = -\frac{S}{n}\left(\frac{\partial T}{\partial S}\right)_{P,n}$ and $\left(\frac{\partial \mu}{\partial P}\right)_{S,n} = \left(\frac{\partial \mu}{\partial V}\right)_{S,n}\left(\frac{\partial V}{\partial P}\right)_{S,n} = \left[-\frac{S}{n}\left(\frac{\partial T}{\partial V}\right)_{S,n} + \frac{V}{n}\left(\frac{\partial P}{\partial V}\right)_{S,n}\right]\left(\frac{\partial V}{\partial P}\right)_{S,n} = -\frac{S}{n}\left(\frac{\partial T}{\partial P}\right)_{S,n} + \frac{V}{n}$. Observe that if either T or P is held constant, only one term results in a partial derivative of μ obtained from the Gibbs-Duhem relation.

[262] However, if there are other potentials in the expression, such as $\left(\frac{\partial T}{\partial U}\right)_{F,n}$, first apply the first law.

Difference in heat capacities: The relationship between c_P, c_V, β, and κ_T for a single-component system with constant n can be derived by expressing dS in terms of T and V and separately in terms of T and P, then equating these two expressions, and then expressing dV in terms of T and P:

$$dS = \left(\frac{\partial S}{\partial T}\right)_{V,n} dT + \left(\frac{\partial S}{\partial V}\right)_{T,n} dV = \left(\frac{\partial S}{\partial T}\right)_{P,n} dT + \left(\frac{\partial S}{\partial P}\right)_{T,n} dP$$

$$dV = \left(\frac{\partial V}{\partial T}\right)_{P,n} dT + \left(\frac{\partial V}{\partial P}\right)_{T,n} dP$$

$$\left(\frac{\partial S}{\partial T}\right)_{V,n} dT + \left(\frac{\partial S}{\partial V}\right)_{T,n}\left(\frac{\partial V}{\partial T}\right)_{P,n} dT + \left(\frac{\partial S}{\partial V}\right)_{T,n}\left(\frac{\partial V}{\partial P}\right)_{T,n} dP = \left(\frac{\partial S}{\partial T}\right)_{P,n} dT + \left(\frac{\partial S}{\partial P}\right)_{T,n} dP$$

$$\left(\frac{\partial S}{\partial T}\right)_{V,n} dT + \left(\frac{\partial S}{\partial V}\right)_{T,n}\left(\frac{\partial V}{\partial T}\right)_{P,n} dT + \left(\frac{\partial S}{\partial P}\right)_{T,n} dP = \left(\frac{\partial S}{\partial T}\right)_{P,n} dT + \left(\frac{\partial S}{\partial P}\right)_{T,n} dP$$

$$\left(\frac{\partial S}{\partial T}\right)_{V,n} dT + \left(\frac{\partial S}{\partial V}\right)_{T,n}\left(\frac{\partial V}{\partial T}\right)_{P,n} dT = \left(\frac{\partial S}{\partial T}\right)_{P,n} dT$$

$$\frac{nc_V}{T} + \left(\frac{\partial S}{\partial V}\right)_{T,n} \beta V = \frac{nc_P}{T}$$

$$\frac{nc_V}{T} + \frac{\beta^2 V}{\kappa_T} = \frac{nc_P}{T}$$

$$c_P = c_V + \frac{\beta^2 T V}{n\kappa_T}$$

The derivative in the next-to-last step was reduced as: $\left(\frac{\partial S}{\partial V}\right)_{T,n} = \left(\frac{\partial P}{\partial T}\right)_{V,n} = -\left(\frac{\partial V}{\partial T}\right)_{P,n}\left(\frac{\partial P}{\partial V}\right)_{T,n} = \beta/\kappa_T$.

Difference in compressibilities: The relationship between κ_T, κ_S, β, and c_P for a single-component system with constant n can be derived through a similar method of deriving a relationship between c_P, c_V, β, and κ_T:

$$dV = \left(\frac{\partial V}{\partial T}\right)_{P,n} dT + \left(\frac{\partial V}{\partial P}\right)_{T,n} dP = \left(\frac{\partial V}{\partial S}\right)_{P,n} dS + \left(\frac{\partial V}{\partial P}\right)_{S,n} dP$$

$$dS = \left(\frac{\partial S}{\partial T}\right)_{P,n} dT + \left(\frac{\partial S}{\partial P}\right)_{T,n} dP$$

$$\left(\frac{\partial V}{\partial T}\right)_{P,n} dT + \left(\frac{\partial V}{\partial P}\right)_{T,n} dP = \left(\frac{\partial V}{\partial S}\right)_{P,n}\left(\frac{\partial S}{\partial T}\right)_{P,n} dT + \left(\frac{\partial V}{\partial S}\right)_{P,n}\left(\frac{\partial S}{\partial P}\right)_{T,n} dP + \left(\frac{\partial V}{\partial P}\right)_{S,n} dP$$

$$\left(\frac{\partial V}{\partial T}\right)_{P,n} dT + \left(\frac{\partial V}{\partial P}\right)_{T,n} dP = \left(\frac{\partial V}{\partial T}\right)_{P,n} dT + \left(\frac{\partial V}{\partial S}\right)_{P,n}\left(\frac{\partial S}{\partial P}\right)_{T,n} dP + \left(\frac{\partial V}{\partial P}\right)_{S,n} dP$$

$$\left(\frac{\partial V}{\partial P}\right)_{T,n} dP = \left(\frac{\partial V}{\partial S}\right)_{P,n}\left(\frac{\partial S}{\partial P}\right)_{T,n} dP + \left(\frac{\partial V}{\partial P}\right)_{S,n} dP$$

$$-\kappa_T V = \left(\frac{\partial V}{\partial S}\right)_{P,n}\left(\frac{\partial S}{\partial P}\right)_{T,n} - \kappa_S V$$

$$-\kappa_T V = \frac{\beta T V}{nc_P}(-\beta V) - \kappa_S V$$

258

$$\kappa_T = \kappa_S + \frac{\beta^2 TV}{nc_P}$$

The derivatives in the next-to-last step were reduced as: $\left(\frac{\partial V}{\partial S}\right)_{P,n} = \left(\frac{\partial V}{\partial T}\right)_{P,n} \left(\frac{\partial T}{\partial S}\right)_{P,n} = \beta TV/(nc_P)$ and $\left(\frac{\partial S}{\partial P}\right)_{T,n} = -\left(\frac{\partial V}{\partial T}\right)_{P,n} = -\beta V$.

Adiabatic index: The adiabatic index, γ, equals the ratio of the molar heat capacity at constant pressure to that at constant volume, and therefore can now also be expressed in terms of the compressibilities:

$$\frac{\beta^2 TV}{n} = \kappa_T(c_P - c_V) = c_P(\kappa_T - \kappa_S)$$
$$\frac{c_P - c_V}{c_P} = \frac{\kappa_T - \kappa_S}{\kappa_T}$$
$$1 - \frac{c_V}{c_P} = 1 - \frac{\kappa_S}{\kappa_T}$$
$$\gamma = \frac{c_P}{c_V} = \frac{\kappa_S}{\kappa_T}$$

Applications of derivative reduction: Since only quasistatic processes are reversible, they are of considerable interest in thermodynamics. A quasistatic process involves making a small change in one or more quantities, while holding other quantities constant. For example, a small addition of heat that increases the volume slightly during an isothermal process can be expressed as $đQ = TdS = T\left(\frac{\partial S}{\partial V}\right)_{T,n} dV$ for a single-component system with constant mole number. From a practical point of view, it is convenient to reduce this derivative in order to express this relationship in terms of quantities that are more accessible in the laboratory. In this case, one of the Maxwell relations can be applied to reduce the derivative: $\left(\frac{\partial S}{\partial V}\right)_{T,n} = \left(\frac{\partial P}{\partial T}\right)_{V,n} = -\left(\frac{\partial V}{\partial T}\right)_{P,n} \left(\frac{\partial P}{\partial V}\right)_{T,n} = \beta/\kappa_T$. Thus, the heat exchange for this isothermal process can be expressed as $đQ = \beta TdV/\kappa_T$. Observe that, by applying the technique of derivative reduction, the heat exchange can be calculated in terms of the small volume change in terms of quantities that are experimentally easy to determine without knowing the fundamental equation. Thus, derivative reduction is a very powerful and useful technique of thermodynamics.

Example. Relate a small pressure change to a small temperature change for a quasistatic adiabatic process for a single-component system with constant mole number.

We need to relate dP to dT through a suitable partial derivative, and then reduce the partial derivative. The key is to determine what is held constant. No heat is exchanged ($đQ = 0$) for an adiabatic process, and since it is quasistatic, $đQ = TdS$, it is implied that the process is also isentropic ($dS = 0$). Therefore, S and n are held constant:

$$dP = \left(\frac{\partial P}{\partial T}\right)_{S,n} dT = -\left(\frac{\partial P}{\partial S}\right)_{T,n}\left(\frac{\partial S}{\partial T}\right)_{P,n} dT = \left(\frac{\partial T}{\partial V}\right)_{P,n} \frac{nc_P}{T} dT = \frac{nc_P}{\beta TV} dT$$

6.7 Phase Transitions

Concavity: A function $f(x)$ is said to be concave up (or convex) if it has a monotonically increasing slope (df/dx), and concave down (or simply concave) if it has a monotonically decreasing slope.[263] For a function that is concave up, the line segment joining any two distinct points, x_1 and x_2, on the curve lies above the curve; whereas for a function that is concave down, such a line segment lies below the curve. The second derivative of a function – d^2f/dx^2 – is non-negative over an interval if it is concave up over that interval, and non-positive over an interval for which it is concave down. Furthermore, if the second derivative is always positive, the function is concave up, and if the second derivative is always negative, it is concave down.[264]

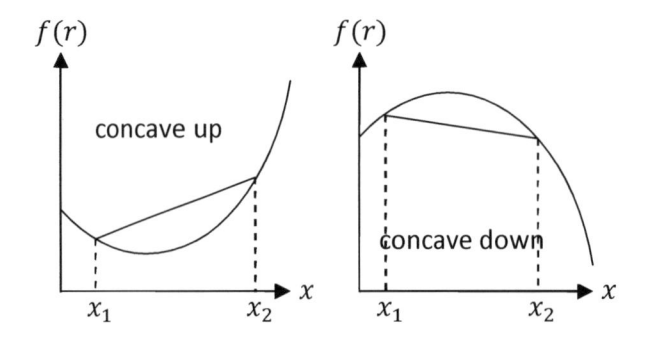

Mechanical stability: Consider an object in classical mechanics for which the potential energy is a function of a single position coordinate, such as $U(r)$. The net force exerted on the object at position r is related to the potential energy via the gradient operator, $\vec{F} = -\vec{\nabla} U$, which in this case reduces to $\vec{F} = -\frac{dU}{dr}\hat{r}$. The object is in an equilibrium state if the net force is zero. Thus, if the slope of the potential energy function is zero $\left(\frac{dU}{dr} = 0\right)$, the object is in equilibrium. Now image that the object receives a small displacement from equilibrium. If the object tends to return to equilibrium regardless of the direction of the small displacement, the equilibrium is termed stable. This occurs if the second derivative of the potential energy function is negative $\left(\frac{d^2U}{dr^2} < 0\right)$ – i.e. if $U(r)$ is concave up in the vicinity of the equilibrium position. In this case, the object tends to oscillate about the equilibrium position. If instead $\frac{d^2U}{dr^2} > 0$, then a displacement in one or both directions – the two 'directions' being increasing or decreasing r – results in an object that tends away, rather than toward, equilibrium; such a position is termed unstable equilibrium. Neutral equilibrium results if $\frac{d^2U}{dr^2} = 0$.

---

[263] Some advanced mathematical works make a distinction between concave and strictly concave, where a function that is concave up has a non-decreasing slope, while strictly concave up means an increasing slope.

[264] Notice the distinction here. For example, for the function $f(x) = x^2$, since $\frac{d^2f}{dx^2} = 2$, it can be concluded that $f(x)$ is concave up. Compare with the function $g(x) = x^4$, where $\frac{d^2g}{dx^2} = 12x^2$ is zero at $x = 0$, yet $g(x)$ is still concave up.

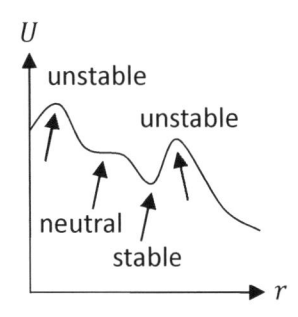

Thus, a state of mechanical equilibrium is stable if a small fluctuation in the system that causes a small displacement from equilibrium results in the system returning to equilibrium. If the potential energy function is concave up $\left(\frac{d^2U}{dr^2} < 0\right)$ in the region around equilibrium, the system will experience stability at the equilibrium position. The line segment joining any two points, r_1 and r_2, in the local neighborhood of a stable equilibrium position lies above the curve, meaning that U is a local minimum at a position of stable equilibrium.

Thermodynamic stability: Any unconstrained parameters of a thermodynamic system approach equilibrium values that result in an extremum for the dependent extensive variable. In the entropy representation, $S(U, V, n_i)$, for example, if the internal energy, volume, and/or mole numbers have freedom, at equilibrium they become those values that maximize the entropy for the given constraints. Setting $dS = 0$ allows us to solve for the equilibrium conditions, which correspond to extreme values (relative minima, relative maxima, and points of inflection) of the entropy. It is then necessary to check that $d^2S < 0$ in order to see if indeed the entropy is maximum. This second differential means:

$$d^2S = \left(dU\frac{\partial}{\partial U} + dV\frac{\partial}{\partial V} + \sum_{i=1}^{r} dn_i \frac{\partial}{\partial n_i}\right)^2 S(U, V, n_i)$$

$$d^2S = \left(dU\frac{\partial}{\partial U} + dV\frac{\partial}{\partial V} + \sum_{i=1}^{r} dn_i \frac{\partial}{\partial n_i}\right)\left(\frac{\partial S}{\partial U}dU + \frac{\partial S}{\partial V}dV + \sum_{i=1}^{r} \frac{\partial S}{\partial n_i}dn_i\right)$$

$$d^2S = \frac{\partial^2 S}{\partial U^2}d^2U + \frac{\partial^2 S}{\partial V^2}d^2V + \sum_{i,j=1}^{r} \frac{\partial^2 S}{\partial n_i \partial n_j}dn_i dn_j + 2\frac{\partial^2 S}{\partial U \partial V}dU dV$$

$$+ 2\sum_{i=1}^{r}\left(\frac{\partial^2 S}{\partial U \partial n_i}dU + \frac{\partial^2 S}{\partial V \partial n_i}dV\right)dn_i$$

Strictly speaking, equilibrium could exist whether entropy is at a relative maximum, minimum, or point of inflection. However, equilibrium is only stable if the entropy is maximum. Unstable equilibrium cannot be expected to persist long due to fluctuations typical of thermodynamic systems – e.g. mechanical fluctuations referred to as Brownian motion. Such fluctuations cause small deviations from equilibrium. A system in stable equilibrium returns to equilibrium after such small departures, whereas a system in unstable equilibrium does not.

Therefore, testing the sign of the second differential, d^2S, is a stability condition: The equilibrium state is stable if $d^2S < 0$ and unstable if $d^2S > 0$. This test relates to the concavity of the multi-dimensional thermodynamic hypersurface formed by the $r+2$ independent extensive parameters in the fundamental relation. Specifically, the entropy of a stable system is concave down. Put another way, the entropy must lie below its tangent hyperplanes. Recall that the fundamental equation relates the possible equilibrium states for a thermodynamic system – i.e. each possible state in the fundamental relation could potentially maximize the entropy, depending upon what constraints actually apply. Thus, the fundamental relation in the entropy representation must lie below its tangent hyperplanes for every point on the hypersurface of the fundamental relation if all of the possible equilibrium states are stable.

However, there can be regions where the entropy is concave up rather than down, corresponding to unstable equilibrium states. These regions are characteristic of phase transitions.

Stability conditions: In the entropy representation, the stability condition is $d^2S < 0$. Geometrically, this means that the entropy hypersurface specified by the fundamental relation, $S(U, V, n_i)$, everywhere lies below its family of tangent hyperplanes. For a system with constant n_i, this translates to:

$$d^2S = \left(\frac{\partial^2 S}{\partial U^2}\right)_{V,n_i} d^2U + \left(\frac{\partial^2 S}{\partial V^2}\right)_{U,n_i} d^2V + 2\frac{\partial^2 S}{\partial U \partial V} dU dV < 0$$

If the system experiences a small fluctuation, corresponding to shifts in internal energy, δU, and volume, δV, which are small fractional changes (i.e. $|\delta U/U| \ll 1$ and $|\delta V|/V \ll 1$), to first-order the stability condition requires that:[265]

$$\left(\frac{\partial^2 S}{\partial U^2}\right)_{V,n_i} (\delta U)^2 + \left(\frac{\partial^2 S}{\partial V^2}\right)_{U,n_i} (\delta V)^2 + 2\delta U \delta V \frac{\partial^2 S}{\partial U \partial V} \leq 0$$

The global criteria that $d^2S < 0$ (i.e. that every point on the entropy hypersurface lie below its family of tangent hyperplanes) can thus be resolved into three local conditions:[266]

$$\left(\frac{\partial^2 S}{\partial U^2}\right)_{V,n_i} \leq 0$$

$$\left(\frac{\partial^2 S}{\partial V^2}\right)_{U,n_i} \leq 0$$

---

[265] This result can be obtained to first-order or higher in δU and δV via a multivariable Taylor series expansion based on the entropy hypersurface being concave down in both U and V.

[266] To see this, multiply all three terms of the original equation by $\left(\frac{\partial^2 S}{\partial U^2}\right)_{V,n_i}$, which reverses the direction of the inequality since this multiplicative factor is non-positive. The inequality can then be restructured as: $\left[\left(\frac{\partial^2 S}{\partial U^2}\right)_{V,n_i}\left(\frac{\partial^2 S}{\partial V^2}\right)_{U,n_i} - \left(\frac{\partial^2 S}{\partial U \partial V}\right)^2\right](\delta V)^2 \geq \left[\left(\frac{\partial^2 S}{\partial U^2}\right)_{V,n_i}\delta U + \left(\frac{\partial^2 S}{\partial U \partial V}\right)\delta V\right]^2 \geq 0.$

$$\left(\frac{\partial^2 S}{\partial U^2}\right)_{V,n_i} \left(\frac{\partial^2 S}{\partial V^2}\right)_{U,n_i} \geq \left(\frac{\partial^2 S}{\partial U \partial V}\right)^2$$

The first condition states that the entropy must be concave down with respect to energy for a fixed volume, and the second makes a similar statement with regard to volume for fixed internal energy. The third condition is necessary if neither internal energy or nor volume are fixed: If the entropy hypersurface is concave down with regard to U for fixed V and with regard to V for fixed U, it is still necessary to check that the entropy hypersurface is not concave up along the in-between directions (i.e. where neither U nor V are fixed).

In the energy representation, the global condition is $d^2 U > 0$, meaning that the internal energy is concave up – i.e. the internal energy hypersurface lies above its tangent hyperplanes. The corresponding local conditions are:[267,268]

$$\left(\frac{\partial^2 U}{\partial S^2}\right)_{V,n_i} \geq 0$$

$$\left(\frac{\partial^2 U}{\partial V^2}\right)_{S,n_i} \geq 0$$

$$\left(\frac{\partial^2 U}{\partial S^2}\right)_{V,n_i} \left(\frac{\partial^2 U}{\partial V^2}\right)_{S,n_i} \geq \left(\frac{\partial^2 U}{\partial S \partial V}\right)^2$$

In the thermodynamic potential representations, the thermodynamic potentials are concave down with regard to extensive parameters (like S and V) and concave up with regard to intensive parameters (like T and P).[269] For example, the Helmholtz free energy is concave down with regard to temperature for fixed volume and concave up with regard to volume for fixed temperature.

Stable systems: Stability places conditions on the second derivatives, which can be expressed in terms of c_P, c_V, κ_T, and/or κ_S. Therefore, the stability conditions can be expressed in terms of the molar specific heat capacities and the compressibilities:

---

[267] In the energy representation, the inequality expressed by the third local condition does not get reversed since in this case we derive the third condition by multiplying through by $\left(\frac{\partial^2 U}{\partial S^2}\right)_{V,n_i}$, which is non-negative, rather than $\left(\frac{\partial^2 S}{\partial U^2}\right)_{V,n_i}$, in the energy representation, which is non-positive and thus flips the inequality.

[268] Because the first derivatives of thermodynamics share a reciprocal relationship – for example, $\left(\frac{\partial S}{\partial U}\right)_{V,n_i} = 1/\left(\frac{\partial U}{\partial T}\right)_{V,n_i}$ – it might be tempting to try to apply this to the second derivatives. However, the second derivatives are not related in this way. For example, $\left(\frac{\partial^2 S}{\partial U^2}\right)_{V,n_i}$ is not the reciprocal of $\left(\frac{\partial^2 U}{\partial S^2}\right)_{V,n_i}$.

[269] The thermodynamic potentials are concave up with regard to extensive parameters because these are the parameters that are left alone in obtaining the thermodynamic potentials from internal energy via Legendre transformations. The conjugate variables that are swapped during the Legendre transformations come with sign changes, however, so a thermodynamic potential is concave down with regard to an intensive parameter.

$$\left(\frac{\partial^2 U}{\partial S^2}\right)_{V,n_i} = \left(\frac{\partial T}{\partial S}\right)_{V,n_i} = \frac{T}{nc_V} \geq 0 \implies c_V \geq 0$$

$$\left(\frac{\partial^2 G}{\partial T^2}\right)_{P,n_i} = -\left(\frac{\partial S}{\partial T}\right)_{P,n_i} = -\frac{nc_P}{T} \leq 0 \implies c_P \geq 0$$

$$\left(\frac{\partial^2 G}{\partial P^2}\right)_{T,n_i} = \left(\frac{\partial V}{\partial P}\right)_{T,n_i} = -V\kappa_T \leq 0 \implies \kappa_T \geq 0$$

$$\left(\frac{\partial^2 H}{\partial P^2}\right)_{S,n_i} = \left(\frac{\partial V}{\partial P}\right)_{S,n_i} = -V\kappa_S \leq 0 \implies \kappa_S \geq 0$$

Conceptually, these stability conditions can be interpreted as follows. If heat is added to a stable system at constant volume, $đQ = nc_V dT$, or constant pressure, $đQ = nc_P dT$, the temperature cannot decrease because the molar specific heat capacity is nonnegative for a stable system. Similarly, if the volume of a stable system is increased at constant temperature, $dV = -V\kappa_T dP$, or at constant entropy, $dV = -V\kappa_S dP$, the pressure cannot increase because the compressibility is non-positive for a stable system.

Le Chatelier's principle: Thermodynamic systems inherently have inhomogeneities and experience constant fluctuations. Any inhomogeneities that develop and any fluctuations that occur spontaneously in a stable system at equilibrium (which is not experiencing any macroscopic changes that would affect the equilibrium conditions – e.g. if a piston is suddenly made movable, this allows the volume to change significantly at the macroscopic level) must induce a process that tends to return the system to its equilibrium state according to Le Chatelier's principle.

For example, microscopically temperature cannot be expected to be perfectly homogeneous, and so there will be small fluctuations in temperature locally throughout a system; but macroscopically, on average the temperature will appear constant and homogeneous at thermal equilibrium. A small local temperature increase will result in a little heat flowing from the higher-temperature inhomogeneity to its lower-temperature surroundings – for a stable system, for which specific heat capacity is nonnegative – which has the effect of making the temperature homogeneous and preserving the average equilibrium temperature. Similarly, due to Brownian motion, small mechanical inhomogeneities and fluctuations in pressure can be expected microscopically, though pressure will appear constant and homogeneous at mechanical equilibrium. A small region of higher pressure tends to expand, lowering its pressure – for a stable system, for which compressibility is nonnegative – which has the effect of making the pressure homogeneous and preserving the average equilibrium pressure.

Phase diagram: Recall that the phases of a substance can be conveniently plotted a P-T diagram (see Sec. 1.5). The solid, liquid, and gaseous phases correspond to combinations of pressure and temperature for which the substance would be in stable equilibrium. The phases are separated by a coexistence curve, which branches in three directions from the triple point. Two phases of the substance coexist for combinations of pressure and temperature that lie on the coexistence curve.

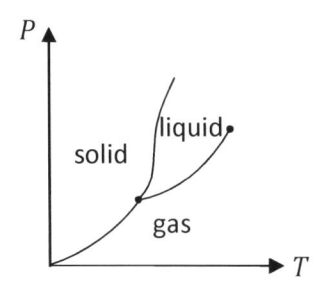

Phase transition: A phase transition occurs if the pressure and/or temperature of the substance are varied such that the coexistence curve is crossed. The stability criteria are not satisfied during a phase transition. The fundamental relation has stable concavity for all combinations of pressure and temperature except for those combinations for which the coexistence curve would be crossed. A phase transition is termed first-order unless it occurs across the critical points (the terminal points) of the coexistence curve (i.e. where the coexistence curve terminates) – a phase transition across the critical points is termed second-order. It is also possible to begin in one phase and use extreme temperatures and/or pressures in order to navigate around the coexistence curve. A phase transition results in an abrupt change of phase, whereas this circumnavigating process does not.

The abrupt change of phase that occurs during a phase transition corresponds to an abrupt change in order in the system – as the solid, liquid, and gaseous phases differ markedly in terms of statistical order. Therefore, the entropy experiences a discontinuity during a phase transition – either a sudden increase or a sudden decrease. This is accompanied by a an exchange of heat. The amount of heat absorbed or released during a phase transition equals $Q = ML$, where M is the mass of the substance and L is the latent heat of transformation. The discontinuity in the entropy and the exchange of heat are related since $đQ = TdS$: $Q = ML = T(S - S_0)$. It is sometimes convenient to work with the molar latent heat, ℓ, which is defined by $Q = ML = n\ell$. That is, ℓ is energy per unit mole, whereas L is energy per unit mass (cf. specific heat, which is per unit mass, $đQ = MC_{P,V}dT$, and molar specific heat, which is per unit mole, $đQ = nc_{P,V}dT$). In terms of the molar latent heat, $\ell = T(s - s_0)$.

The Gibbs representation is particularly well-suited for describing phase transitions, since the Gibbs free energy, $G(T, P, n_i)$, is a function of temperature and pressure (in addition to the mole numbers) – convenient variables as they correspond directly to a P-T diagram. The Gibbs potential is also useful for problems where phase or chemical equilibrium is attained – recalling that for a single-component system, the molar Gibbs potential equals the chemical potential, $g = G/n = \mu$, and for a multi-component system, the molar Gibbs potential equals $g = \sum_{i=1}^{r} \mu_i n_i$. Recall, also, that the chemical potentials are equal in the case of diffusive equilibrium (i.e. equilibrium with respect to the flow of particles), and that $\sum_{i=1}^{r} \mu_i \nu_i = 0$ for chemical equilibrium, where ν_i are the stoichiometric coefficients for the chemical reaction. Furthermore, if the coexistence curve is crossed at either constant temperature or constant pressure – as is easy and convenient to achieve in lab – then the first law simplifies in the Gibbs representation: Since $dG = -SdT + VdP + \sum_{i=1}^{r} \mu_i n_i$, one of the first two terms will be zero for an isotherm or isobar. Thus, the coexistence curve corresponds to points where the substance is in phase equilibrium, for which the two phases have equal chemical potentials, $\mu_1 = \mu_2$, and therefore, for a single-component system, equal molar Gibbs potentials, $g_1 = g_2$.

The enthalpy, $H(S, P, n_i)$, can be obtained as a Legendre transformation directly from the Gibbs free energy, $G(T, P, n_i)$, by exchanging T for its conjugate variable, S. These two thermodynamic potentials are thus related by $H = G + TS$. In molar form, $h = g + Ts$. Therefore, during a phase transition, the change in enthalpy equals: $h - h_0 = T(s - s_0)$. This follows since $g = g_0$ (i.e. the molar Gibbs potential of the final phase equals that of the initial phase) for the phase transition, and since temperature can only change infinitesimally during the infinitesimal path across the coexistence curve that corresponds to a phase transition (so whether or not pressure or temperature are held constant, ΔT and ΔP are approximately zero[270]). We thus see that the change in molar latent heat equals the latent heat of transformation: $\ell = T(s - s_0) = h - h_0$. Whereas P and T scarcely change during a phase transition,[271] S, V, and U change abruptly during a phase transition. In particular, the entropy change is related to the enthalpy change by a simple factor of temperature.

The discontinuity in the entropy, volume, internal energy, and thermodynamic potentials that occur as the coexistence curve is crossed – i.e. during a phase transition – correspond to the different nature of the two phases. The discontinuity is lessened for a phase transition that occurs closer to the critical points[272] – i.e. the endpoints of the coexistence curve. In the limit that the phase transition occurs closer and closer to a critical point, the discontinuity approaches zero. There is a finite discontinuity for a first-order phase transition; a phase transition across a critical point is second-order, where the discontinuity reaches its limit. The two phases, corresponding to two distinct equilibrium states, becomes less and less distinct along the coexistence curve as the critical point is approached; the two phases become indistinguishable at the critical point.

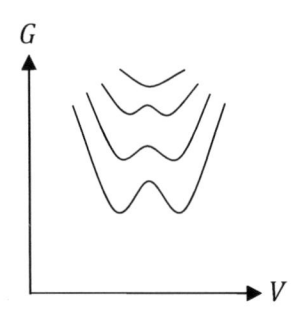

---

[270] This assumes that the coexistence curve is crossed directly, for which the route would be of infinitesimal change in pressure and/or temperature. This would generally be the case for a minute horizontal (isobaric) or vertical (isothermal) shift in the P-T diagram, for example. If you want to be fancy and enter the coexistence curve at one point in the P-T diagram, then adjust pressure and temperature very carefully such that the coexistence curve is followed for quite some time, and finally exit the coexistence curve at some distant point in the P-T diagram, then pressure and temperature may change appreciably.

[271] In holding pressure and temperature relatively constant, the system is essentially in contact with both pressure and thermal reservoirs – i.e. the $-SdT$ and VdP terms of dG are approximately zero. The Gibbs representation is particularly useful for systems in contact with both pressure and thermal reservoirs, for which the Gibbs potential is a minimum at equilibrium.

[272] The terminal points of the coexistence curve are called critical points. The solid-liquid, liquid-gaseous, and gaseous-solid coexistence curves intersect at the triple point. While the triple point is a very significant point, and thus might seem to be 'critical,' the triple point is not a critical point.

Beyond the critical point, there is no clear, well-defined distinction between the two phases. A first-order phase transition results in an abrupt change between the two distinct equilibrium states. Yet the same transformation from one phase to another can be brought about by going around the coexistence curve rather than crossing it. In this latter case, there is not a clearly-defined transition between the two phases.

The coexistence curve corresponds to points of unstable equilibrium. These are regions of parameter space where the fundamental relation does not satisfy the concavity expectations of the stability conditions. The coexistence curve separates two distinct equilibrium states, corresponding to two different phases; the coexistence curve itself is a single equilibrium dividing the two distinct equilibrium states of the two different phases. The substance is in stable equilibrium on either side of the coexistence curve, but in unstable equilibrium on the curve itself. Except for the critical points – i.e. the endpoints of the coexistence curve – the system has two extrema for every point on the coexistence curve. These two extrema (which could be minima or maxima, depending upon the thermodynamic representation – e.g. they would be maxima in the entropy representation, but minima in the energy representation) correspond to the two different phases. The separation between these two extrema corresponds to the discontinuity in the entropy, volume, internal energy, and thermodynamic potentials.

Along the coexistence curve, both phases exist. Just outside of the coexistence curve, one phase is preferred. The two extrema are on an equal footing on the coexistence curve, but there is one true extremum outside of the coexistence curve. This is illustrated with the following example. The temperature of a liquid can actually be cooled so slowly that the liquid is carried into a region of parameter space that would ordinarily correspond to a solid state in a process referred to as supercooling. This supercooled state is an unstable equilibrium state for which there is a preferred extremum, as the spontaneous fluctuations inherent in thermodynamic systems will eventually bring the substance to the preferred extremum, and a phase transition to the solid state will occur.[273] Contrast with the unstable equilibrium state of the coexistence curve, where neither extremum is preferred and so the spontaneous fluctuations do not lead to a pure phase of either variety, but a coexistence of two phases persists.

The two extrema characteristic of the unstable equilibrium between the two phases of the coexistence curve merge into a single extremum at the critical point. A first-order phase transition occurs when the coexistence curve is crossed, and the entropy, volume, internal energy, and thermodynamic potentials experience a discontinuity associated with the separation between these two extrema. A second-order phase transition occurs when the critical point is crossed, which represents the limit where the two extrema, associated with the two phases, have merged together. The discontinuity vanishes in the limit that the coexistence curve is crossed closer to the critical point, yet, as we shall see, there is evidence of the discontinuity at the limit; but beyond the critical point, indications of the discontinuity disappear completely.

---

[273] This is similar to a supersaturated solution. A solution of sugar water is saturated when the maximum amount of solute has dissolved in the water. A saturated solution of sugar water is in stable equilibrium. At higher temperature, more sugar can be dissolved in the water. If the temperature is then slowly cooled – quasistatically – the solution becomes supersaturated – i.e. it contains more dissolved solute than would ordinarily be possible. This supersaturated state is a metastable state: If the solution is disturbed by shaking, for example, the excess sugar will undissolve – as the solution will select the preferred extremum.

One phase can also change into another by going around the coexistence curve, in which case there is no clear, well-defined transition between the two phases, though the initial and final states may be quite distinct; in this case, the system is always in stable equilibrium if the process is quasistatic, and there is always one true extremum.[274]

As the coexistence curve represents states of unstable equilibrium, the region of parameter space corresponding to a phase transition can be found from the stability conditions. If the fundamental relation is known, the concavity conditions for the particular thermodynamic representation employed can be applied to determine for what range of parameters the system is unstable. It is generally more accessible to look at the stability conditions on the molar specific heat capacities and the compressibilities for real systems where the fundamental relation is not so easy to ascertain.

> **Important Distinction.** A first-order phase transition involves a transition between two distinct phases corresponding to two separate extrema, in which the entropy, volume, internal energy, and thermodynamic potentials have finite discontinuities. The two separate extrema merge together and the discontinuities shrink to zero in the limit that the coexistence curve is crossed closer and closer to a critical point. A second-order phase transition occurs when the coexistence curve is crossed at a critical point. Two distinct initial and final phases can also be connected via a route that passes around a critical point, in which case the change is gradual and no sudden, distinct change of phase is observed.

> **Example.** Show that the single-component van der Waals fluid model accommodates a phase transition, but that the single-component ideal gas model does not.
>
> The stability of these models can be explored by checking the signs of the molar specific heat capacities and the compressibilities. The molar specific heat for an ideal gas is a positive number times the universal gas constant. For example, $c_V = 3R/2$ for a monatomic ideal gas. For any c_V, $c_P = c_V + R$. Therefore, the molar specific heat capacity is always positive for an ideal gas: $c_P > c_V > 0$. The isothermal and adiabatic compressibilities for a single-component ideal gas are also always positive:
>
> $$\kappa_T = -\frac{1}{V}\left(\frac{\partial V}{\partial P}\right)_{T,n} = -\frac{1}{V}\left[\frac{\partial}{\partial P}\left(\frac{nRT}{P}\right)\right]_{T,n} = -\frac{nRT}{V}\left[\frac{\partial}{\partial P}\left(\frac{1}{P}\right)\right]_{T,n} = \frac{nRT}{P^2 V} = \frac{1}{P} > 0$$
>
> $$\kappa_S = \kappa_T - \frac{\beta^2 TV}{nc_P} = \frac{1}{P} - \frac{TV}{nc_P}\left[\frac{1}{V}\left(\frac{\partial V}{\partial T}\right)_{P,n}\right]^2 = \frac{1}{P} - \frac{T}{nc_P V}\left[\frac{\partial}{\partial T}\left(\frac{nRT}{P}\right)\right]^2_{P,n}$$
>
> $$\kappa_S = \frac{1}{P} - \frac{nR^2 T}{c_P P^2 V}\left[\frac{\partial}{\partial T}(T)\right]^2_{P,n} = \frac{1}{P} - \frac{R}{c_P P} = \frac{c_P - R}{c_P P} = \frac{c_V}{c_P P} = \frac{1}{\gamma P} > 0$$
>
> Since $c_P > c_V > 0$ and $\kappa_T > \kappa_S > 0$ for all values of P and T for an ideal gas, an ideal gas is always in stable equilibrium and hence the ideal gas model does not accommodate a phase transition.[275]

---

[274] For any given combination of pressure and temperature, there is one true extremum (since the coexistence curve is not crossed in this case), but the value of that extremum gradually changes as the pressure and temperature are gradually changed.

[275] There is good agreement between the ideal gas model and real gases of sufficiently low density, but for higher densities the ideal gas model breaks down. This failure of the ideal gas model to describe higher-density behavior well is related to the problem that the ideal gas model does not accommodate a phase transition. The van der

The molar specific heat capacity at constant volume, c_V, of a van der Waals fluid is one of the parameters of the van der Waals model, along with the constants a and b. So c_V is a parameter of the theory to be determined by experiment. The isothermal compressibility for a van der Waals fluid is:

$$\kappa_T = -\frac{1}{V}\left(\frac{\partial V}{\partial P}\right)_{T,n} = -\frac{1}{V}\left(\frac{\partial P}{\partial V}\right)_{T,n}^{-1} = -\frac{1}{V}\left\{\frac{\partial}{\partial V}\left[\frac{R}{\frac{V}{n}-b} - \frac{ac_V n^2}{V^2\left(\frac{U}{n}+\frac{an}{V}\right)}\right]\right\}_{T,n}^{-1}$$

$$\kappa_T = -\frac{1}{V}\left[\frac{\partial}{\partial V}\left(\frac{nR}{V-bn} - \frac{an^2}{TV^2}\right)\right]_{T,n}^{-1} = -\frac{1}{V}\left[\frac{2an^2}{TV^3} - \frac{nR}{(V-bn)^2}\right]_{T,n}^{-1}$$

The isothermal compressibility is negative if

$$\frac{2an^2}{TV^3} > \frac{nR}{(V-bn)^2}$$
$$2an(V-bn)^2 > RTV^3$$

This shows that the van der Waals model includes regions of parameter space where there is unstable equilibrium in addition to regions where there is stable equilibrium. In this way, the van der Waals model accommodates a phase transition.

Clausius-Clapeyron equation: A relationship corresponding to the coexistence curve can be derived by applying the conditions for phase equilibrium. Consider a points, (P,T), lying on the coexistence curve for a single-component system.[276] At this point, the substance would coexist in two phases, which we may designate by 1 and 2. The two phases must be in phase equilibrium at this point, meaning that their chemical potentials must be equal: $\mu_1 = \mu_2 \equiv \mu$. According to the Gibbs-Duhem relation, $d\mu = -sdT + vdP$. The molar entropies and molar volumes for the two phases may differ substantially: $d\mu = -s_1 dT + v_1 dP = -s_2 dT + v_2 dP$. Solving for dP/dT, we obtain an equation for the slope of the coexistence curve, which is known as the Clausius-Clapeyron equation:

$$\frac{dP}{dT} = \frac{s_2 - s_1}{v_2 - v_1} = \frac{\ell}{T(v_2 - v_1)}$$

The slope of the coexistence curve thus equals the ratio of the discontinuity in the entropy to that of the volume. In practice, very often the volume of one phase is negligible compared to that of the other.

Waals fluid offers an improvement over the ideal gas model that is specifically designed to accommodate higher pressures, well enough that the van der Waals model describes both gases and liquids, including phase transitions.
[276] There is only one chemical component, but there are two phases in the complete system. We may divide this into two subsystems – one for each phase – or we may choose to work with the complete system. Although the complete system may have constant mole number, the mole numbers for each phase may be variable.

Example. Determine the Clausius-Clapeyron equation for a van der Waals fluid.
Using the Clausius-Clapeyron equation and the fundamental relation for a van der Waals fluid,

$$\frac{dP}{dT} = \frac{s_2 - s_1}{v_2 - v_1} = \frac{R \ln\left(\frac{v_2 - b_2}{v_1 - b_1}\right) + \ln\left[\frac{(u_2 + a_2/v_2)^{c_{2V}}}{(u_1 + a_1/v_1)^{c_{1V}}}\right]}{v_2 - v_1} = \frac{R \ln\left(\frac{v_2 - b_2}{v_1 - b_1}\right) + \ln\left[\frac{(n_2 c_{2V} T)^{c_{2V}}}{(n_1 c_{1V} T)^{c_{1V}}}\right]}{v_2 - v_1}$$

where the relation $n_{1,2} c_{1,2;V} T = u_{1,2} + a_{1,2}/v_{1,2}$, which is easily obtained from one of the equations of state, has been used. Note that for any point on the coexistence curve, $T_1 = T_2$ because the two phases have the same pressure and temperature (but, of course, the pressure and temperature each vary along the curve). The molar volumes of the two phases, v_1 and v_2, are not constant, but depend upon the pressure and temperature. In particular, the difference $v_2 - v_1$ diminishes to zero as the critical point is approached. There are two different phases with two different molar volumes for each point, (P,T), on the coexistence curve.

Each molar volume can now be expressed in terms of the variables P and T, the constants $a_{1,2}$, $b_{1,2}$, and $c_{1,2;V}$, and the mole numbers $n_{1,2}$. Integration of the Clausius-Clapeyron would then, in principle, yield the equation of the coexistence curve (as a definite integral from initial pressure and temperature to final pressure and temperature). However, since a quadratic equation must be solved to express v_1 and v_2 each in terms of pressure and temperature, the differential equation is rather cumbersome. We will see an alternative method of obtaining the equation of the coexistence curve when we learn about the line of equal areas.

Stable and underlying thermodynamic relations: The fundamental relation may not be stable for all regions of parameter space; in fact, in order to describe phase transitions, there must be regions where the stability conditions are not satisfied. It is conceptually and mathematically useful to work with two fundamental relations for such a system: (1) The 'underlying' fundamental relation includes regions of instability, while (2) the 'stable' fundamental relation is a modification of the underlying fundamental relation, which is stable throughout the parameter space.

The stable fundamental relation is obtained as the set of 'superior' or 'inferior' tangent hyperplanes – i.e. in the region of instability, the concavity is 'corrected' by drawing a 'superior' or 'inferior' tangent hyperplane. For example, recall the entropy is concave down for stable regions, where the entropy hypersurface lies below its tangent hyperplanes. In a region of instability – where the entropy is instead concave up – the stable fundamental relation is obtained by patching over the concave up region with a 'superior' tangent hyperplane. The superior tangent hyperplane that patches over the concave up region lies above the entropy hypersurface. The complete set of tangent hyperplanes for a stable fundamental relation in the entropy representation everywhere lies above the entropy hypersurface, because all of the tangent hyperplanes that did not were replaced with superior tangent hyperplanes. In a representation where concave down is stable and concave up is unstable, the unstable regions are patched up with 'inferior' tangent hyperplanes, which lies below the potential hypersurface.

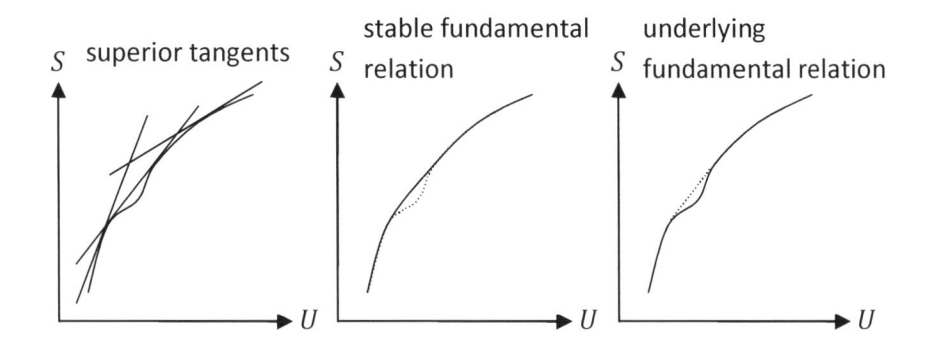

Conceptually, the stable fundamental relation is important because it agrees with the macroscopic thermodynamic averages of the extensive parameters – like volume and entropy. The underlying fundamental relation is important because it describes where the system is stable or unstable, and therefore determines the behavior of the system due to the presence of statistical fluctuations. Both fundamental relations are conceptually and mathematically useful.

> **Important Distinction.** The underlying fundamental relation includes regions where the system is unstable as well as where it is stable, and thus determines the behavior of the system due to statistical fluctuations. The stable fundamental relation is obtained from the underlying fundamental relation by patching up the concavity problems with superior or inferior tangent hyperplanes, and agrees with the macroscopic averages of the extensive parameters. Both fundamental relations are conceptually and mathematically useful.

Line of equal areas: Consider the van der Waals fluid, which was found to have a region of instability characteristic of a phase transition. Specifically, we saw that the isothermal compressibility, $\kappa_T = -\frac{1}{V}\left(\frac{\partial V}{\partial P}\right)_{T,n}$, was negative for some regions of parameter space. This means that $\left(\frac{\partial P}{\partial V}\right)_{T,n}$ is positive for some regions of parameter space. For high enough temperatures, $\left(\frac{\partial P}{\partial V}\right)_{T,n}$ is negative for any combination of P and V, meaning that the system is stable. At a low enough fixed temperature, $\left(\frac{\partial P}{\partial V}\right)_{T,n}$ is negative for small V as well as large V, but positive in between.

This behavior can be conveniently depicted in a P-V diagram. For a low enough fixed temperature, the P-V curve has a relative minimum and a relative maximum.[277] For higher fixed temperatures, the relative minimum and maximum soften and approach one another, until eventually they meet at a point of inflection. The point of inflection marks the critical point.

---

[277] Intuitively, one may want to point to the relative minimum and say that it corresponds to the liquid phase and likewise associate the relative maximum with the gaseous phase, and to measure the horizontal separation, ΔV, between the two to use in the Clausius-Clapeyron equation. However, this would be incorrect. The discontinuity in volume is actually larger than the horizontal separation between the relative minimum and maximum. As we will see, what we really want to do is find the 'line of equal areas,' which properly separates the curve into liquid, coexistence, and gaseous regions.

A first-order phase transition features a relative minimum and maximum of the P-V curve, while a second-order phase transition features a point of inflection of the P-V curve. The horizontal tangent at the point of inflection shows a momentary instability at the critical point. For temperatures above the critical temperature, the system is completely stable. No distinct, sudden transition is observed when one phase passes into another at such high temperatures.

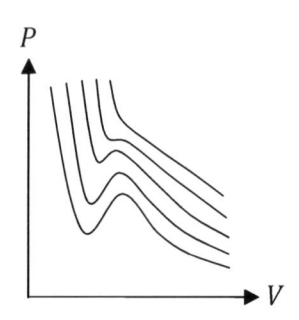

Let us consider one of the low-temperature isotherms in a P-V diagram for a single-component van der Waals fluid, for which there is a first-order phase transition. Recalling that the first law in the molar Gibbs representation is the same as the Gibbs-Duhem relation, $dg = d\mu = -sdT + vdP$, the equation of the isotherm can be expressed as $\mu - \mu_0 = \int_{P=P_0}^{P} vdP$. The integral $\int_{P=P_0}^{P} vdP$ is interpreted as the area under the P-v curve and μ_0 is an unknown function of temperature (since temperature is constant for an isotherm, the 'constant' of integration is actually a function of a variable that is held constant).

Observe that a horizontal line intersects the P-v curve in three places (since we are still concerned with a low enough temperature corresponding to a first-order phase transition). These three points have the same pressure and temperature, but different molar volume. Thus, all three of these points correspond to the same point on of a P-T diagram.

We can now see that it is convenient to choose the initial and final points of the integral to correspond to the region of coexistence between the two phases: That is, P_0 and μ_0 correspond to the maximum volume for which the phase would be pure liquid for the given temperature, and P and μ correspond to the minimum volume for which the phase would be pure gas for the given temperature. The region between v_0 and v then represents the coexistence between the two phases. In this case, the three points of intersection between the P-v curve and the horizontal line, $P = P_0$, all lie on the coexistence curve. Therefore, $\mu = \mu_0$ because the two phases must be at equilibrium for any point (P, T) on the coexistence curve. Conceptually, fort his choice, it is convenient to relabel the initial point as (v_ℓ, P_ℓ) and the final point as (v_g, P_g).

For this choice, we see that $\int_{P=P_\ell}^{P_g} vdP = 0$. Geometrically, this represents the area between the P-v curve and the horizontal line, $P = P_\ell = P_g$. The total area is zero, meaning that the positive area and negative area cancel perfectly. Hence, the line $P = P_\ell = P_g$ that makes $\int_{P=P_\ell}^{P_g} vdP = 0$ is called the 'line of equal areas.' It is convenient to split this integral into two pieces, using the middle point that has the same pressure as the initial and final points: $\int_{P=P_\ell}^{P_m} vdP + \int_{P=P_m}^{P_g} vdP = 0$. Although $P_\ell = P_m = P_g$, the two separate integrals are nonzero, which is easy to see in the figure below.

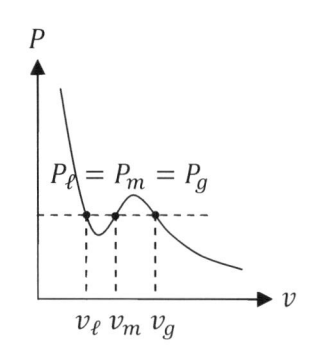

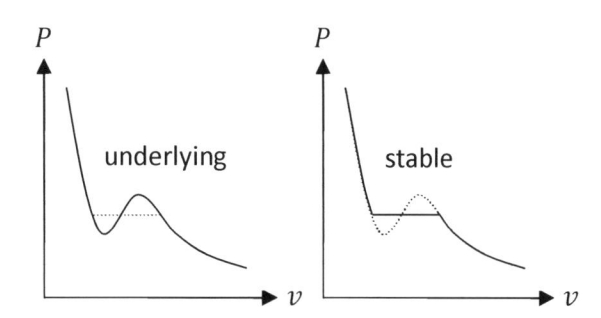

The line of equal areas separates the pure liquid phase, the coexistence phase, and the gaseous phase. It is the horizontal line segment connecting (v_ℓ, P_ℓ) to (v_g, P_g). The liquid phase corresponds to $v < v_\ell$, the liquid and gaseous phases coexist together for $v_\ell \leq v \leq v_g$, and the gaseous phase corresponds to $v_g < v$.

Conceptually, the line of equal areas is interpreted as follows. For a low enough fixed temperature, the P-v curve features a region where $\left(\frac{\partial P}{\partial v}\right)_{T,n}$ is positive, meaning that κ_T is negative, which corresponds to a region of instability. This P-v isotherm corresponds to the underlying fundamental relation. The horizontal line of equal areas replaces the upward sloping region of the P-v curve in the stable fundamental relation. The stable P-v isotherm, which involves the line of equal areas, is a physical isotherm in the sense that it is macroscopically observed. The underlying P-v isotherm determines the behavior of the system due to statistical fluctuations as it describes the stability conditions for the system.

The horizontal location, v_m, of the middle pressure, P_m, can be found from the lever rule: Namely, $v_m = \frac{n_\ell v_\ell + n_g v_g}{n}$ since $nv_m = V_m$ and $n_\ell v_\ell + n_g v_g$ both represent the total volume of the coexisting middle state.[278] Conceptually, the middle point (v_m, P_m) represents what fraction of the system exists in the liquid and gaseous phases.[279]

---

[278] It is termed the 'lever rule' because it has the same structure as balancing torques.

[279] Whereas the fractions may be unequal, the two areas are equal.

The integral, $\int_{P=P_\ell}^{P_m} v dP + \int_{P=P_m}^{P_g} v dP = 0$, along with the lever rule, provides an alternate means of obtaining the equation of the coexistence curve. This entails solving the equation of state (which is quadratic in v for the van der Waals fluid), and holding temperature constant during the integral (since the P-v curve corresponds to an isotherm).

This technique can be generalized to other systems and other stability conditions, as well as to transitions between different phases. In the case of an isobar, instead of an isotherm, the Gibbs-Duhem relation gives $\mu - \mu_0 = \int_{T=T_0}^{T} s dT$, in which case a similar technique applies to the discontinuity in entropy rather than the volume. The behavior of the T-s isobar is analogous to that of the P-v isotherm that we have considered. The P-v isotherm is useful for analyzing instabilities associated with $\kappa_T = -\frac{1}{V}\left(\frac{\partial V}{\partial P}\right)_{T,n}$, whereas the T-s isobar is useful for analyzing instabilities associated with $c_P = \frac{T}{n}\left(\frac{\partial S}{\partial T}\right)_{P,n}$.

Critical points: The coexistence curve ends with terminal points that are referred to as critical points. They are labeled as (P_{cr}, T_{cr}). As we have seen, the two extrema that represent the two coexisting phases merge together at the critical point. The discontinuity in the entropy, volume, internal energy, and thermodynamic potentials vanishes in the limit that a phase transition approaches the critical point. The critical point is part of the coexistence curve as it still corresponds to a point of stable equilibrium. We observed this in the case of the instability associated with the isothermal compressibility in the case of the van der Waals fluid: The upward slope of the P-v isotherm at sufficiently low temperatures turned into a point of inflection for the critical point, and both the upward slope and the point of inflection represent unstable equilibrium. The nature of the phase transition is different for the two cases; a phase transition that occurs across the critical point is deemed second-order, while for any other part of the coexistence curve it is deemed first-order.

For an unstable region in the underlying thermodynamic relation that has the form of the van der Waals instability in the isothermal compressibility, the critical point corresponds to a point of horizontal inflection on the P-v isotherm. In this case, $\left(\frac{\partial P}{\partial v}\right)_{T,n} = 0$ because the slope is zero and $\left(\frac{\partial^2 P}{\partial v^2}\right)_{T,n} = 0$ because it is a point of inflection. This provides a means of calculating the critical pressure and temperature. A point of inflection on any other diagram, such as an isobar on a T-s diagram, can be found analogously.

Example. Determine the pressure and temperature for the critical point of a van der Waals fluid.

Set the first and second partial derivatives of the pressure with respect to the molar volume equal to zero in order to determine the critical pressure and temperature:

$$\left(\frac{\partial P}{\partial v}\right)_{T,n} = \left[\frac{\partial}{\partial v}\left(\frac{RT}{v-b} - \frac{a}{v^2}\right)\right]_{T,n} = \frac{2a}{v^3} - \frac{RT}{(v-b)^2} = 0$$

$$\left(\frac{\partial^2 P}{\partial v^2}\right)_{T,n} = \left[\frac{\partial}{\partial v}\left(\frac{2a}{v^3} - \frac{RT}{(v-b)^2}\right)\right]_{T,n} = \frac{2RT}{(v-b)^3} - \frac{6a}{v^4} = 0$$

It is convenient to first solve for molar volume by eliminating temperature from the two equations:

$$\frac{a}{RT} = \frac{v_{cr}^3}{2(v_{cr}-b)^2} = \frac{v_{cr}^4}{3(v_{cr}-b)^3}$$
$$2v_{cr} = 3(v_{cr}-b)$$
$$v_{cr} = 3b$$

The critical temperature of the van der Waals fluid is then found to be:

$$T_{cr} = \frac{2a(v_{cr}-b)^2}{Rv_{cr}^3} = \frac{8a}{27bR}$$

The critical pressure of the van der Waals fluid is:

$$P_{cr} = \frac{RT_{cr}}{v_{cr}-b} - \frac{a}{v_{cr}^2} = \frac{4a}{27b^2} - \frac{a}{9b^2} = \frac{a}{27b^2}$$

Conceptual Questions

The selection of conceptual questions is intended to enhance the conceptual understanding of students who spend time reasoning through them. You will receive the most benefit if you first try it yourself, then consult the hints, and finally check your answer after reasoning through it again. The hints and answers can be found, separately, toward the back of the book.

1. Describe how you could top a pizza so that it is: homogeneous and isotropic; homogeneous, but not isotropic; isotropic, but not homogeneous; and neither homogeneous nor isotropic.

2. Which would have the same value for any subset of the system as for the whole system – extensive or intensive parameters? Which would have a fractional value compared to the value it has for the whole system – extensive or intensive parameters?

3. Indicate whether each of the following quantities is extensive or intensive: weight, molar mass, number per unit volume, surface area, chemical potential, enthalpy, Helmholtz free energy, and molar entropy.

4. The thermodynamic properties associated with stretching a rubber band can actually be modeled to good approximation with a simple thermodynamic formulation. Experimentally, for example, the temperature of a rubber band increases if it is stretched. In this case, the variables of interest are the internal energy, entropy, temperature, length, and tension of the rubber band. Which of these variables are extensive and which are intensive? What partial derivative can be applied to the fundamental relation for a rubber band that is analogous to mechanical pressure?

5. Magnetic systems involve two additional parameters: the magnetic dipole moment[280] and the external magnetic field. Indicate whether each of these is extensive or intensive. Express the magnetic partial derivative that is analogous to temperature and pressure.

6. Consider the coefficient of thermal expansion, the compressibilities, and the molar heat capacities. Do these behave as extensive or intensive parameters, or as something different? In particular, examine their additivity. Are they homogeneous first-order, homogeneous zero-order, or something different?

7. Develop a conceptual argument to justify that if the entropy did not have its maximum value for the specified internal energy, then the internal energy could not be minimum (in equilibrium). That is, conceptually justify that the energy minimum principle requires maximization of the entropy for the given internal energy.

8. Imagine a sphere rolling along a curved surface for which the Brownian motion produces visibly noticeable, yet reasonably small, motions. Make analogies between positions of mechanical equilibrium and the positions of stable and unstable thermodynamic equilibrium of thermodynamic systems. Also discuss how the Le Chatelier principle applies to the mechanical case of the marble.

Practice Problems

The selection of practice problems primarily consists of problems that offer practice carrying out the main problem-solving strategies or involve instructive applications of the concepts (or both). You will receive the most benefit if you first try it yourself, then consult the hints, and finally check your answer after working it out again. The hints to all of the problems and the answers to selected problems can be found, separately, toward the back of the book.

Formulation of equilibrium thermodynamics

1. Determine whether each of the following functions is a homogeneous, first-order or homogeneous, zero-order function of its arguments: (A) $f(x,y,z) = \frac{xyz}{xy+yz+zx}$; (B) $g(x,y,z) = (z^2xy)^{\frac{1}{4}}$; (C) $h(x,y,z) = \frac{x}{y} + \frac{y}{z} + \frac{z}{x}$; and (D) $i(x,y,z) = x^{1/2}y^{-1/4}z^{-1/4}$.

2. Under what circumstances would the relation $U = aS^b V^c/n$, where a, b, and c are constants, be a viable fundamental relation for a thermodynamic system? In any case(s) in which the relation would not be viable, identify the problem(s). Under the circumstances in which the relation would be viable, show that it satisfies all of the requirements; also, express the intensive parameters in terms of S, V, and n.

---

[280] Recall that the magnetic moment of a particle equals its charge-to-mass ratio times one-half its angular momentum, and that moving charges (which a magnetic moment inherently is, even if the particle appears to be stationary – in which case its spin is said to be intrinsic) interact with external magnetic fields.

3. Show explicitly whether or not the fundamental relation for the ideal gas satisfies the requirements for a viable fundamental relation for a thermodynamic system.

4. The fundamental relation of a magnetic system is $u = ae^{bs+cm^2}$, where m is the molar magnetic dipole moment (see Conceptual Question 5 and the corresponding footnote) and a, b, and c are constants. (A) Find the equations of state for this system, including one for the external magnetic field B. (B) By analogy, what is the expression for the magnetic work done by the system?

5. Two equations of state for a system are $P = U/(2V)$ and $T = (2a)^{-1}U^{3/5}V^{-1/5}n^{-2/5}$, where a is a constant. (A) Show that these equations of state are compatible. (B) Find the third equation of state for this system. (C) Determine the fundamental relation in the energy representation for this system.

6. Find an equation of state that is compatible with $P = au/v^2$, where a is a constant.

7. Integrate the following equation: $\frac{y^2 z}{w} dx + \frac{xyz}{w} dy = \frac{w}{xz} dz - \frac{2}{x} dw$.

8. Two subsystems with equal mole numbers are separated by an immovable, diathermal wall that divides a perfectly insulating container exactly in half. The fundamental relations for the two subsystems are $S_1 = a(U_1 V_1 n_1)^{1/3}$ and $S_2 = a(U_2 V_2 n_2)^{1/3}$, where a is a constant. The initial temperatures of the two subsystems are $T_{10} = 200$ K and $T_{20} = 300$ K. What is the final equilibrium temperature for the system?

9. Two subsystems with equal mole numbers are separated by a movable, diathermal wall that divides a perfectly insulating container exactly in half. The fundamental relations for the two subsystems are $S_1 = a(U_1 V_1 n_1)^{1/3}$ and $S_2 = a(U_2 V_2 n_2)^{1/3}$, where a is a constant. The initial temperatures of the two subsystems are $T_{10} = 200$ K and $T_{20} = 300$ K. The total volume of the container is 1.00 m³. (A) What is the temperature of each subsystem when equilibrium is reached? (B) What is the pressure of each subsystem when equilibrium is reached? (C) Where is the piston when equilibrium is reached? (Just give the ratio of the volumes.)

Simple thermodynamic systems

10. An ideal monatomic gas initially at STP in a thermally insulated container expands isentropically as a movable piston slides until its volume increases by a factor of 8. Use the fundamental relation in the entropy representation to answer these questions. (A) What is the temperature of the gas after the expansion – compared to its initial temperature? (B) What is the pressure of the gas after the expansion – compared to its initial pressure?

11. An ideal monatomic gas initially at STP in a thermally insulated container expands isenthalpically as a movable piston slides until its volume increases by a factor of 8. (A) What is the temperature of the gas after the expansion – compared to its initial temperature? (B) What is the pressure of the gas after the expansion – compared to its initial pressure? (C) How much work is done during the expansion?

12. A monatomic ideal gas and diatomic ideal gas with equal mole numbers are separated by an immovable, diathermal wall that divides a perfectly insulating container exactly in half. The initial temperature of the monatomic ideal gas is 200 K and that of the diatomic ideal gas is 300 K. What is the final equilibrium temperature?

13. A monatomic ideal gas and diatomic ideal gas with equal mole numbers are separated by a movable, diathermal wall that divides a perfectly insulating container exactly in half. The initial temperature of the monatomic ideal gas is 200 K and that of the diatomic ideal gas is 300 K. The total volume of the container is 1.00 m^3. (A) What is the temperature of each gas when equilibrium is reached? (B) What is the pressure of each gas when equilibrium is reached?

14. A van der Waals fluid expands isentropically. Derive an equation that is analogous to the equation for the adiabat of an ideal gas, but not necessarily in terms of P and V (e.g. P and T, or some other combination of easy-to-measure variables may be more convenient).

15. A van der Waals fluid expands isobarically until its volume doubles. (A) What is the temperature of the fluid after the expansion – compared to its initial temperature? (B) What is the pressure of the fluid after the expansion – compared to its initial pressure?

16. A photon gas expands isentropically until its volume doubles. (A) What is the temperature of the photon gas after the expansion – compared to its initial temperature? (B) What is the pressure of the photon gas after the expansion – compared to its initial pressure? (C) Derive an equation for the work done in terms of its initial pressure and volume.

17. A photon gas expands isenthalpically until its volume doubles. (A) What is the temperature of the photon gas after the expansion – compared to its initial temperature? (B) What is the pressure of the photon gas after the expansion – compared to its initial pressure? (C) Derive an equation for the work done in terms of its initial pressure and volume.

Thermodynamic potentials

18. Given the function $U(S) = ae^{bS}$, where a and b are constants, determine the Legendre-transformed potential $A = A(T)$.

19. Given the Legendre-transformed potential $A(T) = c/T + b$, where b and c are constants, determine the function $U = U(S)$.

20. The fundamental relation for a system is $S = aU^{1/4}V^{1/2}n^{1/4}$, where a is a constant. Find the fundamental relation in the Helmholtz, Gibbs, and enthalpy representations.

21. The fundamental relation in the Helmholtz representation for a system is $a(T^5n^2V)^{1/3}$, where a is a constant. Find the fundamental relation in the energy and Gibbs representations.

22. Express the fundamental relation for the van der Waals fluid in the Helmholtz representation.

23. Express the fundamental relation for the magnetic system of Problem 4 in the Helmholtz representation and also in the magnetic potential representation in which the two conjugate magnetic variables are exchanged in the Legendre transformation.

The Maxwell relations and the extremum principles

24. Derive expressions for c_P and c_V for a system with the fundamental relation $S = aU^{1/4}V^{1/2}n^{1/4}$, where a is a constant.

25. Derive expressions for β, κ_T, and κ_S in terms of T or P for a single-component ideal gas.

26. Derive an expression for κ_T in terms of P and v for a van der Waals fluid.

27. Derive an expression for β for a photon gas.

28. Derive expressions for the isothermal and adiabatic molar susceptibilities, $\chi_T = \frac{\mu_0}{n}\left(\frac{\partial m}{\partial B}\right)_{T,n}$ and $\chi_S = \frac{\mu_0}{n}\left(\frac{\partial m}{\partial B}\right)_{S,n}$, respectively, where μ_0 is the permeability of free space, for the magnetic system of Problems 4 and 23.

29. Read off the Maxwell relation for each of the following derivatives: $\left(\frac{\partial T}{\partial n}\right)_{S,V}$, $\left(\frac{\partial S}{\partial n}\right)_{V,\mu}$, and $\left(\frac{\partial \mu}{\partial P}\right)_{T,n}$.

30. Reduce the following derivatives: (A) $\left(\frac{\partial P}{\partial V}\right)_{T,n}$; (B) $\left(\frac{\partial V}{\partial S}\right)_{T,n}$; (C) $\left(\frac{\partial V}{\partial \mu}\right)_{S,n}$; (D) $\left(\frac{\partial P}{\partial T}\right)_{\mu,n}$; (E) $\left(\frac{\partial U}{\partial P}\right)_{F,n}$; and (F) $\left(\frac{\partial P}{\partial S}\right)_{G,n}$.

31. Show that the quasistatic heat flux, đQ, can be expressed as $nc_V dT + \frac{T\beta}{\kappa_T} dV$ or $nc_P dT - \beta TV dP$.

32. For the magnetic system of Problems 4, 23, and 28, show that the derivatives of thermal variables with respect to magnetic variables, such as $\left(\frac{\partial T}{\partial m}\right)_{S,n}$, are equal to zero. Also, explain why these derivatives equal zero. Are the Maxwell relations useful for this model? (This simple model is a special case; there is mixing between the thermal and magnetic variables in general magnetic systems.)

33. Relate a small pressure change to a small temperature change for a quasistatic isochoric process for a single-component system with constant mole number.

34. Relate a small internal energy change to a small temperature change for a quasistatic adiabatic process for a single-component system with constant mole number.

Phase transitions

35. Determine the region(s) where each of the following functions are concave up and concave down: (A) $f(x) = e^{-x}$, (B) $g(x) = x^4 - 2x^2$, (C) $h(x) = \sqrt{x}$, and (D) one period of a sine wave.

36. Determine the region(s) where the fundamental relation in the entropy representation for an ideal gas is concave up or concave down with respect to the extensive variables.

37. Determine the region(s) where the fundamental relation in the entropy representation for a van der Waals fluid is concave up or concave down with respect to the extensive variables.

38. Does $S = a(UVN)^{1/3}$, where a is a positive constant, satisfy the conditions for stability?

39. Does $S = a(U^3V^{-2}N^3)^{1/4}$, where a is a positive constant, satisfy the conditions for stability?

Hints to Conceptual Questions and Practice Problems

Hints to Chapter 0 Conceptual Questions

1. Review the definitions of temperature and heat, especially the distinction between the two (also, see the related conceptual examples). Imagine dipping your hand into both pots so that in either case the total amount of surface area of your hand that is covered with water would be the same in each case. What can you say about the temperature of the water in each pot? Which pot has the greater potential transfer heat? For both questions, heat is transferred – in one case, to your hand, and in the other case, to the ice. Is the amount of heat transferred the same or different in each case?

2. Think about the transformation of energy that occurs. Is there a transfer of thermal energy (heat) involved? If so, does this raise the internal energy of the washer's molecules?

3. Temperature and internal energy have similar definitions, but there is a very important adjective that is different with regard to the kinetic energies of the molecules: average versus total.

4. Of course, you can look up some values in a table, but then you should also try to think about the reason for the difference. Here are some ideas that you might consider: How much the molecules' average kinetic energy changes in each case, how much the volume changes in each case, and/or how does the latent heat figure into the first law of thermodynamics?

5. Imagine that the steam condenses on your hand to give you the burn, so that in both cases you are ultimately burned by liquid. In this case, looking at the average kinetic energies of the molecules will not help directly. Instead, think about the latent heat.

6. What will happen if you put one of these in a bathtub full of water?

7. Air is a fluid. See the discussion of buoyant force and Archimedes' principle in the hint to Question 8.

8. There is an upward buoyant force that results because there is greater pressure at greater depth. Because ice is less dense than water, the upward buoyant force exceeds its weight if it is wholly submerged. The upward buoyant force is reduced when part of the ice sticks out of the water. Apply Newton's second law and the equations for density – plus Archimedes' principle, which states that the weight of the displaced water is equal to the buoyant force – in order to determine the fraction. For the last question, note that as the ice melts, it actually becomes more dense in the form of liquid water.

9. Do objects of different volume necessarily have similar surface area? If you're not familiar with Gabriel's horn, you should look it up. Here is a similar idea, except in terms of perimeter (or circumference) and area. Consider an island that is roughly circular. If you look at the island closely, and the coast is very squiggly like the fjords of Norway, its perimeter is significantly enlarged compared to the circumference of a circle, yet its area is roughly unaffected. In this way, a shape can have finite area, yet infinite perimeter. You can draw this limiting case rather easily using the concept of a fractal.

10. Think about the area of contact in each case and the definition of pressure.

11. First, see the hint to Question 9. Next, realize that there is an important difference – a trigonometric factor appearing in the definition of pressure, that is not analogous to the 'alternative density' of Question 9.

12. Compare the two definitions carefully. By comparison, see what you need to multiply and/or divide by in order to turn one definition into the other.

13. Compare the two definitions carefully. By comparison, see what you need to multiply and/or divide by in order to turn one definition into the other.

14. The lid affects the pressure at which the water boils. Think about how pressure (review its definition, if necessary) affects the boiling process (also, review the definition of temperature, if needed).

15. What happens to the density of liquid water when it freezes to ice? See also Questions 6 and 8.

16. Look at the equation in which the coefficient of thermal expansion is used. If you still don't see it, try working out a numerical example in both sets of units and you should see what happens.

17. Think about the molecular bonding. There is a lot of oxygen in the air – which form?

Hints to Chapter 0 Practice Problems

1. Use a periodic table to find the molar mass of each element, in order to find the molar mass of baking soda. From this and the total mass, you can figure out the number of moles of baking soda, from which you can figure out the number of atoms corresponding to each element.

2. The first question you can figure out without knowing the mass of water. See the hints to Question 1. Also, review the definitions of molar mass and molar volume.

3. See the hints to Question 1. Also, review the definition of mole fraction and the related example.

4. Write out the balanced chemical equation for the chemical reaction. See the hints to Questions 1 and 2. Note that liters are not SI units.

5. It should be easy to deduce that the volumes are the same. From the volumes, you can find the dimensions of each object, from which you can compute the surface areas (symbolically).

6. Simply convert the given temperature to SI units.

7. Set the two temperatures equal to each other in the conversion formulas and proceed to solve for the temperature. The algebra will tell you if a solution is not possible (which you can also see conceptually).

8. See the hints to Question 7, but express that one is twice the other (instead of setting them equal).

9. Simply look up the conversion factors. You can convert force and volume separately – i.e. you should not try to look up a direct conversion, but work it out in steps. Note that exponents are very important. For example, although 1 cm = 10 mm, in terms of area 1 cm^2 = 100 mm^2 since 1 cm^2 = (10 mm)^2.

10. Simply plug in the SI units of each and reduce. You will need to know how to express various units in multiple forms – e.g. 1 N = 1 kg·m/s^2 – in order to simplify the answer.

11. Simply plug in the SI units of each and reduce. You will need to know how to express various units in multiple forms – e.g. 1 N = 1 kg·m/s^2 – in order to simplify the answer. Note that the units of distance get squared. Note that it's distance-squared times pressure, and not a second differential of pressure.

12. Simply plug in the SI units of each variable and solve for the SI units of the constant. You will need to know how to express various units in multiple forms – e.g. 1 N = 1 kg·m/s^2 – in order to simplify the answer.

13. Use algebra to combine the two equations with the common coefficient eliminated. Plug in the SI units of each variable to see what values of the exponents will make this work.

Hints to Chapter 1 Conceptual Questions

1. Try adding these vectors tip to tail. The answer is **not** simply the missing line segment that extends from the origin to the unused corner (but it's close!). Just by looking, if this were an unbalanced force table, it should be obvious in what direction the resultant points (which is why the previous statement has the word 'not' underlined).

2. This relates to the zeroth law of thermodynamics.

3. There is a single word that explains what a thermos bottle does that explains both cases. Comparing a coffee mug and a thermos bottle, if you pour hot coffee into a coffee mug, it will cool rather quickly compared to a thermos bottle, and if you pour ice water into a coffee mug, it will warm rather quickly compared to a thermos bottle (with an underlying assumption here that should be apparent). What is different between the thermos bottle and coffee mug that explains this?

4. Does the thermos bottle allow a significant amount of heat to be exchanged between the system and surroundings? Look at the first law of thermodynamics and consider the energy transformations. Also review the definition of temperature and the second law of thermodynamics.

5. This is one way that the coffee in a coffee mug cools over time.

6. Don't neglect friction. This will figure into the first and second laws.

7. Think about how you could design a perpetual motion machine if the second law did not exist. In this case, what is the role of the first law? What is different about the machine compared to the universe?

8. Here is an interesting idea to ponder: Is thought a closed system? Or does the increased order of thought tie into increased entropy of the universe as the brain proceeds to order the thoughts?

9. The P-V curve could instead cross itself. Look at the first law to relate work and heat. Internal energy change is given by the situation – it is not an assumption you have to make. However, there is an assumption that we frequently make when we apply the first law. That is, in the most general case, the first law would consist of more than just three terms.

10. Does the work integral depend on the shape of the path or just the area enclosed?

11. For example, if the diagram looks like a triangle, would it make a difference which vertex you start at?

Hints to Chapter 1 Practice Problems

1. In what quadrants is this possible? Bear in mind that magnitudes must be positive, but components may be negative. In 3D, in which octants is this possible?

2. What shape must these vectors make if you add them tip-to-tail?

3. The components should be obvious, from which it should be easy to find the magnitude or direction. Make sure that your dot product results in a scalar, not a vector. The scalar product comes in two forms, one of which makes it easy to find the angle between two vectors. Note that adding and subtracting vectors is particularly simple when the vectors are expressed in terms of unit vectors.

4. Apply successive derivatives. Note that Cartesian (but not polar) unit vectors are constant in time. Of course, you must take the derivatives before you plug in time.

5. Apply calculus. Note that Cartesian (but not polar) unit vectors are constant in time.

6. Follow the order of operations. Note that the addition or subtraction of two vectors results in a vector, whereas the dot product results in a scalar.

7. Use the chain rule.

8. Use the chain rule to express the derivative with respect to position in terms of a derivative with respect to time (as in an example).

9. When you take a partial derivative with respect to one variable, treat the other independent variable as a constant.

10. Use the chain rule when taking these partial derivatives.

11. Note that the gradient operator results in a vector, not a scalar.

12. Apply the partial derivatives first, and plug in the coordinates after.

13. Follow the prescription and example for implicit differentiation.

14. Follow the prescription and example for implicit differentiation.

15. This is largely conceptual, as the math is rather trivial. Mostly, you need to study the definitions of these terms, especially the distinction between similar concepts.

16. The net displacement should be very easy, while you'll need to apply calculus in order to find the arc length.

17. Use the component form of the dot product to separate this into two integrals. Note that each integral is over a single variable, while the integrand is multi-variable. Use the path to express the un-integrated variable in terms of the other (or a constant) over each section of the path. You must do this before integrating. You'll want to understand what a conservative field is.

18. Use the mean-value theorem from calculus.

19. Note that the polar unit vectors are not constant. Use the chain rule to express derivatives of polar unit vectors with respect to time in terms of derivatives with respect to angle.

20. Integrate the volume. Whether or not you use cylindrical coordinates, one or more of the limits will be variable. The limit tells the integral which region of the trough you want to include in or exclude from the volume.

21. Integrate the volume. Whether or not you use spherical coordinates, one or more of the limits will be variable. The limit tells the integral which region of the trough you want to include in or exclude from the volume.

22. Set this up as a surface area integral using spherical coordinates. The solid angle integral is very similar.

23. Express these mass integrals in terms of volume integrals, writing $dm = \rho dV$. You can't pull density out of the integral, of course, since it's non-uniform, but you can pull β out. These masses are directly proportional to the weights.

24. Express this mass integral as a volume integral using spherical coordinates (actually, 2D polar coordinates will suffice). Compare two integrals to answer the question – one for the entire mass and the other for the specified range.

25. Since $F_n = F \cos \theta$ at the finite scale, at the differential level it becomes $dF_n = dF \cos \theta - F \sin \theta \, d\theta$. The first term is zero, assuming F to be distributed uniformly.

26. The net work corresponds to the area of the triangle, while the work for each process is given by a trapezoid or rectangle (the area between the line segment and the horizontal axis). The direction of the arrow determines the sign of the work.

27. See the footnote to figure out how to compute the area of the semi-ellipse, which will give the net work. For the work done during each process, you'll also need to account for the rectangle below the horizontal line segment. The direction of the arrow determines the sign of the work.

28. You can perform these work integrals by noting what is constant in each case. You can't pull a variable out of the integrand, but you can express it in terms of constants and variables over which you are integrating. See the examples in the section on work.

29. Use the given equation to express pressure in terms of volume (similar to the example of an isotherm or adiabat) in order to compute the work integral.

Hints to Chapter 2 Conceptual Questions

1. The thickness of the ring will change negligibly compared to the change in arc length.

2. Notice the difference in dimensionality between the nature of the expansions in the first two questions. Thermal expansion applies directly to the solid material, which indirectly affects the hole.

3. What is probably different for the two metals? Why doesn't the strip stay straight as the metals get longer when heated or shorter when cooled?

4. The last question is a little tricky. Consider not only possible differences in the thermal properties of the metals, but also the different dimensionality between the nature of the expansions of the ring and the cylinder.

5. What happens to the temperature of the tires when a car is driven compared to when it is parked? (Does it matter if it's a cold winter day or a hot summer day?) Do you want the tires to expand or contract with this temperature change?

6. If the pond has a temperature gradient, where do you expect the most and least dense layers of water (or ice) to be? Note that all of the water must reach 4° C before the pond can freeze. Why?

7. Think about the conceptual meaning associated with heat capacity and/or thermal conductivity.

8. Think about walking on a tile floor in the middle of a cool night, for example, and then compare this to a wooden floor, which has a lower thermal conductivity.

9. Think about the conceptual meaning associated with heat capacity and/or thermal conductivity and/or emissivity.

10. Try to reason this out conceptually, though it may be useful to let the math guide your reasoning.

11. Each one primarily corresponds to one of the three methods of heat transfer.

12. Does heat flow from high to low temperature, or vice-versa?

13. Think about the conceptual meaning associated with heat capacity and/or thermal conductivity.

14. Does boiling occur through conduction or a different method of heat transfer?

15. What can you say about the absorption and emission rates for an object that is in equilibrium? Conservation of energy requires the percentage absorbed plus the percentage reflected to equal 100% (instantaneously).

16. See the hint to Question 15 for the effect of the silvered cover. For the last question, does the same silvered cover help keep the car from getting very cold on a cold winter day?

17. How does intensity depend upon distance? Work out the distances of the reflected images carefully. It might help to draw it out from a side view and trace the reflected rays.

18. Compare the equations and determine which variables are analogous. It should be obvious whether thermal conduction is static or dynamic.

19. Heat flows from high temperature to low temperature, even though all (absolute) temperatures are positive, whereas electric field flows from positive to negative charge. Compare with gravitational flux. All masses are positive and attract, so a gravitational field map of the earth and moon would look like the heat flow map, which also looks like the electric field map of an electric dipole. In the case of gravity and heat, unlike electricity, all of the sources are positive. (Of course, an electric field map with all positively charged sources looks much different – showing repulsion rather than attraction.)

Hints to Chapter 2 Practice Problems

1. The solid cube expands three-dimensionally. Find the expansion of the volume, and find the new edge length from this. The wire cube expands one-dimensionally along each edge. Find the expansion of each wire, and find the new volume from this.

2. Note that it's the tape measure and not the pole that expanded significantly. Note also that the pole is longer than two meters because of the tape measure's expansion.

3. Notice the different dimensionality between the nature of the expansions of the ring and cylinder.

4. The thickness of the ring will change negligibly compared to the change in arc length.

5. Notice the difference in dimensionality between the nature of the expansions in this problem and the previous problem. Thermal expansion applies directly to the solid material, which indirectly affects the hole.

6. Set this up as a system of equations with equal number of equations and unknowns.

7. You must first account for the heat needed to change the temperature to the freezing or boiling point, and then the heat associated with the phase change.

8. Apply the calorimeter strategy.

9. If the ice does not receive enough heat, it won't melt. If it does melt, you must account for the heat exchange associated with the melting. Check your signs carefully.

10. Set the formulas for power across each conductor equal at the interface.

11. Apply the strategy from Problem 10 at each interface.

12. Going from the left wall to the right wall, two materials placed sequentially from one wall to the other are in series, whereas two materials side-by-side are in parallel (but not side-by-side in the same sense as the sequential case, of course). For example, the set of white cubes is in parallel with the set of checkered cubes, while the set of gray cubes is in series with the set of black cubes. The gray and black cubes make an equivalent thermal conductor, and similarly with the white and checkered cubes; then decide if these two equivalent thermal conductors are in series or parallel, and combine them as well.

13. One case is no different from a rectangular conductor. In the other case, you will need to integrate. Study the example and integrate by analogy using cylindrical coordinates.

14. Of course, heat flows from high temperature to low temperature. Recall that ice is less dense than water when it freezes. Where will the layer of ice form? So where are the highest and lowest temperatures on the temperature gradient of the tub? One of these symbols is the latent heat. A simple alteration of the equation given in (B) will provide an equation for power, for which you also know another equation for heat conduction (see Question 13 – the case that corresponds to a rectangular conductor). Set these two power equations equal to one another, eliminating power. Separate variables and integrate both sides.

15. Separately compute the work done and the heat exchange, and plug these into the first law, being careful with the signs. You'll need to treat the thermal expansion before you can determine how much work is done.

16. The total heat supplied includes the heat needed to raise the liquid's temperature to the boiling point plus the heat associated with the phase transformation – realizing that only a percentage of the total mass changes phase. As a result of boiling, the volume of the system increases. Compute this volume in order to determine how much work is done. Lastly, use the first law, being careful with the signs.

17. The net work can be found from the area. You should know what the net internal energy change is for this complete cycle. Being careful with the signs, use the first law to find the heat exchanged.

18. Use the first given equation to express pressure in terms of volume to perform the work integral. Use the second given equation to determine the internal energy change. Being careful with the signs, use the first law to find the heat exchanged.

19. The internal energy change is path-independent. So find the work done along the adiabat, and use the first law to determine the internal energy change for the adiabat. The internal energy change is the same for the alternate path. Perform the work integral for the isobar. Being careful with the signs, use the first law to find the heat exchanged.

20. Use the appropriate form of the specific heat equation to determine the heat exchange for each case. Perform the work integral for each case. Being careful with the signs, use the first law to find the internal energy change for each case. Use the given equation to express the answer in the specified form.

Hints to Chapter 3 Conceptual Questions

1. Does this intermolecular attraction enhance or reduce the net force exerted on the walls of the container?

2. Try setting one of the variables constant.

3. The text discusses how the speed of sound compares to the average speed of the molecules, to give you some idea of how fast the molecules travel, on average. You should have experience with how the speed of smell compares to that of sound. Think about how smell propagates.

4. It will help to look at a periodic table. Also, look at the formulas for the average speed of the molecules and the kinetic energy to figure out the effect of whether the molecules are monatomic or diatomic.

5. See the hints to Question 4. Here the difference is between diatomic and polyatomic.

6. The molecules gain kinetic energy when they absorb solar radiation. Also, there are very few molecules in the outermost layers.

7. You can see which is larger in a formula. Try to reason it out conceptually, though.

8. You can work the answer out mathematically, combining the equation for an adiabat with the ideal gas law. Try to explain the answer conceptually.

Hints to Chapter 3 Practice Problems

1. Review the kinetic theory of the pressure that an ideal gas exerts on a plane surface. Apply the same mathematics to calculate the rain pressure.

2. Apply the ideal gas law to determine the new temperature. Then review the kinetic theory of ideal gases in relation to kinetic energy to determine the average speed of the molecules from the temperature. Remember to use absolute temperature and SI units for mass.

3. Review the kinetic theory of ideal gases in relation to kinetic energy to determine the average speed of the molecules from the temperature. See the hints to the Conceptual Questions regarding the nature (e.g. diatomic) of the bonding. Remember to use absolute temperature and SI units for mass.

4. A formula for the speed of sound in an ideal gas was derived in an example. Use this result. Remember to use absolute temperature and SI units for all quantities.

5. Use the definition of the adiabatic index and the relation between the heat capacities at constant pressure and at constant volume. Just play with the algebra.

6. Start with the equation of an adiabat at constant pressure in terms of initial and final pressure and volume. Use the ideal gas law to eliminate the desired quantities.

7. Review the derivation of the adiabatic bulk modulus. How will this be different in the isothermal case?

8. Use the volume changes to determine how much work is done in each process. For each process, determine how much heat is exchanged. Remember that you can also use equations that pertain to ideal gases – such as the ideal gas law or the equation for internal energy. The first law may also be useful. It may also be useful to review examples where work, heat, internal energy, and the first law are applied to ideal gases.

9. See the hints for Problem 8.

10. See the hints for Problem 8. Remember that work can be found in terms of area. Determine the temperatures to deduce which value of the adiabatic index is appropriate.

11. See the hints for Problem 8.

12. See the hints for Problem 8.

Hints to Chapter 4 Conceptual Questions

1. Compare the two definitions.

2. Think about the plausible levels of agreement or disagreement between the actual outcomes versus the expected (probable) outcomes over the course of a small number of events compared to the course of a large number of events.

3. You will always win $1, eventually, even if black and/or green come up several times in a row, unless what?

4. What's the distinction between total distance traveled and net displacement? What quantity similar to average velocity completes the analogy?

5. If an event is repeated thousands of times, one of these outcomes is most likely to occur, based on its definition. Review the kinetic theory of ideal gases to see which of these was applied to find the average speed of the molecules. The explanation lies in the derivation (where this quantity first appeared).

6. Look at the definition of the pressure of a fluid.

7. Look at the definition of the pressure of a fluid.

8. A mirage can form above hot desert sand, but not anywhere above the surface of the earth. The mirage is related to this question.

9. Compare the initial speed of one ball to the final speed of the same ball. What can you expect in general?

Hints to Chapter 4 Practice Problems

1. Use the definition of a permutation and consult the examples.

2. This time you need to permute the letters into a subset. Look at the rules for how to compute various types of permutations and review the examples.

3. Look at the rules for how to compute various types of permutations and review the examples.

4. Apply the definitions of permutations and combinations. The difference is that you can't draw the same card from the same deck twice (assuming that you don't put it back in the deck before drawing the second card).

5. Look at the rules for how to compute various types of permutations and combinations, and review the examples.

6. Review the section on discrete probability distributions and review the examples therein.

7. Review the section on discrete probability distributions and review the examples therein. You probably already know that blackjack is a face card or a ten plus an ace.

8. Review the section on discrete probability distributions and review the examples therein. See the distinction described in the hints to Problem 4.

9. Review the section on discrete probability distributions and review the examples therein. You probably already know that a full house is three of one card plus two of another card (e.g. three queens plus two fives).

10. Review the section on discrete probability distributions and review the examples therein. Also look at the definitions of these three terms.

11. Which type of continuous probability distribution applies to the ideal gas?

12. Review the definitions of these terms for a continuous probability distribution.

13. Look at how we calculated the average value and the average squared value. Apply a similar calculation to find the average inverse value.

14. Review what is meant by the mean-free path.

15. This derivation is analogous to the derivation for the distribution of molecules in the atmosphere, except that the formula for centripetal force, rather than the dependence of the pressure in a fluid on depth, dictates the probability distribution.

Hints to Chapter 5 Conceptual Questions

1. Review the quasistatic process that was described qualitatively using the sand technique. Construct an analogous technique that you can apply to an isobar.

2. What local changes could you make, and what would be the global consequences?

3. Think about the statistical order of the water molecules. If the order increases, what other effect must go along with this in order to satisfy the second law?

4. What type of energy change is involved? Is any of the energy converted into heat (e.g. by a resistive force that will produce heat)? The refrigerator is related to a heat engine, from which you should be able to apply the second law directly.

5. Is the efficiency of the heat engine related to the outside temperature?

6. Don't neglect friction!

7. Compare with similar processes in P-V diagrams. Obviously, the same processes will look much different on the T-S diagrams. Think about how the various physical quantities are related, in general. You might use an ideal gas as a guide, for which you have more equations to guide your expectations.
8. Use the results of Question 7.

Hints to Chapter 5 Practice Problems

1. Assume the process to be approximately quasistatic, and even reversible, if necessary. What is the (macroscopic) quantitative definition of entropy?

2. If necessary, also assume the process to be approximately reversible. What is the (macroscopic) quantitative definition of entropy?

3. Start with the (macroscopic) equation that defines entropy for a reversible path. The integral is analogous to work integrals, but in terms of different variables.

4. See the hints for Problem 3. Show that $dS = \frac{nc_V dT}{T} + \frac{nRdV}{V}$. Use the given equation and the ideal gas law to express temperature and its differential in terms of volume (as the only variable), and then integrate.

5. Use both the microscopic and macroscopic definitions of the entropy. For the macroscopic case, review the discussion of the entropy change for the adiabatic free expansion of an ideal gas.

6. Review the derivations of efficiency in the heat engine examples.

7. Review the derivations of efficiency in the heat engine examples. There are two ways to draw this.

8. Review the derivations of efficiency in the heat engine examples.

9. Begin with the efficiencies of the air-standard diesel engine and Otto engine and use applicable equations for each engine to express both in terms of two common variables (such as the high and low temperatures). Review the derivations of efficiency in the heat engine examples. Show that the diesel engine is less efficient than an Otto engine with the same compression ratio;[281] since we have already shown that the Otto engine is less efficient than the Carnot engine, this will imply that the diesel engine is less efficient than the Carnot engine. In your final expression, for fixed γ, look at the limit as r_c approaches zero; note that $\gamma > 1$ and use l'Hôpital's rule.

10. Use the strategy of Problem 9.

11. Begin with the efficiency of the endoreversible Carnot-like engine and use applicable equations to express it in terms of the appropriate temperatures. Review the derivations for the endoreversible Carnot-like engine.

Hints to Chapter 6 Conceptual Questions

1. Read the definitions of these terms and review the distinction between them.

---

[281] A typical diesel engine, however, generally has a higher compression ratio than a typical Otto engine, so in practice the diesel engine may be more efficient than the Otto engine. Nonetheless, we may compare diesel and Otto engines with the same compression ratio, theoretically, to show that the diesel engine is less efficient than a Carnot engine operating between the same extreme temperatures.

2. For example, consider a specific extensive or intensive parameter, and imagine measuring its value for just half the system, compared to that of the whole system.

3. Review the distinction between extensive and intensive parameters.

4. Length and tension are analogous to use usual variables. Which ones?

5. If necessary, review your understanding of the magnetic dipole moment and external (not to be confused with the magnetic field that the magnetic dipole itself makes) magnetic field. One is extensive, one is intensive.

6. Look at the defining equations and explore the last question first.

7. This is the inverse of an argument that was made in the text. Review that and try to formulate the analogous inverse argument.

8. This relates directly to stable/unstable/neutral equilibrium points on a gravitational potential energy curve, which is analogous to thermodynamic systems.

Hints to Chapter 6 Practice Problems

1. Follow the examples in the text.

2. You should get conditions on the constants (such as sign constraints or mathematical relations among them) from imposing requirements like additivity, the sign of a particular partial derivative, and related requirements on a viable fundamental relation. Review the example in the text where the viability of a fundamental relation is checked.

3. See the hints for Problem 2.

4. Review how to derive equations of state from a fundamental relation, including related examples. Review the interpretation of mechanical and chemical work in thermodynamics to work out the analogy.

5. There is a compatibility check in the text. There is also a strategy, including examples, of how to find the third equation of state and the fundamental relation.

6. Review the text, including an example, regarding how to find a compatible equation of state.

Hints

7. This is related to the technique that is described – and applied to thermodynamic systems – for how to integrate thermodynamic equations like the first law, which feature three or more variables. Separate pairs of variables on both sides, and think of using the product rule backwards on both sides, implicitly, like examples in the text.

8. What is the equilibrium condition? Apply it. It may also be useful to write out equations for variables that add up to a constant – such as energy or volume, if applicable.

9. See the hints to Problem 8. This problem may be a little more involved, but it's the same strategy with different equilibrium conditions.

10. Plug the volume change into the fundamental relation and use the fact that the process is isentropic. It may also be useful to use other equations that apply to the ideal gas.

11. This is similar to Problem 10, but you should instead work in the enthalpy representation.

12. See the hints for Problem 8.

13. See the hints for Problem 9.

14. Review the derivation of the equation of an adiabat for an ideal gas (in a previous chapter), and work out an analogous derivation for the van der Waals fluid. You may find it convenient to work with a different pair of easy-to-measure variables.

15. Play with the equations that apply to the van der Waals fluid, plugging in the specified volume change.

16. See the hints to Problem 10, but apply this to the photon gas.

17. See the hints to Problem 11, but apply this to the photon gas.

18. Review the discussion of the single-variable Legendre transformation and the example in the text.

19. Review the discussion of the inverse single-variable Legendre transformation and the example in the text.

20. Perform these Legendre transformations, following examples in the text.

21. Perform this inverse Legendre transformation, following examples in the text.

22. Perform this Legendre transformations, following examples in the text, starting with the fundamental relation for the van der Waals fluid.

23. Perform this Legendre transformations, following examples in the text, starting with the specified fundamental relation for this magnetic system.

24. Apply the mathematical definitions of these heat capacities.

25. Apply the mathematical definitions of the coefficient of thermal expansion and the compressibilities.

26. Apply the mathematical definition of the isothermal compressibility.

27. Apply the mathematical definition of the coefficient of thermal expansion.

28. This is the magnetic analog of the previous Problems.

29. Review the technique for working out the Maxwell relations.

30. Review the strategy for how to reduce derivatives and study the examples.

31. This is related to derivative reduction. You can find a useful example of applying the derivative reduction technique in the text.

32. Apply the Maxwell relations to this magnetic system.

33. You will need to reduce an appropriate derivative. See the examples of how to apply the derivative reduction technique.

34. You will need to reduce an appropriate derivative. See the examples of how to apply the derivative reduction technique.

35. Review the mathematical test for the concavity of an algebraic function and related examples.

36. Review the mathematical test for the concavity for a thermodynamic system and related examples.

37. Review the mathematical test for the concavity for a thermodynamic system and related examples.

38. Review the stability conditions for a thermodynamic system and related examples.

39. Review the stability conditions for a thermodynamic system and related examples.

Answers to Conceptual Questions and Selected Practice Problems

Answers to Chapter 0 Conceptual Questions

1. No. Larger pot.

2. It increases. Mechanical energy of hammer swing is converted into internal energy.

3. Same. Iceberg's is greater. Iceberg has more internal energy to convert to heat.

4. It's smaller. The average distance between molecules changes the most in the solid ↔ liquid transition, meaning that more work must be done.

5. Thermal energy is associated with the phase change from steam to water.

6. They float. Submerge them mid-level and see if they rise or sink. About the same. Change its density slightly.

7. It actually depends upon what you mean by weight. Its actual weight – the gravitational force that the earth exerts on it – would be the same in either case. However, its apparent weight, which would be read by a scale, would be less in air due to the buoyant force.

8. Wholly submerged, the buoyant force would exceed its weight. It sticks out to reduce the volume of displaced fluid, thereby reducing the buoyant force according to Archimedes' principle. Its fractional extent is such that the buoyant force and its weight are equal and opposite at equilibrium. 1/19. Nothing! You can show mathematically that the reduced volume associated with the density increase as the ice melts compensates the fractional volume of the ice that sticks out.

9. Yes. Like Gabriel's horn, which has finite volume, but infinite surface area. Not unless it could have infinite volume, but finite surface area.

10. Edge. Greater pressure with less contact area.

11. Force is applied directly to a portion of the surface of an object, creating a tangible physical quantity that we call pressure. Force is generally not applied throughout a volume, except for forces created by external fields – but even in this case, it's the pressure that is most convenient, since an object wholly submerged in a fluid in an external gravitational field results in buoyancy (which we understand in terms of pressure). Not unless it could have infinite volume, but finite surface area.

12. Mass of solute divided by volume of solution is not equal to the reciprocal of molar volume. Multiply the reciprocal of molar volume by the ratio of the total volume to the volume of the solution and also multiply by the ratio of the mole number for the solute to the total mole number.

13. Mole number of solute divided by mass of solvent is not equal to the reciprocal of molar mass. Multiply the reciprocal of molar mass by the ratio of the total mass to the mass of the solvent and also multiply by the ratio of the mole number for the solute to the total mole number.

14. At higher (lower) air atmospheric pressure, the molecules must move faster (slower) on average to balance the higher (lower) pressure exerted on the surface of the water by the air molecules. The temperature at which boiling occurs must result in faster (slower) moving steam molecules, meaning that the boiling point is higher (lower) than it is at standard atmospheric pressure.

15. The density of water is maximum at 4 °C. As water cools to 4 °C, it falls to the bottom of the lake because it is most dense. When all of the lake has reached 4 °C, parts of the lake may then cool to a lower temperature. As this happens, the water is actually less dense and rises to the top. The ice that forms at 0 °C thus forms at the top of the liquid.

16. Change in temperature is the same in degrees Kelvin or Celsius.

17. How the molecules are bonded together. Diatomic oxygen.

Answers to Selected Chapter 0 Practice Problems

1. 1.3×10^{22}, 1.3×10^{22}, 1.3×10^{22}, 3.9×10^{22}.

3. 0.00262, 0.997.

5. 1 : 1.24 : 1.14.

7. $-40°$, would be less than absolute zero, not possible.

10. m/s.

12. $J/(K^4 m^3)$

13. $b = -\frac{1}{2}, c = \frac{3}{2}, d = -1$.

Answers to Chapter 1 Conceptual Questions

1. 1 m to the left.

2. Same. Zeroth law of thermodynamics.

3. Thermal insulation slows down the rate at which heat is exchanged between the system and the surroundings. The difference between cooling and heating is only the direction in which heat is exchanged. The coffee mug is not thermally insulated, so heat exchange is significant for the coffee mug.

4. Assuming that the thermos bottle is an excellent thermal insulator, no significant amount of heat is transferred from the surroundings to the contents of the thermos bottle. The mechanical work done to shake the thermos bottle is ultimately transferred to internal energy of the liquid, causing the liquid's molecules to move faster on average, meaning that its temperature has increased. This is an irreversible process, so the overall entropy increases. Assuming that the thermos bottle is an excellent thermal insulator, only the entropy of the system increases noticeably, and the entropy of the surroundings is not significantly altered.

5. Simply remove the fastest molecules from the system and the average molecular speed (and hence temperature) will be reduced. One way that the hot coffee in a cup cools over time is that the fastest coffee molecules escape into the air, which you can smell with your nose.

6. No. Mechanical energy is converted into heat energy due to friction between the two surfaces. Heat energy cannot be converted back into mechanical energy without some exhaust (second law applied to heat engines). The heat will also affect the internal energies of the ground and box (associated with an increase in their temperatures).

7. The first law permits perpetual motion, provided that resistive or dissipative forces like friction and air resistance can be avoided – but it is not possible to completely avoid such forces. The second law prevents resistive forces like friction from spontaneously converting heat into mechanical energy (otherwise, the first law would permit forces like friction to aid, rather than hinder, motion). The second law also prevents perpetual motion by driving the conversion of heat into mechanical energy, since in this case there is also the production of exhaust. Celestial objects and microscopic particles appear to perpetually be in motion, quite naturally, though – where such forces do not play a significant role.

8. Humans tend to strive to order their thoughts. At the same time, the amount of information pouring into human brains is also increasing. The number of human brains is increasing as well. The brains that store and think the thoughts are physical, and the second law of thermodynamics states that if the brains become more ordered, the entropy must increase elsewhere to compensate.

9. For example, consider a figure eight. It's zero, assuming no change in mole numbers.

10. Area dependent. Just the area. No.

11. No. The same area will result regardless of the starting point, so the net work is unaffected by this choice. The net internal energy change is zero. The first law then shows that the heat exchange is also unaffected by the choice of starting point.

Answers to Selected Chapter 1 Practice Problems

1. (A) They lie in Quadrants II and III. $\vec{A} = -\frac{A}{2}\hat{i} \pm \frac{A\sqrt{3}}{2}\hat{j}$. (B) $\vec{A} = -\frac{A}{2}\hat{i} \pm \frac{A\sqrt{11}}{4}\hat{j} - \frac{A}{4}\hat{k}$.

3. $\sqrt{5}$, $\tan^{-1}(2/3)$, 4, $\cos^{-1}(4/\sqrt{65})$, $\sqrt{26}$, $\tan^{-1}(1/5)$, $\sqrt{10}$, $\tan^{-1}(3)$, 7, $\pi/2$ rad.

5. 1 m/s, $2\sqrt{10}$ m/s^2, $\sqrt{41}$ m/s, $\frac{64}{3}$ m $\hat{i}$ − 20 m $\hat{j}$.

7. $2\omega \sin(3\theta) + 6\theta\omega \cos(3\theta)$.

8. 2.5 s^{-1}.

9. $8x^3 - 2xy^2 + 3y^3$, $-2x^2y + 9xy^2$.

12. $0.29\,\hat{i} + 0.38\,\hat{k}$.

14. $\sin x\, dx + x \cos x\, dx = 4y^3 dy - 4dy$.

15. $\Delta \vec{r} = -2\sqrt{3}$ m $\hat{i}$ + 2 m $\hat{j}$, 2π m, 4.0 m/s, $\vec{v}_{ave} = -\frac{4\sqrt{3}}{\pi}$ m/s $\hat{i}$ + $\frac{4}{\pi}$ m/s $\hat{j}$.

18. 0.5.

19. $\vec{v} = R_0\omega_0\hat{\theta}$, $\vec{a} = -R_0\omega_0^2\hat{r}$.

22. 33%, 4.18 steradians. (But you need to do the double integrals; don't simply divide by 3.)

23. $W/4$, $3W/4$.

24. $M/16$.

25. 2.65 Pa.

26. 0.38 MJ, 88 kJ, −0.35 MJ, 0.11 MJ.

27. 1.0 MJ, −0.40 MJ, 0.63 MJ.

28. $3P_0V_0/4$, $3P_0V_0$, 0.

Answers to Chapter 2 Conceptual Questions

1. The gap length decreases in arc length and angle from the center as the arc length of the metal ring expands.

2. It increases as the shape expands radially outward.

3. The two metal strips expand/contract by different amounts. The metal that expands more when heated – or less when cooled – will correspond to the outer arc length.

4. The ring expands when heated and the rod contracts when cooled. It might not be possible to heat or cool the system so that the ring may be removed without damaging either the rod or the ring (and it's not possible to heat one, but not the other, since they are in thermal contact) because they expand according to different formulas, as the dimensionality of the expansion is different for the two geometries.

5. When rolling rubber tires heat up due to friction, the rubber tires contract rather than expand and become loose.

6. The density of water is maximum at 4 °C. As water cools to 4 °C, it falls to the bottom of the lake because it is most dense. When all of the lake has reached 4 °C, parts of the lake may then cool to a lower temperature. As this happens, the water is actually less dense and rises to the top. The ice that forms at 0 °C thus forms at the top of the liquid.

7. The coffee mug with the smallest thermal conductivity is the poorest conductor of heat, and so heat does not flow as rapidly from the coffee to the mug, which is in thermal contact with it. With less heat going to the mug, it will also be cooler to touch.

8. On a cool night, a tile floor feels cooler to bare feet than a wooden floor because tile has a higher thermal conductivity than wood. The temperature of the floors is about the same; it's how well they exchange heat when your feet make thermal contact that makes the difference.

9. Specific heat capacity determines how well each material absorbs heat when it is hot or releases heat when it is cool. During the daytime, emissivity determines how well each material absorbs thermal radiation from the sun. These factor into the temperatures of the surfaces themselves. The thermal conductivity, however, is directly responsible for the exchange of heat between the walkway and your feet. Whether the two walkways are very hot or very cool, you would like the thermal conductivity to be small (so they are poor conductors of heat) so that less heat is transferred to warm or cool your feet.

10. Solving for the equilibrium temperature mathematically, you can see that in the limit that C_s approaches zero, the equilibrium temperature approaches T_ℓ, which is the lowest equilibrium temperature that can result. So the highest equilibrium temperature does not result by making C_s as low as possible. Similarly, dividing both sides by C_ℓ, you can see that in the limit that the ratio C_s/C_ℓ becomes very small, the equilibrium temperature approaches T_ℓ. So the highest equilibrium temperature also does not result by making C_ℓ very large. Therefore, you need a larger C_s or a smaller C_ℓ in order to have a high equilibrium temperature.

11. Radiation, conduction, convection, convection, conduction, radiation.

12. Internal energy from your fingers is transformed into heat, which flows from your fingers to the ice. The heat is transferred via conduction along the metal rod.

13. Wood; more heat flows from your body to the metal since the metal has a higher thermal conductivity. Same; no heat is exchanged. Metal; more heat flows from the metal to your body.

14. Boiling occurs through convection.

15. In equilibrium, the absorption and emission rates must be equal. Conservation of energy requires that the more radiation it reflects, the less it absorbs. If it reflects well, it absorbs little, and therefore in equilibrium it also emits little.

16. It blocks sunlight so that the car absorbs less thermal radiation. It reflects thermal radiation, so that the car will absorb less. If the temperature inside the car is decreasing because the car emits more thermal radiation than it absorbs, radiation heading out of the car can be blocked to reduce the amount of radiation that the car emits (while also blocking some of the radiation heading into the car).

17. It is four times greater. Light from the brightest image seen travels a distance equal to the width of the room (halfway twice). Light from the second brightest image seen travels twice as far.

18. They add in reciprocal in series, but add in parallel. No, thermal conduction involves a dynamic flow of heat. A steady state can be achieved, but it is not thermostatic.

19. Heat flows from high temperature to low temperature, whereas electric field lines do not flow from a more positive charge to a less positive charge (but instead from positive charge to negative charge). The relative highs and lows.

Answers to Selected Chapter 2 Practice Problems

2. 0.44 mm longer than 200 cm.

4. 0.40 cm, 0.027 rad.

6. 25 °C.

8. 24 °C.

9. 25%, 0 °C, 100%, 11 °C.

10. 50 °C.

11. 47 °C, 60 °C.

13. $P = \pi k(b^2 - a^2)\Delta T/L$, $P = 2\pi k L \Delta T \frac{1}{\ln(b/a)}$.

14. $x = \sqrt{\frac{2k\Delta T t}{L\rho}}$.

16. 6.0 MJ, 0.40 MJ by the system, 5.6 MJ increase.

17. 0.32 MJ, absorbed.

19. 0.13 MJ, absorbed.

Answers to Chapter 3 Conceptual Questions

1. It's reduced.

2. The ideal gas law reduces to Boyle's law along an isotherm, Charles's law along an isobar, and Gay-Lussac's law along an isochor.

3. It's considerably less. The molecules of the substance that travel through the air to transport the odor are very slow compared to the average speed of the air molecules, and hence are also very slow compared to the speed of sound propagating through air.

4. $v_{rms}^{Cl_2} < v_{rms}^{S} < v_{rms}^{N_2} < v_{rms}^{Ne} < v_{rms}^{He}$. Regardless of the structure of the molecules, $v_{rms} = \sqrt{3k_B T/m_m}$. Whether the molecules are monatomic, diatomic, or polyatomic does figure into this formula, even though there is no adiabatic index: It must be incorporated into the molecular mass. So if the temperatures of various gases are equal, the mass of one mole determines the average speed of its molecules.

5. $v_{rms}^{O_3} < v_{rms}^{O_2}$. See the answer to Question 4.

6. The increase in temperature with altitude in the stratosphere can be attributed to the absorption of ultraviolet radiation by ozone molecules. The increase in temperature with altitude in the thermosphere has to do with the fact that there are very few air molecules in this layer, and any solar radiation that these molecules absorb increases their average kinetic energy significantly. (The temperatures of the thermosphere and exosphere can exceed a thousand degrees Kelvin, yet you would freeze in these outermost layers. Why?)

7. The molar specific heat capacity at constant pressure. For an ideal gas, the change in internal energy is directly proportional to the temperature change. Suppose we have a particular temperature change in mind, so that the internal energy change is fixed. At constant volume, all of this internal energy change would be associated with heat, since no work is done along an isochor. At constant pressure, this internal energy change would be related to both heat and work – with a relative minus sign between them. Solving for the heat transferred, work comes to the other side, becoming a plus sign. Therefore, in the case of an isobar, some of the heat goes toward doing work and some goes into changing temperature; while in the case of an isochor, all of the heat goes toward the temperature change. The heat exchanged equals the molar specific heat capacity times the mole number times the temperature difference. For the same heat transfer and mole number, a greater temperature change will result at constant volume, compared to constant pressure, which means that the molar specific heat capacity at constant pressure must be larger.

8. Its temperature decreases. No heat is exchanged, so the internal energy change – which is proportional to the temperature change – is proportional to the negative of the work done – which is positive since volume increased. Since there is a minus sign (in "the negative of the work done"), the internal energy – and hence temperature – decrease.

Answers to Selected Chapter 3 Practice Problems

2. 1.4 km/s, 1.7 km/s.

4. Neon is the closest monatomic gas.

7. $B = P$.

8. 36 kPa, 4.5 m³, −5.0 kJ, 0, 5.0 kJ, 0.11 MJ, 0.11 MJ, 0, 30 kJ, 25 kJ, −5.0 kJ.

9. $\Delta U = \frac{c_V}{R}\left(\frac{3^{5/3}}{6} - 1\right) P_0 V_0$.

11. $1/2^{2/3}$.

Answers to Chapter 4 Conceptual Questions

1. The number of permutations. A single combination can be reordered to make a different permutation.

2. The odds at which the casino pays bets are based on 36 slots, but the actual probabilities – and hence expectation values – are based on 38 slots. Assuming a fair roulette wheel, you may win money in the short run because actual outcomes may deviate from expected outcomes in the short run, but in the long run the actual outcomes fill out the expected probability distributions very closely.

3. You need and infinite supply of both money and time (and a casino that doesn't impose a maximum bet). Otherwise, eventually you'll bust. If you're lucky to quit before you bust, you probably won't make much money – it won't take too long before you have to make a huge wager just to make one dollar.

4. Zero. Nonzero. The average velocity of the molecules is zero, but the average speed is not.

5. Hence the name most probable value, which corresponds to the top of the probability distribution.

6. Very small, but deep. Pressure in a fluid depends on depth, but not area.

7. Neither. Pressure in a fluid depends on depth, which is the same for both, gravitational field, which is the same for both, and the density of the liquid, which is the same for both.

8. No. Air is less dense just above very hot sand and increases in density for some region above this as the air cools. The general atmospheric trend is for the air to be most dense near the surface and to decrease in density up to space.

9. They will change. No.

Answers to Selected Chapter 4 Practice Problems

2. 840, 120, 20.

3. 720.

5. 1,024.

7. 32/663.

9. 6/4165.

10. 7.0, 7.4, 7.

12. $2\sqrt{b/3}$, $16\sqrt{b}/15$, $2\sqrt{b/3}$.

13. $\overline{1/v} = \sqrt{\dfrac{2m}{\pi k_B T}}$, $1/\bar{v} = \sqrt{\dfrac{\pi m}{8 k_B T}}$.

Answers to Chapter 5 Conceptual Questions

1. Use the sand method to carry out the process quasistatically, ensuring that pressure remains constant as volume slowly changes and as heat is slowly exchanged.

2. Yes. Make the air around it cold enough to freeze. This does not contradict the second law of thermodynamics. Rather, the second law of thermodynamics tells you that the entropy must increase – enough that the overall entropy does not decrease – elsewhere in order to drive this process.

3. It decreases. It is. The entropy of its surroundings will increase such that the overall entropy does not decrease.

4. Increase. Electrical energy is converted into mechanical energy. There are resistive forces involved in the rotation of the fan, which means that nonconservative work is done. This results in heat. Since the room is perfectly insulated, there will be a slight overall increase in temperature. The reason that a fan makes you feel cooler has to do with helping to evaporate sweat. Opening a refrigerator door may make you feel cooler while you stand by the refrigerator, but will actually raise the temperature of the (insulated) kitchen. In the case of the refrigerator, the temperature rises directly in accordance with the second law of thermodynamics applied to heat engines.

5. A lower exhaust temperature provides for a greater maximum possible Carnot efficiency.

6. No. Mechanical energy is converted into heat. According to the second law of thermodynamics applied to heat engines, this can't be reversed to convert heat into mechanical energy without also producing some exhaust.

Answers to Selected Chapter 5 Practice Problems

1. 0.30 kJ/K.

3. $\Delta S = nc_P \ln 2$.

4. $\Delta S = nc_V \ln\left(V^{1-\frac{b}{a}} - V_0^{1-\frac{b}{a}}\right) + nR \ln\left(\frac{V}{V_0}\right)$.

7. $e = 1 - \dfrac{1}{\frac{1}{\gamma} + \frac{R}{c_P} \frac{r}{r-1} \ln r}$. There are two ways to draw this, and your answer may vary accordingly.

8. $e_S = \dfrac{1}{\frac{c_V}{R \ln r} + \frac{T_h}{T_h - T_c}}$.

Answers to Chapter 6 Conceptual Questions

1. A homogeneous pizza must have its toppings evenly distributed throughout, while an isotropic pizza must have its toppings radially symmetric. If you add toppings like horizontal stripes, they are evenly spaced, but there is a preferred direction. If you add toppings like concentric rings that are unevenly spaced, they are the same in all directions from the center, but not homogeneous.

2. Intensive. Extensive.

3. Extensive: weight, surface area, enthalpy, Helmholtz free energy. Intensive: molar mass, number per unit volume, molar entropy, chemical potential.

4. Extensive: internal energy, entropy, length. Intensive: temperature, tension. Tension equals the partial derivative of the internal energy with respect to length for constant entropy. (Tension is analogous to the negative of mechanical pressure.)

5. Magnetic dipole moment is extensive. External magnetic field is intensive. The external magnetic field equals the partial derivative of the internal energy with respect to magnetic dipole moment, holding entropy, volume, and mole numbers constant.

6. Intensive. Homogeneous zero-order.

8. Regions where the gravitational potential energy function is concave up are stable; regions where it is concave down are unstable; a point of inflection is neutral. A slight disturbance from a stable equilibrium position results in oscillation about the equilibrium position, whereas slight disturbances from other types of equilibrium positions do not result in oscillation about equilibrium.

Answers to Selected Chapter 6 Practice Problems

1. First-order, first-order, zero-order, zero-order.

2. $b > 1, c < 1, b + c = 2, a^{1/b} b > 0, T = abS^{b-1}V^c/n, P = -acS^b V^{c-1}/n, \mu = -aS^b V^c/n^2$.

4. $T = abe^{bs+cm^2}$, $B = 2acme^{bs+cm^2}$, $đW_{mag} = BdM$, where M is the (non-molar) magnetic dipole moment.

5. $\mu = -u$, $U = (5a)^{5/2} S^{5/2} V^{-1/2} n^{-1}$.

7. $x^2 y^2 - x_0^2 y_0^2 = \frac{2w_0^2}{z_0} - 2\frac{w^2}{z}$.

8. 253 K.

10. 75 K, $\frac{1}{32}$ atm.

12. 263 K.

14. $T(v-b)^{R/c_V} = const$.

16. $T = T_0/2^{1/3}$, $P = P_0/2^{4/3}$, $W = 3(1 - 2^{-1/3})P_0 V_0$.

18. $A(T) = \frac{T}{b}\left[1 - \ln\left(\frac{T}{ab}\right)\right]$.

20. $F = \left(\frac{1}{4^{4/3}} - \frac{1}{4^{1/3}}\right)(a^4 T^4 V^2 n)^{1/3}$.

21. $U = \dfrac{S^{5/2}}{\left[\left(\frac{2}{5}\right)^{5/3} - \left(\frac{2}{5}\right)^{2/3}\right]^{3/2} a^{1/2} nV^{1/2}}$.

22. $F = n\left(c_V - \frac{S_0}{n_0}\right)T - \frac{an^2}{V} + nT \ln\left[\left(\frac{\frac{V}{n} - b}{\frac{V_0}{n_0} - b}\right)^R \left(\frac{T}{T_0}\right)^{c_V}\right]$.

309

Answers

25. $1/T, 1/P, 1/(\gamma P)$.

26. $\dfrac{1}{\dfrac{2a}{v^2} - \dfrac{Pv + \dfrac{a}{v}}{v-b}}$.

29. $\left(\dfrac{\partial \mu}{\partial S}\right)_{V,n}$; $-\left(\dfrac{\partial \mu}{\partial T}\right)_{S,V}$; $\left(\dfrac{\partial V}{\partial n}\right)_{T,P}$.

30. $-\dfrac{1}{\kappa_T V}$, $\dfrac{\kappa_T}{\beta}$, $\dfrac{-n}{\dfrac{\beta TS}{nc_V \kappa_T} - \dfrac{1}{\kappa_S}}$, $\dfrac{S}{V}$, $\dfrac{\beta TSV - nc_P \kappa_T PV - \kappa_T PSV}{S + \beta PV}$, $\dfrac{TS}{\beta TSV - nc_P V}$.

33. $dP = \dfrac{\beta}{\kappa_T} dT$.

35. Always concave up, concave up for $|x| > 1/\sqrt{3}$, always concave down, concave up for the second half of the cycle.

References

The accepted numerical values of a large variety of constants can be accessed through many online resources, such as The National Institute of Standards and Technology at www.nist.gov and Material Property Data at www.matweb.com.

If you're looking for an introductory calculus-based textbook, the classic is Halliday and Resnick's *Physics*, Part 1. I recommend the third edition, published in 1977, if you can find a used copy. Also, if you're curious about the meaning of negative temperatures or the entropy of thought, for example, which were alluded to in this text, you can find useful references to these fascinating notions therein. The newer edition, with Walker as a coauthor, is a popular textbook used in courses these days, and is also good.

I also recommend Serway's *Physics for Scientists and Engineers*, Volume 1. If you don't need the current edition for a course, there are many old editions that you can find a good deal on. These calculus-based physics textbooks will have a few chapters devoted to thermal physics.

If you are looking for a friendly-to-read introduction to very basic notions, consider Hewitt's *Conceptual Physics*. Some first-year students who thrive on the math, but struggle with the concepts, often ask me where they might find supplemental material to help them understand the conceptual aspect of the course better. They invariably enjoy the book, and often show improvement in the course. Like the calculus-based textbooks, conceptual physics books usually only have a couple of chapters devoted to thermal physics.

If you are taking thermodynamics, there's a good chance that you're using Callen's *Thermodynamics and an Introduction to Thermostatistics*. If not, and you want an additional reference on this subject, start with this. At the end of Callen's text, you can also find other recommended readings on thermodynamics and statistical mechanics.

My list is short because my policy is not to overwhelm the reader with too many references, and to focus on recommending the textbooks that were most useful when I was learning the material and which have served as the most useful references when I needed to consult or review something.

Index

additivity	204-205
adiabatic	
adiabat	21,147-149
adiabat for an ideal gas	92-97
adiabatic expansion	150-152
adiabatic index	90,259
adiabatic wall	214
cf. isothermal	21
work done along an adiabat	47,49,71
air conditioner	159-161
arc length	37
atmosphere	
distribution of molecules	121-127
law of atmospheres	124-127
layers of earth's atmosphere	97
automobile engine	182
average	
average value	116-117
expectation value	116-117
root-mean-square (rms) value	117
Avogadro's number	7
binomial distribution	109
binomial expansion	106
blackbody radiation	231-234
boiling point	18
Boltzmann constant	15
Boltzmann distribution law	127
Boyle's law	45,81
Brayton cycle	191-193
Brownian motion	136
bulk modulus	95-97
calorimetry	60,152-154

Carnot	
Carnot cycle	32,163-182, 216
Carnot efficiency	174-182,216
Carnot temperature	181-182
Carnot theorem	177-180
centrifuge	139
chain rule	35
Charles's law	45,81
chemical composition	19
chemical equilibrium	20,220
chemical potential	14,206-207, 221
chemical work	207
Clausius form of the 2nd law	31,177
Clausius-Clapeyron equation	52,269
closed system	214
coefficient of performance (COP)	
heat pump	162,176,181
Carnot cycle	173
coefficient of thermal expansion	15,58-59, 249-250
coexistence curve	51,264-275
collision frequency	131
collision probability	132
combinations	105-106
compressibility	251,258-259
compression ratio	185-187, 189-190
combustion engine, internal	159,180
concavity	260
condensation point	18

conduction		
electrical conduction	66-68	
heat conduction	45,62-68	
thermal conductivity	15,67	
thermal current	67-68	
conjugate variables	208	
conservation of energy	60-61	
see also *first law*		
conservative field	38	
convection	62	
conversions		
energy conversions	17	
force conversions	16	
mass conversions	16	
power conversions	17	
pressure conversions	17	
temperature conversions	16	
volume conversions	16	
convexity	260	
coordinate systems		
2D polar coordinates	39-40	
cylindrical coordinates	41	
spherical coordinates	40	
critical point	51,264-275	
current density	67	
cutoff ratio	186,189-190	
cycle		
Brayton cycle	191-193	
Carnot cycle	32,163-182	
cyclic process	22,47,71,158	
diesel cycle		
Otto cycle		
Stirling cycle		
cylindrical coordinates	41	
degrees of freedom	87-88	
density	14,43	
deposition point	18	
derivative reduction	255-257,259	
dew point	18	
diathermal wall	214	
diatomic	87-90	

diesel engine	159,188-190	
diffusive equilibrium	19,219-220	
displacement vector	37	
displacement volume	186	
distribution		
see probability distribution		
dot product	34	
driven processes	175	
Dulong and Petit	46	
efficiency		
Brayton cycle	192-193	
Carnot cycle	174-182,216	
diesel cycle	189-190	
endoreversible cycle	199	
heat engine	14,30-32,46,158,166-167,171,174-182,216	
Otto cycle	185-188	
electric fan	199	
electrical conduction	66-68	
current density	67	
electric charge	68	
electric conductivity	67	
electric current	68	
electric field	67	
electric flux	66	
electric potential	66	
Gauss's law	67	
Kirchhoff's rules	68	
resistance	67	
electromagnetic radiation	45,62,231-234	
emissivity	15,68-69	
energy	86-88	
see also first law		
see also conservation of energy		
see also internal energy		
see also heat		
energy conversions	17	
energy minimum principle	205,218,247-248	

Index

energy representation	205-206
engine	
automobile engine	182
Brayton engine	191-193
Carnot engine	32,163-182, 216
diesel engine	159,188-190
endoreversible engine	193-199
engine displacement	186
engine efficiency	14,30-32,46, 158,166-167,171, 174-182,216
engine power	186
engine torque	186
gas turbine engine	190-193
heat engine	14,30-32,46, 140-201,244
internal combustion engine	159-180
Otto engine	183-188
steam	159
Stirling engine	201
enthalpy	
enthalpy representation	13,227
extremum principle	248
for a photon gas	234
for an ideal gas	236
entropy	13,28-32,43, 142-154,158
entropy maximum principle	205,215, 217,218, 220,242, 245,247-248
entropy representation	206
equal areas, line of	271-274
equations of state	208,228-230

equilibrium	
chemical equilibrium	20,220
diffusive equilibrium	19,219-220
equilibrium	203,214-221
mechanical equilibrium	19,216-218
phase equilibrium	20
radiative equilibrium	20,69,221
thermal equilibrium	19,60,214-215,218,220
thermodynamic equilibrium	20,221
equipartition of energy	46,86
Euler equation	211
evaporation	18
exact differentials	7,49,203
excited state	20
exhaust	157
expectation value	116-117
extensive parameters	202-203
extremum principles	205,215-218,220, 241-248
first law of thermodyanmics	26-27,42,70-73,145,166, 175-176, 206,228
flux	
electric	66-67
heat	7-8,13,26-27,42,60-61,66-67,70-73,88-97, 144-145, 166-171, 175-180, 206-207, 216,228, 246-247
force conversions	16
free expansion	
adiabatic free expansion	150-152
definition of	22
entropy change for	150-152
work done during	47,49,71

freezing point	17
fundamental relation	204
gas centrifuge	139
gas constant	15
gas turbine engine	190-193
Gauss's law	67
Gaussian distribution	112-115, 118-121
Gay-Lussac's law	45,82
Gibbs free energy	
extremum principle	249
for an ideal gas	236
Gibbs representation	226-227
Gibbs-Duhem relation	212-213
gradient	36-37
grand canonical potential	227
ground state	20
heat	
cf. internal energy	13
cf. temperature	8,144
heat capacity	9,41,60,88-90,246-247,250,258
heat conduction – see cond.	
heat engine – see heat engine	
heat flux	7-8,13,26-27,42,60-61,66-67,70-73,88-97,144-145,166-171,175-180,206-207,216,228,246-247
heat of transformation	8,42,61
heat pump	162,172-173,176,181
heat source	157,242
latent heat	8,42,61
mechanical equivalent of heat	72-73,216-218

heat engine	
automobile engine	182
Brayton cycle	191-193
Carnot cycle	32,163-182,216
diesel cycle	159,188-190
efficiency	14,30-32,46,158,166-167,171,174-182,216
endoreversible engine	193-199
gas turbine engine	190-193
general heat engines	30-32,140-201,244
internal combustion engine	159-180
Otto cycle	183-188
steam engine	159
Stirling cycle	201
heater	161
Helmholtz free energy	225-226
extremum principle	248
for an ideal gas	236-237
for a photon gas	234
Helmholtz representation	225-226
histogram	108
homogeneous	202
homogeneous first-order	204-205
homogeneous zero-order	208
ideal gas	
fundamental relation for	234-237
ideal gases	45,79-99,234-237
ideal gas law	45,79,81-83
kinetic theory	79-81,83-88
imperfect differentials	203
implicit differentiation	36
inexact differentials	203
intensity	69,231-232
intensive parameters	202-203
internal combustion engine	159-180

internal energy	12-13, 26-27, 42, 70-73, 88-89, 91-97, 145, 166, 171, 175-176, 178-180, 206
inverse Legendre transformation	224-225
irreversible	22, 27-31, 140-41, 147, 149-154, 221, 241-242
isenthalpic	22
isentropic	
cf. adiabatic	147
isentrope	22, 147-148
isobaric	
isobar	21, 148, 150
work along an isobar	47, 49, 71
isochoric	
isochor	21, 148-149
work along an isochor	47, 49, 71
isothermal	
cf. adiabatic	21, 148, 150
for an ideal gas	90-92
isotherm	21
work along an isotherm	47, 49, 71
isotropic	58, 202
isovolumetric	
see isochoric	
Joule's experiment	72-73
Kelvin-Planck form of the 2nd law	31, 177
kinetic theory	79-81, 83-88, 131-136
Kirchhoff's rules	68
latent heat	8, 42, 61
law of atmospheres	124-127
law of cosines	34
law of conservation of energy	
see first law of thermodynamics	
see conservation of energy	
law of Dulong and Petit	46
law of equipartition of energy	46, 86

law of large numbers	102
le Chatelier's principle	264
Legendre transformation	222-227
line of equal areas	271-274
line integral	37-38
linear expansion	
see thermal expansion	
macroscopic coordinates	202
macrostate	19, 142
magnetic systems	276-277, 279
mass	
conversions	16
definition of	9
molar mass	19
maximum work theorem	242-245
Maxwell-Boltzmann distribution	121-130
Maxwell relations	249-259
mean-free path	131-133
mean-free time	131
mean value	114
mean-value theorem	38
mechanical equilibrium	19, 216-218
mechanical equivalent of heat	72-73
melting point	17
metastable states	52
microstate	19, 142
molality	23
molar heat capacity	
see heat capacity	
molar mass	19
molar scaling	205
molarity	23
mole	
definition of	17
mole fraction	10
mole number	9, 43
molecular flux	134-136
monatomic	86-90
most probable value	118
Nernst postulate	208
non-equilibrium states	20
normalization	113-114

number density	123-124
Otto cycle	183-188
partial derivative	35
path-dependence	38,49,70, 144-145, 148-152, 203,210
perfect differentials	7,49,203
permutation	104-105
phase	
phase curve	51,264-265
phase diagram	51,264-265
phase equilibrium	20,220
phase transition	51-52,149, 260-275
photon gas	45,62, 231-234
polyatomic	87-90
position vector	33
potentials	222-227
power	
conversions	17
definition of	12,42
for an endoreversible engine	195-198
radiancy	231-232
rate of energy/heat flow	62-69
pressure	
atmospheric pressure	121-123
conversions	17
definition of	11,42
for an ideal gas	83-86
pressure ratio	192-193
standard (STP)	18
thermodynamic	206-207,218
probability	
probability and statistics	100-139
probability density	111-112
probability distributions	101-102, 104-121
P-T diagram	51,264-265
P-V diagram	47-48

quasistatic	22,140-141, 221,241-242
R value	65
radiancy	231-232
radiation	15,45,62, 231-234
radiative equilibrium	20,69,221
reduction of derivatives	255-257,259
refrigerator (see also COP)	159-161
relaxation time	242
reservoir	157,243
resistance	67
reversible	22,27-31, 140-141, 147-151, 154,221, 241-242
reversible heat source	157,242
reversible work source	242
room temperature	18
root-mean-square (rms) value	84-86,96, 117
rubber band	275
scalar product	34
scaling	205,211
scattering	132
second derivatives	249-259
second law of thermodynamics	
Clausius form	31,177
Kelvin-Planck form	31,177
second law	27-32,145-146,166-167,176-177,245-247
simple systems	203,228-241
solid angle	44,231
specific heat	
see heat capacity	
speed of sound	96-97
spherical coordinates	40
spontaneous processes	22,174
stability	260-275

Index

standard deviation	115
standard temp. and pressure (STP)	18
state variables	154
statistics	102-103
steady state	20
steady-state	21
steam engine	159
Stefan-Boltzmann constant	15,68-69
Stefan's law	45,68-69,232
Stirling cycle	201
stoichiometric coefficients	220-221
sublimation point	18
superior tangent	270-271
surface area	11,39,43-44
temperature	
cf. heat	8,144
cf. internal energy	13
conversions	16
definition	7,144
room temperature	18
standard (STP)	18
thermodynamic	206-207,216
thermal conduction	
see conduction	
thermal conductivity	15,67
thermal current	67-68
thermal energy	
see heat	
thermal equilibrium	19,60,214-215,218,220
thermal expansion	15,41,58-59,249-250
thermal radiation	15,45,62,231-234
thermal reservoir	157,243
thermodynamic cross-polytope	254

thermodynamic equilibrium	20,221
thermodynamic octahedron	253-254
thermodynamic potentials	222-227
thermodynamic square	252-253
thermodynamics	202-280
thermometer	58,82
third law of thermodynamics	32,181-182,208
total derivative	35
transient state	20
triple point	18,52
two-dimensional polar coordinates	39-40
unit vectors	33
universal gas constant	15
van der Waals fluid	238-241
variance	118
virial expansion	237-238
virial coefficients	238
volume	10,16,39,44
volume conversions	16
work	12,26-27,36,42,47-50,70-73,91-97,145,158,166-171,175-176,178-180,206-207,218,228
work source	242
zeroth law of thermodynamics	26,208

Notes

Notes

Made in the USA
Monee, IL
23 December 2019